Molecular Sieves
Science and Technology

Editors: H. G. Karge · J. Weitkamp

Volume 1

Springer-Verlag Berlin Heidelberg GmbH

Synthesis

With contributions by
J. S. Beck, G. Bellussi, E. N. Coker, P. Cool,
S. Ernst, H. Gies, J. C. Jansen, C. T. Kresge, B. Marler,
S. B. McCullen, R. Millini, G. Perego, W. J. Roth,
S. A. Schunk, F. Schüth, R. Szostak, R. W. Thompson,
E. F. Vansant, J. C. Vartuli, U. Werthmann

Springer

Editors:

Dr. Hellmut G. Karge
Fritz-Haber-Institut der Max-Planck-Gesellschaft
Faradayweg 4–6
D-14195 Berlin
Germany

Professor Dr.-Ing. Jens Weitkamp
Institut für Technische Chemie I
Universität Stuttgart
D-70550 Stuttgart
Germany

Cataloging-in-Publication Data applied for

Die Deutsche Bibliothek – CIP-Einheitsaufnahme

Synthesis / with contributions by J. Beck ... – Berlin ; Heidelberg ; New York ; Barcelona ; Budapest ; Hong Kong ; London ; Milan ; Paris ; Santa Clara ; Singapore ; Tokyo : Springer, 1998
(Molecular sieves ; Vol. 1)

ISBN 978-3-662-30838-7 ISBN 978-3-540-69615-5 (eBook)
DOI 10.1007/978-3-540-69615-5

Ursprünglich erschienen bei Springer-Verlag Berlin Heidelberg New York 1998.
Softcover reprint of the hardcover 1st edition 1998

Typesetting: Dataconversion by Fotosatz-Service Köhler OHG, Würzburg
Coverdesign: Friedhelm Steinen-Broo, Estudio Calamar, Pau/Spain

SPIN: 10127286 2/3020 – 5 4 3 2 1 0 – Printed on acid-free paper

Preface to the Series

Following Springer's successful series *Catalysis – Science and Technology*, this series of monographs has been entitled *Molecular Sieves – Science and Technology*. It will cover, in a comprehensive manner, all aspects of the science and application of zeolites and related microporous and mesoporous materials.

After about 50 years of prosperous research, molecular sieves have gained a firm and important position in modern materials science, and we are witnessing an ever increasing number of industrial applications. In addition to the more traditional and still prevailing applications of zeolites as water softeners in laundry detergents, as adsorbents for drying, purification and separation purposes, and as catalysts in the petroleum refining, petrochemical and chemical industries, novel uses of molecular sieves are being sought in numerous laboratories.

By 1998, the Structure Commission of the International Zeolite Association had approved approximately 100 different zeolite structures which, altogether, cover the span of pore diameters from about 0.3 nm to 2 nm. The dimensions of virtually all molecules (except macromolecules) chemists are concerned with fall into this same range. It is this coincidence of molecular dimensions and pore widths which makes zeolites so unique in adsorption and catalysis and enables molecular sieving and shape-selective catalysis. Bearing in mind that each zeolite structure can be modified by a plethora of post-synthesis techniques, an almost infinite variety of molecular sieve materials are nowadays at the researcher's and engineer's disposal. In many instances this will allow the properties of a zeolite to be tailored to a desired application. Likewise, remarkable progress has been made in the characterization of molecular sieve materials by spectroscopic and other physico-chemical techniques, and this is particularly true for structure determination. During the last decade, we have seen impressive progress in the application of quantum mechanical ab initio and other theoretical methods to zeolite science. The results enable us to obtain a deeper understanding of physical and chemical properties of zeolites and may render possible reliable predictions of their behavior. All in all, the science and application of zeolites is a flourishing and exciting field of interdisciplinary research which has reached a high level of sophistication and a certain degree of maturity.

The editors believe that, at the turn of the century, the time has come to collect and present the huge knowledge on zeolite molecular sieves. *Molecular Sieves – Science and Technology* is meant as a handbook of zeolites, and the term "zeolites" is to be understood in the broadest sense of the word. While, throughout the handbook, some emphasis will be placed on the more traditional alumosilicate zeolites with eight-, ten- and twelve-membered ring pore openings, materials with other chemical compositions and narrower and larger pores (such as sodalite, clathrasils, $AlPO_4$–8, VPI-5 or cloverite) will be covered as well. Also included are microporous forms of silica (e.g., silicalite-1 or -2), alumophosphates, gallophosphates, silicoalumophosphates and titaniumsilicalites etc. Finally, zeolite-like amorphous mesoporous materials with ordered pore systems, especially those belonging to the M41S series, will be covered. Among other topics related to the science and application of molecular sieves, the book series will put emphasis on such important items as: the preparation of zeolites by hydrothermal synthesis; zeolite structures and methods for structure determination; post-synthesis modification by, e.g., ion exchange, dealumination or chemical vapor deposition; the characterization by all kinds of physico-chemical and chemical techniques; the acidic and basic properties of molecular sieves; their hydrophilic or hydrophobic surface properties; theory and modelling; sorption and diffusion in microporous and mesoporous materials; host/guest interactions; zeolites as detergent builders; separation and purification processes using molecular sieve adsorbents; zeolites as catalysts in petroleum refining, in petrochemical processes and in the manufacture of organic chemicals; zeolites in environmental protection; novel applications of molecular sieve materials.

The handbook will appear over several years with a total of ten to fifteen volumes. Each volume of the series will be devoted to a specific sub-field of the fundamentals or application of molecular sieve materials and contain five to ten articles authored by renowned experts upon invitation by the editors. These articles are meant to present the state of the art from a scientific and, where applicable, from an industrial point of view, to discuss critical pivotal issues and to outline future directions of research and development in this sub-field. To this end, the series is intended as an up-to-date highly sophisticated collection of information for those who have already been dealing with zeolites in industry or at academic institutions. Moreover, by emphasizing the description and critical assessment of experimental techniques which have been used in molecular sieve science, the series is also meant as a guide for newcomers, enabling them to collect reliable and relevant experimental data.

The editors would like to take this opportunity to express their sincere gratitude to the authors who spent much time and great effort on their chapters. It is our hope that *Molecular Sieves – Science and Technology* turns out to be both a valuable handbook the advanced researcher will regularly consult and a useful guide for newcomers to the fascinating world of microporous and mesoporous materials.

Hellmut G. Karge
Jens Weitkamp

Preface to Volume 1

Obviously, the preparation of molecular sieve materials stands at the origin of their use in science and technology. Since the pioneering work of Barrer and his co-workers and the fascinating achievements of Milton, Breck, Flanigen and others in the Union Carbide laboratories, a wealth of zeolites and related microporous and mesoporous materials have been synthesized, and novel materials of this class will continue to be discovered. In almost all instances, hydrothermal synthesis is the method of choice for preparing zeolites, and structure-directing auxiliaries, often referred to as templates, frequently play a vital role. The techniques for hydrothermal synthesis of molecular sieves and the search for novel and more efficient structure-directing agents have reached a high level of sophistication, yet the scientific understanding of the very complex series of chemical events en route from the low-molecular weight reagents to the inorganic macromolecule remained rather obscure.

Consequently, Chapter 1 written by *R. W. Thompson* gives a modern account of our present understanding of zeolite synthesis. The fundamental mechanisms of zeolite crystallization (primary and secondary nucleation and growth) in hydrothermal systems are highlighted.

Chapter 2 by *H. Gies, B. Marler* and *U. Werthmann* critically reviews the methods for synthesizing porosils, the all-silica end members of zeolites. Depending on their pore or cage apertures the porosils are subdivided into clathrasils (at most six-membered ring windows) and zeosils (at least eight-membered ring windows), the latter being valuable adsorbents with hydrophobic surface properties.

In Chapter 3, *S. Ernst* gives an overview on more recent achievements in the syntheses of alumosilicates with a pronounced potential as catalysts or adsorbents. Examples are zeolites MCM-22, NU-87 and SSZ-24, zeolites with intersecting ten- and twelve-membered ring pores and the so-called super-large pore alumosilicates.

Chapter 4 authored by *J. C. Vartuli, W. J. Roth, J. S. Beck, S.B. McCullen* and *C. T. Kresge* is devoted to the synthesis and properties of zeolite-like amorphous materials of the M41S class with ordered mesopores. These mesoporous solids are currently being scrutinized in numerous laboratories for their potential as adsorbents and catalysts.

Apart from the pore width and pore architecture, the crystal size of a zeolite is often very important. In Chapter 5, *E.N. Coker* and *J. C. Jansen* present a systematic evaluation of the attempts to synthesize either ultra-small (i.e., much smaller than 1 μm) or ultra-large (i.e., much larger than 1 μm) zeolite crystals.

The second most important class of molecular sieves besides the alumosilicates are without any doubt the alumophosphates and their derivatives containing elements other than aluminum and/or additional elements in the framework. Chapter 6 authored by *R. Szostak* is a review covering the synthesis of these molecular sieve phosphates.

The subsequent Chapter 7 is devoted to the synthesis and characterization of molecular sieve materials containing transition metals in the framework. Authored by *G. Perego, R. Millini* and *G. Bellussi*, this Chapter focuses on titanium-silicalite-1 which has recently been found to be a unique catalyst for selective oxidations with hydrogen peroxide. Also covered in this Chapter is the synthesis of vanadium- and iron-containing molecular sieves.

In Chapter 8, *S.A. Schunk* and *F. Schüth* are going one step further by reviewing the literature on microporous and mesoporous materials which are traditionally less familiar to the zeolite community, but rather scattered over the literature on solid-state chemistry. The main intention of this Chapter is to bring this wealth of knowledge to the attention of researchers who routinely look for applications of molecular sieves.

Last but not least, a class of porous materials closely related to zeolites is addressed in Chapter 9: *P. Cool* and *E.F. Vansant* discuss the basic principles of preparing pillared clays, and methods for the proper characterization of these fascinating materials are outlined.

Thus Volume 1 of *Molecular Sieves – Science and Technology* covers the synthesis methods for a broad variety of porous solids. In addition to the critical discussion of the synthesis procedures, the reader will find numerous references to the original literature. May we express our hope that Volume 1 of the series helps the community of scientists to prepare all those microporous and mesoporous materials they need for their purposes.

Hellmut G. Karge
Jens Weitkamp

Contents

Recent Advances in the Understanding of Zeolite Synthesis

Robert W. Thompson

Department of Chemical Engineering, Worcester Polytechnic Institute, 100 Institute Road, Worcester, Massachusetts 01609, USA. *E-mail: rwt@wpi.edu*

1 Introduction

The objective of this chapter is to review the open literature on molecular sieve zeolite synthesis, highlighting information regarding the fundamental mechanisms of zeolite crystallization in hydrothermal systems. The text, therefore, focuses on the three primary mechanistic steps in the crystallization process: nucleation of new populations of zeolite crystals, growth of existing populations of crystals, and the role played by existing zeolite crystal mass in the subsequent nucleation of new crystals or the growth of zeolite crystals in the system.

The perspective taken in this work, based on research results from the literature, has been that molecular sieve zeolite crystals are formed from the species dissolved in the caustic solution medium, and that formation of zeolites by solid-solid transformations does not occur. As such, classical treatments of

Molecular Sieves, Vol. 1

crystallization systems should adequately describe molecular sieve zeolite crystallization processes. However, it is suggested that this absolute perspective may have to be modified to qualify our future thinking, as noted in this review. Some recent work has investigated the very early transformations occurring in several of these alumino-silicate systems, and revealed that colloidal assemblages may form just prior to the creation of crystal nuclei, and may be precursors to nucleation. Consequently, the source of nuclei may be revealed to be associated with species entering the system from well-defined origins.

Growth of molecular sieve zeolites in hydrothermal systems has been shown to occur from sub-micron sizes to macroscopic sizes in a continuous fashion. While agglomeration of crystals is known to occur, it does not appear to be a predominant growth mechanism, nor is it an essential feature of these systems. Assimilation of material from the solution phase has been speculated to involve "secondary building units", that is the myriad alumino-silicate oligomers known to exist in the solution. However, it has been argued convincingly that such relatively large units, while they do exist in the medium, probably have little to do with the actual growth of zeolite crystals, other than to provide a reservoir of material. It is more likely that the growth units are monomers, dimers, or other small alumino-silicate units which also are known to persist in these basic environments.

The addition of zeolite seed crystals to hydrothermal synthesis media have long been known to accelerate the crystallization process, and even direct the outcome of syntheses in certain circumstances. The mechanism by which this occurs has been shown to involve very small alumino-silicate fragments in the seed crystal sample, either actually adhering to the seed crystal surfaces, or simply co-existing in the sample. These "initial-bred nuclei", as they have been labeled, do not appear to prohibit the nucleation of zeolite crystals which would form in their absence in some cases. However, there are several examples reported in the open literature in which the phase formed by the unseeded solution did not form when seeds of another crystalline phase were added to the solution. An interpretation of these results is provided.

1.1 Background

Molecular sieve zeolites are crystalline alumino-silicates in which the aluminum atoms and the silicon atoms are present in the form of AlO_4 and SiO_4 tetrahedra. Consequently, the crystalline framework has net negative charge due to the presence of the alumina tetrahedra, which must be compensated by associated cations, e.g., Na^+, K^+, Ca^{2+}, H^+, NH_4^+, etc. The silica tetrahedra have no net charge, and, therefore, need not have any compensating cations associated with them. The alumina tetrahedra in the lattice must be adjacent to silica tetrahedra, while the silica tetrahedra may have adjacent alumina or silica tetrahedra as neighbors. The tetrahedra may be oriented in numerous arrangements, resulting in the possibility of forming some 800 crystalline structures, less than 200 of which have been found in natural deposits or synthesized in laboratories around the world. Synthetic zeolites are used commercially more often than mined natural zeolites, due to the purity of the crystalline products, the uniformity of particle sizes, which usually can be accomplished in manufacturing facilities, and the relative ease with which syntheses can be carried out using rather inexpensive starting materials.

The synthesis of most molecular sieve zeolites is carried out in batch systems, in which a caustic aluminate solution and a caustic silicate solution are mixed together, and the temperature held at some level above ambient (60–180 °C) at autogenous pressures for some period of time (hours-days). It is quite common for the original mixture to become somewhat viscous shortly after mixing, due to the formation of an amorphous phase, i.e., an amorphous alumino-silicate gel suspended in the basic medium. The viscous amorphous gel phase normally becomes less viscous as the temperature is raised, but this is not universally true, as in the case of some NH_4OH-based systems which remain viscous throughout the synthesis. The amorphous gel can be filtered from the solution and dehydrated by conventional drying methods.

As the synthesis proceeds at elevated temperature, zeolite crystals are formed by a nucleation step, and these zeolite nuclei then grow larger by assimilation of alumino-silicate material from the solution phase. Simultaneously, the amorphous gel phase dissolves to replenish the solution with alumino-silicate species. In short, the two phases have different solubilities, with the solubility of the amorphous gel being higher than that of the crystalline zeolite phase. Thus, during a zeolite synthesis, one might imagine that the alumino-silicate concentration in solution lies somewhere between the solubility levels of the gel and crystal phases, as shown along the vertical dashed line in Fig. 1. During the synthesis, then, the amorphous gel has a thermodynamic tendency to dissolve, while the thermodynamic driving force is toward formation of the crystalline zeolite

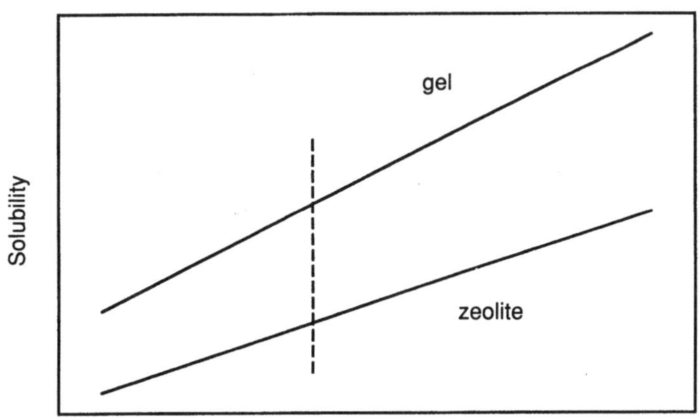

Fig. 1. Illustration of the classical thermodynamic driving force for zeolite crystallization. As crystallization occurs, the solution composition falls between the gel solubility and the crystal solubility. Zeolite crystal growth stops when sufficient material has been deposited to reduce the solution concentration to the zeolite "equilibrium" level

phase. The first step in this transformation process usually involves the formation of the smallest entity having the identity of the new crystalline phase, the crystal nucleus. That event is normally followed by the subsequent assimilation of mass from the solution and its reorientation into ordered crystalline material via crystal growth. The particular rates at which zeolite crystals form by nucleation, or grow instead of nucleating more new crystals, are more difficult to predict, however.

As a consequence of the transformation of amorphous gel to crystalline zeolite, by transport of material through the solution phase, the amount of zeolite relative to amorphous gel increases as the synthesis proceeds. In fact, if one either takes representative samples from a large batch system, or divides the large batch into smaller self-contained vessels to be quenched intermittently, the fraction of crystalline zeolite material in the solid sample (the remainder being the amorphous solid) normally increases slowly at first, then more rapidly, and finally slows down as reagents are depleted, giving a typical S-shaped profile when plotted as a function of time. Kerr [1] illustrated that, when plotted on semi-logarithmic coordinates, this "crystallization curve" increased linearly, characteristic of an autocatalytic process, then slowed down once the reagent supply became rate-limiting. A great deal more has been made of the "crystallization curve" than is warranted, since it has been shown [2] that it is impossible to generate information regarding zeolite crystallization mechanisms from it, in spite of many attempts to do so. Activation energies for "nucleation" and "growth," for example, based on analysis of the induction time and slope of the "crystallization curve" are almost completely unrelated to those processes, and, therefore, the numerical values obtained are all but meaningless.

1.2 Crystallization Mechanisms

Crystallization is conventionally agreed to proceed through two primary steps: nucleation of discrete particles of the new phase, and subsequent growth of those entities. (Agglomeration is viewed, perhaps naively, as undesirable, and, therefore, will not be dealt with to a great extent in this discussion.) The first, and most intriguing, process can be broken down further in the following way [3]:

Nucleation

1. Primary nucleation
   a) Homogeneous nucleation
   b) Heterogeneous nucleation
2. Secondary nucleation
   a) Initial breeding
   b) Micro-attrition
   c) Fluid shear-induced nucleation

Primary nucleation mechanisms occur in the absence of the desired crystalline phase, i.e., they are solution-driven mechanisms. In the case of homogeneous nucleation, the mechanism is purely solution-driven, while heterogeneous nucleation relies on the presence of extraneous surface to facilitate a solution-driven nucleation mechanism. The extraneous surface is thought to reduce the energy barrier required for the formation of the crystalline phase, but this mechanism has not received a great deal of rigorous study in the crystallization literature.

Secondary nucleation mechanisms require the desired crystalline phase to be present to catalyze a nucleation step. Initial breeding stems from microcrystalline "dust" being washed off the surface of seed crystals into the growth medium, thereby providing nuclei directly to the solution. In the absence of seed crystals added to the solution, however, agitation can sometimes promote nuclei formation by micro-attrition, i.e., by causing microcrystalline fragments to be broken off of existing growing crystals in the medium. These fragments arise from crystal contacts with the stirrer, other crystals, or the walls of the container, and may become growing entities in a supersaturated solution. Lastly, it has been speculated that nuclei can be created by fluid passing by the surface of a growing crystal with sufficient velocity to sweep away quasi-crystalline entities (clusters, embryos, ...) adjacent to the surface, which were about to become incorporated in the crystalline surface. If these clusters are swept away into a sufficiently supersaturated environment, they will have the thermodynamic tendency to grow, and become viable crystals. Thus, in the event that high shear fields in the neighborhood of growing crystal surfaces exist, nucleation can sometimes be promoted. Further details of crystal nucleation mechanisms, with numerous primary references, may be found in the text by Randolph and Larson [3].

A more detailed review of these mechanisms, and their relevance to zeolite crystallizations may be found elsewhere [4, 5]. Briefly, however, it is not expected that fluid shear-induced nucleation will be relevant to zeolite syntheses, due to the viscosity of the solutions, and because it is not believed to be important except at quite high agitation rates, or quite high fluid velocities relative to crystals in the medium [6]. Whereas many zeolite syntheses are carried out with no agitation, or very mild agitation, micro-attrition breeding also may be viewed as not universally important in zeolite crystallizations (systems using intense agitation being the exception). Thus, in this review, those mechanisms will be understood to be relevant only in special circumstances.

One should understand the differences between secondary nucleation and seeding, i.e., the common strategy of promoting the "rate of crystallization" by adding crystals of the desired phase to a precipitating system. Secondary nucleation is, strictly speaking, the promotion of crystal nucleation due to the physical presence of crystals of the desired phase, while seed crystals may promote crystallization by providing additional surface area for dissolved material to grow onto. However, a seed crystal sample may contain sub-micron-sized fragments which eventually grow to macroscopic sizes, that is, in some cases it may appear that a newly formed population was created, when, in fact, it was the result of growth on very small seed crystal pieces.

It is tempting to simply refer the reader to prior analyses in which convincing arguments have been made quite eloquently, and with adequate references, e.g., [7–10], rather than attempt to restate what has been said previously. Therefore, while the reader will most definitely benefit from reading those, and other, prior works, it is hoped that some new insights and interpretations of existing data may be provided in this chapter.

2
Thermodynamic Considerations

Prior to discussing the kinetics of zeolite crystal nucleation and growth it is beneficial to consider several thermodynamic aspects which have bearing on the phase transformation process. Reports on this topic are not abundant in the zeolite crystallization literature, but several papers will serve to illustrate various issues which should be important in these systems.

Lowe et al. [11–13] have considered the change of pH during the synthesis of high-silica zeolites, EU-1 and ZSM-5, in hydrothermal systems, and developed a model to interpret these changes. In the first of these papers [11], it was demonstrated that, during the synthesis of EU-1, the pH of the solution increased at about the same time that the crystallization curve began to show that a significant level of crystalline material was forming. The suggestion was made that measuring the pH of the synthesis solution would be a reasonable way to monitor a zeolite synthesis in progress, since it was quick, easy, and did not require that the crystals be separated from the mother liquor, washed dried, or handled in any special way. However, changes in pH were not significant during the early stages of the process.

In the second report [12], an equilibrium model was developed which accounted for changes in pH during zeolite crystallization. It was demonstrated that the largest pH changes would be expected to be associated with the most stable zeolite produced. Yields of specific phases were shown to be dependent on the starting batch composition, and especially the amount of base in the batch. Further, it was shown that pH changes were smaller for systems buffered by amines, and that yields would be expected to be higher in those systems. It also was noted that the pH of the synthesis solution should be governed by the solubility of the most soluble solid present in the system, and that, therefore, the pH would be expected to remain essentially constant until the amorphous gel had dissolved. Thus, the increase in pH should mark the nearly complete conversion of amorphous gel to substantial amounts of crystalline zeolite. This prediction was corroborated by the prior results with EU-1 [11]. Clearly, these changes in the solution, occurring predominantly late in the synthesis, should not provide information about the nucleation behavior.

A comprehensive evaluation of the effects of alkalinity on the synthesis of silicalite-1 (aluminum-free ZSM 5) also was conducted [13]. The study was carried out using the batch composition:

$$x\ Na_2O : 2\ TPABr : 20\ SiO_2 : 1000\ H_2O$$

where TPABr represents tetrapropylammonium bromide, and the alkalinity, x, was varied between 0.25 and 6.5 moles. The initial pH of the solution was increased with increasingly higher values of x. Syntheses were carried out at 95 °C, and were evaluated by changes in pH (measured at ambient conditions), powder X-ray diffraction, and electron microscopy.

Changes in the pH of the mother liquor during the syntheses were demonstrated to change in a systematic way depending on the starting alkalinity, x. Unlike in the previous studies [11, 12] there were occasions in which the pH decreased during synthesis rather than increased, and it was predicted, by interpolation of the results, that at a starting level of $x = 0.75$ there would be no change in pH during the synthesis. At the highest base level used, $x = 6.5$, the amorphous gel dissolved "completely," and precipitation of silicalite-1 occurred from the clear solution rather slowly. It also was noted that the highest yield of silicalite-1 was obtained at the lowest value of x, and that the yield decreased with increased alkalinity levels. Extrapolation of the data indicated that at $x = 6.7$ the yield would fall to zero. Therefore, the solubility of both the amorphous gel and silicalite-1 increased with increasing alkalinity, and the thermodynamic yield decreased accordingly. It is noteworthy that even though the authors later showed that the final crystal size became smaller as the alkalinity was increased (their Fig. 7), that result could very well have been due to the combined effects of reduced yield and enhanced nucleation.

The presence of silicate ions in solution buffers the solution during much of the synthesis. Near the end of the synthesis, when the silicate ion concentration begins to decrease, the buffering capacity decreases, and the pH rises because there is a smaller rate of decrease of the base concentration in the solution, since relatively small amounts of base are incorporated into the crystalline phase. Synthesis solutions with lower initial alkalinities have a lower buffering capacity to compensate for the loss of base from the solution during synthesis due to the lower concentration of silicate ions in solution. Therefore, in those systems the pH decreased in the early stages of synthesis, followed by a rapid increase in pH, due to the same changes noted for the systems with higher base content. For the system with $x = 0.25$ the removal of base from the solution had a dominant effect in reducing the pH, but the unusually low final pH value (ca. 8.3) was attributed to incorporation of CO_2 from air.

The rate of formation of zeolite mass was correlated with the slope of the curve expressing the percent zeolite in the solid phase against time. The rates estimated this way increased with increasing values of x, and then approached a constant. While the rates appeared to become essentially constant at higher values of x, because the yield decreased at high x values, there actually was a maximum in the growth of zeolite mass at around $x = 3$. The reason for the optimum was explained to be the low concentration of silicate oligomers at low alkalinities, and the high solubility of the zeolite phase at high alkalinities.

3 Nucleation

As previously noted, most zeolite syntheses of commercial value occur in systems clouded with an amorphous gel phase due to higher product yields, admitting to the possibility of homogeneous nucleation due to solubility differences, or to heterogeneous nucleation due to the abundance of foreign surface in the medium. Seeding these mixtures, or agitating the solutions, could induce nucleation by any of the secondary nucleation mechanisms. However, zeolite syntheses also have been conducted successfully in dilute clear alumino-silicate media, i.e., in the absence of any amorphous gel phase [14–26]. In fact, one of the early papers by Kerr [1] reported on a technique whereby dried gel was placed on a filter membrane, and hot caustic solution was circulated over it to induce crystallization on a second filter membrane connected by a pump. A second pump recirculated the filtrate back to the dried gel on the first membrane to continue the process. A "clear" solution is only clear insofar as the technique used to monitor the solution (the naked eye, laser light scattering, small angle neutron scattering, etc.). Kerr's conclusion that the experiment proved that zeolite crystallization occurred from the solution phase must be accepted in the context of the filter membranes used in the experiments, since some colloidal material may have passed through. This point will be revisited below.

Recently, several clear solution syntheses have been monitored by quasi-elastic laser light scattering spectroscopy (QELSS) techniques [19–26], which have demonstrated that the solutions contained essentially no colloidal material prior to the onset of nucleation, at least not present in sufficient concentration, or of sufficient size, to be observed by the light scattering techniques. In one report, the solution was concentrated at early times [22], and no mention was made of amorphous material being present prior to the onset of crystal growth. Therefore, the evidence from these reports suggests that zeolite nucleation may be driven purely from dissolved species present in the liquid phase, even though other mechanisms also may be important in more concentrated systems. Thus, it is tempting, from the evidence cited, to assert that the fundamental zeolite nucleation mechanism involves species coming together in solution to create a metastable entity, which grows spontaneously after reaching a critical size, very much in the classical way. While the clear solution systems may not have much commercial significance, they are informative from a fundamental perspective in revealing information regarding mechanisms of nucleation and growth. However, any analysis of zeolite nucleation in hydrothermal systems must consider nucleation events in all of the media noted, as well as by the more recent analytical techniques used to evaluate particles in "clear" solutions, discussed below.

Consider the results from Zhdanov et al. reproduced in Fig. 2 [27], which were reported previously by Zhdanov and Samulevich [28], based on a technique reported by Zhdanov in 1971 [29]. In that figure, the apparent nucleation history of the synthesis system was determined by monitoring the growth of several of the largest crystals in the system over time, determining the crystal size distribution of the final crystalline zeolite product, and using both sets of data to estimate when each class of particles had been nucleated during the synthesis. The same technique has been used by others [7, 30–32] with very similar results.

The results in Fig. 2 indicate that nucleation began after some time had passed, most likely due to a transient heat-up time and some time required for dissolution of the amorphous gel to achieve some threshold concentration. However, it is most noteworthy that the nucleation event in zeolite crystallization systems always has been determined to have ended when only about 10–15% of the alumino-silicate material had been consumed. That is, it is remarkable that with 85–90% of the alumino-silicate reagents left in the system, the nucleation process was somehow caused to cease, while crystal growth proceeded for the duration of the synthesis. This must be noted in the context of the amorphous gel dissolving sufficiently fast that the solution phase concentration was essentially constant up to almost 80% conversion in some cases [33–35],

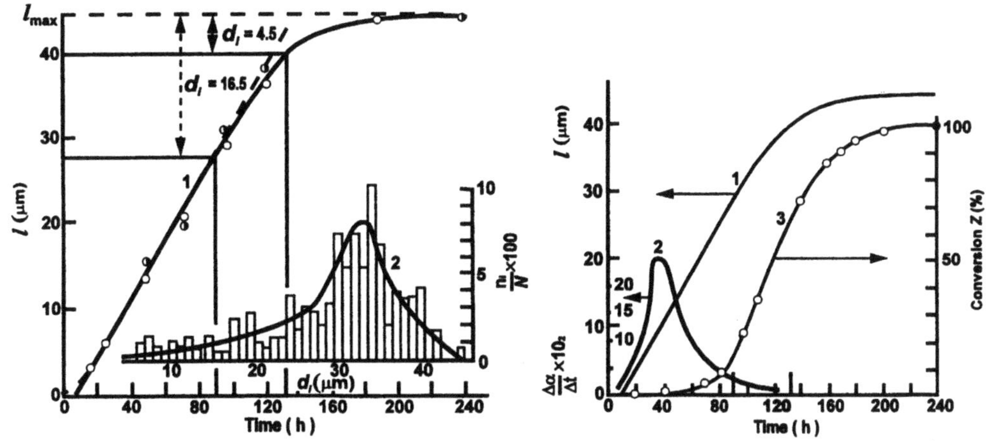

Fig. 2. **a** Linear crystal dimension of the largest crystals of zeolite NaX (curve *1*), and the crystal size distribution of the final product (curve *2*). **b** The crystal dimension of the largest crystals (curve *1*), the apparent nucleation rate derived from the curves in a (curve *2*), and the calculated and measured degree of crystallinity for the NaX synthesis. Figure redrawn with permission from [27]. Original data and computational technique reported in [28]

which means that the driving force measured in terms of concentration really did not change very much at all. Figure 3 shows the results of a population balance analysis (based on the development in [36]) of a silicalite synthesis carried out by Golemme et al. [32], which indicates that the classical homogeneous (or heterogeneous) nucleation mechanism for that crystallization did not represent the nucleation profile at all, even though the "crystallization curve" was fit very well. The predictions of the classical homogeneous nucleation theory are that crystal nucleation should have occurred over a much longer time than observed, because of the relatively constant supersaturation, or driving force for nucleation. However, we should bear in mind that the concept of a "supersaturation" in zeolite synthesis solutions is rather ambiguous, since the concentration of more than a single species is normally involved, and changes in these concentrations as the synthesis proceeds may be affected by the framework Si/Al ratio, and pH changes during synthesis. Furthermore, making note of silicate and aluminate concentrations in solutions may be inadequate to describe the driving force for zeolite nucleation and growth, since silicate ions in basic solutions form myriad oligomers, and even more complex structures with aluminate ions. Therefore, the notion of a "supersaturation" which can be correlated with nucleation and crystal growth rates may be superficial at best.

Thus, one has the dilemma of explaining why the nucleation process should cease in the course of a typical hydrothermal zeolite synthesis when there appears to be an abundance of material remaining in the system, ca. 85–90%, from which nucleation could be sustained. Additionally, solutions of mathematical models based on fundamental principles [36] have demonstrated that nucleation should continue over a much longer time than observed, if classical nucleation concepts applied to these systems.

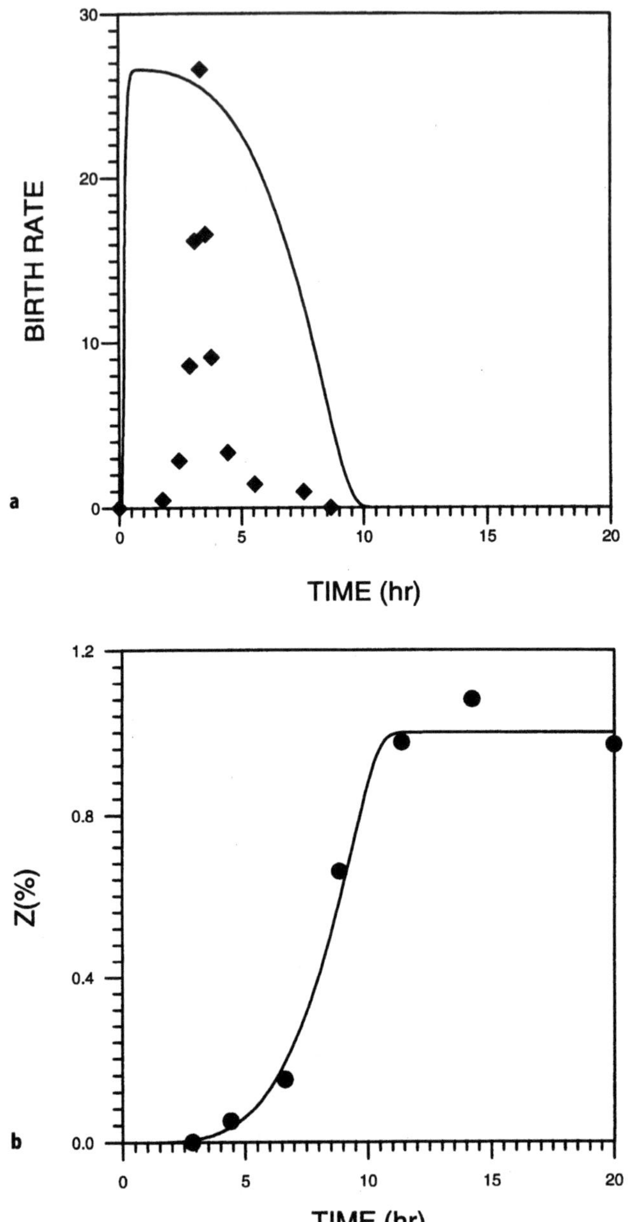

Fig. 3. **a** Predicted (curve) and experimental (points) values of the nucleations rate vs. time. Theoretical values based the classical homogeneous nucleation model and the population balance model developed in [36]. Data for silicalite synthesis replotted from [32]. **b** Predicted (curve) and experimental (points) values of the per cent zeolite in the solid phase. Model calculations and data from same sources as **a**

It also should be mentioned at this point that, while there are known to be myriad oligomeric species which exist in caustic silicate (or alumino-silicate) solutions [37], their rearrangement to the equilibrium distribution of oligomers occurs in seconds [38] or milliseconds [9]. Equilibration of oligomeric species was noted to be rapid at room temperature, and was even faster at elevated temperatures [38]. This observation means that the rate-limiting step in the nucleation process may not be the build-up of sufficiently large structures (at least relative to many simpler inorganic salts, for example) to create metastable nuclei, as the classical nucleation mechanism would suggest, since that process would be expected to occur rapidly.

In a series of papers [32, 34, 35, 39–41] Subotic and his co-workers have proposed and studied a so-called autocatalytic nucleation mechanism. According to their concept, suggested by the proposal of Zhdanov somewhat earlier [29], there are three possible sources of nuclei: homogeneous nuclei formed by a classical mechanism, heterogeneous nuclei formed in conjunction with foreign particulate matter, and autocatalytic nuclei from within the amorphous gel phase (although later in the series of papers the homogeneous nucleation mechanism seems to have been rejected as unimportant). According to the conceptual model, the autocatalytic nuclei lie dormant in the amorphous gel phase until they are released into the solution by dissolution of the gel phase and become active growing crystals. As the cumulative zeolite crystal surface area increases due to crystal growth, the rate of solute consumption increases, which, in turn, increases the rate of gel dissolution, resulting in increased rate of dormant nuclei activation, etc. It is clear why the mechanism was labeled autocatalytic.

Mathematical models to describe the crystallization process in batch systems, assuming uniform distribution of the autocatalytic nuclei in the amorphous gel particles, were developed and solved. However, it was shown recently [36] that the model predicts that nucleation should continue much longer in the process than has been observed in several studies previously (e.g., [28, 30–32]). It was shown to be more realistic, therefore, to assume that the dormant nuclei were located preferentially near the outer edges of the gel particles, and, thus, became activated much earlier in the process. Alternatively, there may be some other mechanism by which dormant nuclei are activated as the process evolves, which will be discussed below.

Following upon the pioneering work of Freund [42] and Lowe et al. [43], Hamilton et al. [44] reported on a study in which several different powdered silica sources were used in the synthesis of molecular sieve zeolite NaX. In the two previous works [42, 43] the authors had determined that "active silicates" were those which had relatively high levels of aluminate impurities. The more recent study sought to correlate "active silicates" with those from which an increased number of nuclei formed in the hydrothermal system, that is, the "activity" was associated with the number of crystals which were formed, which added to the cumulative crystal surface area available to assimilate material from solution. In that study, the batch composition was the same in all experiments, and given by:

$$4.76\ Na_2O : Al_2O_3 : 3.5\ SiO_2 : 454\ H_2O : 2.0\ TEA$$

where TEA represented triethanolamine used to stabilize the sodium aluminate solution and produce slightly larger particles than otherwise would form. All syntheses were carried out in Teflon-lined autoclaves at 115 °C and autogenous pressure. The sodium aluminate solution used for each experiment was from the same preparation, while the various silicate solutions all had the same composition and pH. The aluminate solution and all of the silicate solutions were clear, and, additionally, filtered through Gelman membrane filters to remove any particulate matter larger than 0.20 mm in dimension. While 0.20 mm is large compared to the size of crystal nuclei, and material smaller than 0.20 mm could have served as heterogeneous nuclei, light scattering analyses of several filtered solutions failed to correlate nuclei formation with particulate material in the filtrate. The amounts of each solution added to each final mixture were the same.

The rather startling results obtained, in spite of everything being identical, except the source of the silica powders, were that the synthesis times for each experiment were quite different and the ultimate particle sizes from each solution were quite different. Some of their results are summarized in Table 1, but the original paper contains more results and more details [44]. Each system created different numbers of nuclei, which consumed material from the solution at different rates, due to the different cumulative surface areas, and, therefore, converted the amorphous gel to crystalline zeolite NaX in different time periods. The results were interpreted in terms of inherent differences in the silicate solutions formed from the various silica powders, because all the silica powders were completely dissolved and filtered prior to combining with the aluminate solutions. At the time, the strongest correlation to explain the results appeared to be with the impurity levels contained in the silica powders, however it also was noted that the correlation of the number of nuclei formed with impurity levels was equally good with Al^{3+}, Fe^{3+}, Mg^{2+}, or Ca^{2+}. Similar impurities added to the silicate solutions did not have any observable effect on the outcome. At that time no convincing argument could be found persuasive to identify any particular impurity as the key ingredient in promoting nucleation in that system. In fact, it is possible that some other impurity or ingredient in the silica powders was essential in that role. This conclusion is different from those in the papers of

Table 1. Selected results on the effect of silica sources on the crystallization of zeolite NaX [44]

Silica source	Maximum crystal size [μm]	Synthesis time [days]
$Na_2SiO_3 \cdot 9H_2O$	7.5	1.0
$Na_2SiO_3 \cdot 0H_2O$	20	1.0
$Na_2SiO_3 \cdot 5H_2O$	50	3.0
Cab-O-Sil	85	4.0
Silicic Acid	90	4.0

Product using Cab-O-Sil contained 50% zeolite NaX and 50% chabazite.
Product using Silicic contained 40% zeolite NaX and 60% chabazite.
Particle sizes reported are only the zeolite NaX component.

Freund [42] and Lowe et al. [43] in which the "activity" of the silica powders was concluded to reside with aluminate impurities in the silica powders.

3.1 Clear Solution Studies

There have been recent reports of clear solution syntheses of zeolites which were monitored by in-situ methods, either optical microscopy or quasi-elastic laser light scattering spectroscopy, QELSS [19–26]. In each of the cases to be discussed [7, 10, 22, 25], silicalite, or aluminum-free ZSM-5, was the zeolite of study, therefore, valid comparisons can be made. It will be insightful to consider the results of these studies both in relation to nucleation mechanisms and crystal growth mechanisms. Table 2 contains summary information on the synthesis conditions and selected results from the studies.

In view of the similarities in these studies, it is interesting to put two items in perspective, using the results of Twomey et al. [25] in this example. The batch composition used in that study is shown in Table 2 and illustrates that the batch system contained four times the stoichiometric amount of TPA^+ required to fill the pores at complete conversion, so it was not a limiting species. Both dynamic and static light scattering techniques were utilized to monitor the progress of the synthesis in-situ. The fact that the particle size distribution was very narrow, that is, that the crystals were all about the same size, permitted the determination of the total number of crystals during the experiments using static light scattering data. The number of crystals remained essentially constant during each experiment, suggesting that one nucleation event occurred in most experiments.

The results of those experiments indicated that there was a lag time, or induction time, prior to the onset of crystal formation, that fewer nuclei were formed at higher temperatures, but in much shorter times, and that crystal growth was essentially constant during each experiment as long as the observations could be made with the light scattering system. It also was demonstrated that the crystal size distribution was quite narrow compared to the size distribution usually obtained from more concentrated systems containing amorphous gel. These results were consistent with several of the observations of the other groups as

Table 2. Synthesis conditions for clear solution Al-free ZSM-5 experiments

Reference	Batch compositions	Temperature [°C]	Linear growth rate [μm h^{-1}]
[7]	Na_2O : 20 SiO_2 : 2 TPABr : 1960 H_2O : 80 EtOH	95	0.038–0.04
[10]	Na_2O : 60 SiO_2 : 3 TPABr : 1500 H_2O : 240 EtOH	95	0.022
[22]	0.1 Na_2O : 25 SiO_2 : 9 TPAOH : 480 H_2O : 100 EtOH	98	0.00379
[22]	same as above, but with 1500 H_2O	98	0.0101
[25]	Na_2O : 25 SiO_2 : 9 TPAOH : 450 H_2O	96	0.036
[31]	Na_2O : 38.5 SiO_2 : 3.8 TPABr : 954 H_2O	150	0.81

well [7, 10, 22]. In particular, Schoeman et al. [22] also observed that the total number of crystals formed decreased with increasing temperature, that the induction time decreased with increasing temperature, that the linear growth rate increased with temperature, and that the crystal growth rate was constant during each experiment.

The fact that a very narrow crystal size distribution was formed permits one to assume that nucleation occurred as a single event, starting and ending rather abruptly, causing a shower of nuclei to be formed, and that they grew uniformly from that moment. If one presumes that the nuclei themselves were on the order of 50 Å in diameter (approximately the detection limit of the instrument), and that the final crystal size of the silicalite particles was 0.95 mm (as observed), then one can estimate that the nucleation event consumed 1.46×10^{-5}% of the silica finally incorporated into each particle, or 1.46×10^{-7}, expressed as a fraction. This small fraction represents an imperceptible reduction in the silica present in the solution, and could not be modeled with reasonable values of the activation energy and frequency factor for classical nucleation [45].

Second, considering that on the order of 10^{12} particles cm^{-3} were nucleated by this spontaneous nucleation event, the density of nuclei can be estimated. In that case, assuming that nucleation occurred throughout the medium uniformly, the volume of the medium (i.e., 1 cm^3) can be imagined to be divided into 10^{12} separate boxes having volumes of 1 μm^3. That means that, on average, each new nucleus occupied a volume of 1 μm^3, i.e., a box 1 µm on a side, which, in turn, means that each nucleus was, on average, 1 µm away from its nearest neighbor. The chance that their adjacent diffusion fields should interact or affect the growth of their neighbors during the very early stages is quite low, due to the comparatively large distance between the new growing centers. In other experiments reported, where fewer particles were nucleated, the distance between growth centers would be greater, and vice versa.

These two simple calculations illustrate the dilemma regarding why nucleation in these systems ceases. The nucleated growth centers were relatively far apart, and their formation should not noticeably have changed the concentration of the limiting material in the medium which, according to classical considerations, should have promoted nucleation, i.e., the silicate anions (since the tetrapropyl ammonium ions were present in excess). Recalling that changes in the silicate anion oligomer distribution in the solution due to the onset of nucleation should be momentary, at best, since rearrangements occur in seconds [9, 38], the cause of the onset of nucleation and its cessation needs to be investigated further. Let us consider the possibility that some other limiting reagent may be involved.

It has long been known that adding triethanolamine (TEA) to zeolite NaA or NaX systems results in larger crystal formation, due to the reduction in nucleation [46–54]. It has been suspected that the reduction in nucleation is due to the fact that the TEA complexes with free aluminate anions in the solution [52–54], thereby reducing the concentration of species which could otherwise participate in crystal nucleation. However, it also has been reported that the TEA can complex with Fe^{3+}, effectively reducing the amount of iron incorporated into the crystals [55]. This observation illustrates the fact that additives, TEA for example, can interact with numerous species in synthesis solutions, some of which may be important, while others may not be. Therefore, if a limiting reagent existed in those systems, which affected the nucleation behavior, it was not the SiO_2 species, since TEA was shown to not interact with them [53], but could have been the aluminate present, or any of several trivalent T-atom species, present as impurities in the reagents. It is clear that additions of triethanolamine resulted in the reduction of the number of nuclei formed in the high aluminate synthesis solutions, i. e., for zeolites NaA and NaX, and that TEA complexed reversibly with aluminate species. It also is apparent that TEA complexed with other species, but not silicates. It is not obvious from the results reported that aluminate species necessarily were the limiting reagent responsible for nucleation. However, it is difficult to determine what other species might have been the key limiting ingredient, because of the abundance of aluminate in those systems, and the fact that the ^{13}C NMR spectra had no peaks other than those associated with the TEA-aluminate complexes. That is, any other species which might have been complexed with the triethanolamine were in sufficiently low concentration that they could not be observed in the NMR spectra.

The relevance of the results of Freund [42], Lowe et al. [43] and Hamilton et al. [44] to this discussion should not be forgotten. That is, the onset and cessation of nucleation may be coupled with other materials in the starting reagents, e.g., the silica sources, and may be associated with impurities unavoidably present.

Two recent reports have been published which shed new light on the possible structure of precursors to zeolite nucleation and crystal growth. The first of these [56] reported on ^{1}H-^{29}Si and ^{1}H-^{13}C cross-polarization MAS-NMR observations of a pure-silica ZSM-5 synthesis mixture (0.5 TPA_2O : 3 Na_2O : 10 SiO_2: 2.5 D_2SO_4 : 380 D_2O; 110 °C). The results of that study revealed that TPA-silicate structures form prior to the formation of observable long-range crystalline structure, and have short-range interactions on the order of 3.3 Å, indicative of van der Waals interactions. The proposed structure for these inorganic-organic entities, and their role in the synthesis process are shown in Fig. 4. The authors argued that the observed layered intergrowth behavior noted in several high silica zeolite systems (e.g., ZSM-5/ZSM-11, beta, etc.) supported the hypothesized model of nucleation and growth by the TPA-silicate species suggested by their results.

An excess of 2.4 times the maximum amount of TPA that could be occluded in the final product was used in the syntheses noted above [56]. In the first day after heating there was evidence of both Q_3 and Q_4 silicate interactions (where Q_n species are those tetrahedrally coordinated Si atoms having n bonded SiO_4 neighbors). However, even after two days of heating the $^{1}H-^{13}C$ CP MAS-NMR spectra suggested that a small fraction of the TPA was associated with the silicate species, as noted by the small peak at 10.1 ppm in their Fig. 7c. Additionally, comparing the results of Figs. 5B and 5D, reproduced from their Figs. 8b and 8d, one notes that the amount of TPA^+ per solid is much smaller in the 1-day heated sample than in the final product, as pointed out by the authors. If all of the silica had been associated with TPA-silicate structures of the type described in Fig. 4, the signal in Fig. 5B probably would have been sharper, because the

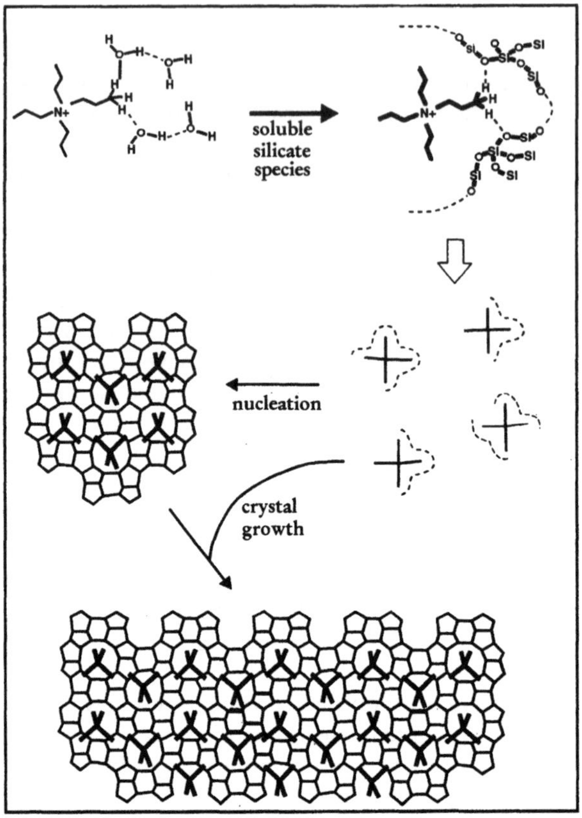

Fig. 4. Schematic illustration of the proposed conceptual model for the TPA-facilitated nucleation and crystal growth of all-silica ZSM-5. Figure redrawn with permission from [56]

stoichiometric amount of TPA^+ would have been present in the sample, as it was in Fig. 5D. And, if the stoichiometric amount of TPA^+ had been incorporated in those structures, then as much as 42% (based on the excess TPA^+ used by the authors) would have been associated with these structures; their Fig. 7b, c do not appear to bear this out. Taking into consideration the previous observation that silicate species, up to groups of 12 T-atoms, were shown to re-equilibrate in seconds [38], or milliseconds [9], and that the authors indicated that the observed structures were perhaps as large as 24 T-atoms, it would appear that some silicate species, and even more TPA^+ (due to the excess), were not associated with the structures proposed by the authors. This conclusion would lead one to admit to the possibility that nucleation of the ZSM-5 structure might involve some of these inorganic-organic species, or perhaps other species not associated with the TPA-silicate entities.

It is possible that the inorganic-organic structures noted by the results after heating for 1 day were, in fact, the nuclei, or small fragments of crystalline material, too small to be detected by X-ray diffraction (i.e., smaller than about

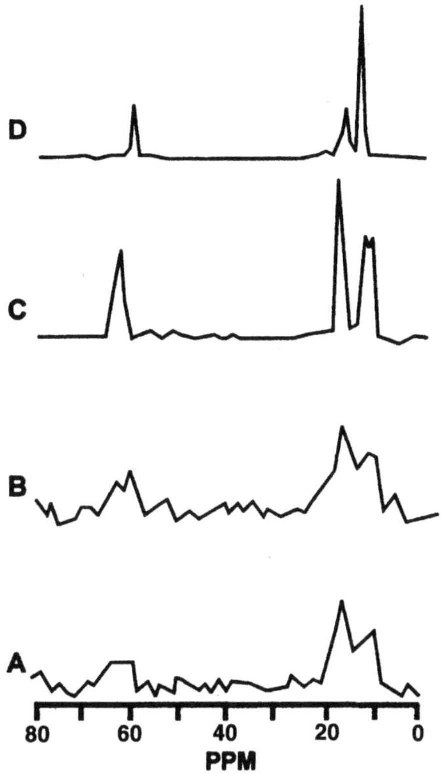

Fig. 5. ^{1}H-^{13}C CP MAS NMR spectra of freeze-dried and washed samples from the TPA-facilitated all-silica ZSM-5 synthesis. **A** Unheated amorphous gel, **B** gel heated 1 day at 110 °C, **C** TPA trapped in all-silica ZSM-5, and **D** pure TPABr. Figure redrawn with permission from [56]

80–100 Å, noted as the detection limit by the authors). TPA^+ was present in excess in these experiments [56], as in the previous study discussed [25], and it, therefore, was not a limiting reagent. One has the same question in this case, then, of why only a fraction of the TPA^+ would be expected to participate in nucleation. However, it is worth considering that the proposed structures, proven to exist for the first time by these authors [56], could be participants in zeolite crystal nucleation and growth.

The second recent work which must be mentioned is the in-situ combined small-angle X-ray scattering/wide-angle X-ray scattering (SAXS/WAXS) monitoring of an all-silica ZSM-5 crystallization [57, 58]. The combined technique allows one to simultaneously observe particles in the system, to determine their fractal dimension, and to determine the level of crystallinity within the particles, and the crystalline phase(s) present. Based on the data collected, some of which is reproduced in Fig. 6, the authors proposed the nucleation mechanism depicted in Fig. 7. In essence, the authors suggested that primary silica particles formed quite early in the process, perhaps of the nature described by

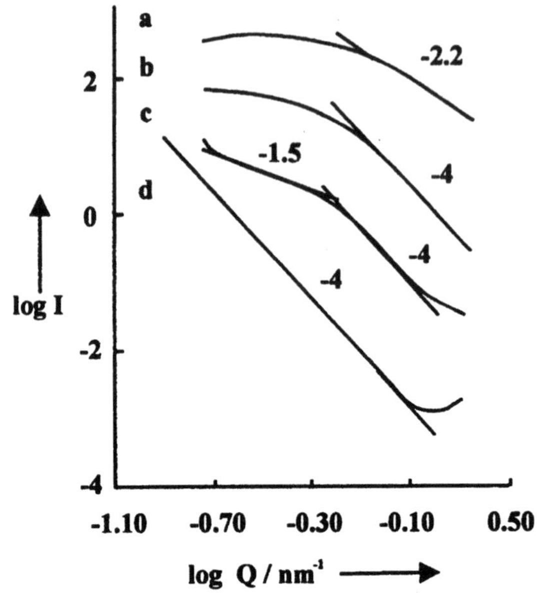

Fig. 6. Plot of log I versus log Q from the small-angle X-ray scattering spectra of a clear silicalite synthesis mixture after various reaction times: *a* 5 minutes, *b* 35 minutes, *c* 75 minutes, and *d* 105 minutes. Figure redrawn with permission from [58]

Burkett and Davis [56], which then underwent a series of aggregation and densification steps to ultimately form growing crystals of ZSM-5. The initial primary particles, were proposed to aggregate into clusters having surface fractal dimension with slope of −2.2, corresponding to a fairly open aggregate, as depicted by Fig. 7b. Densification and subsequent aggregation of those densified aggregates ultimately led to crystalline mass, and crystal growth occurred in the normal way. At this time, it is not clear why the densification occurs in this way, or what mechanism of re-orientation occurs within the amorphous particles to initiate crystal formation.

Cundy et al. [7] also proposed that silicalite nucleation occurred on, or "in," amorphous gel "rafts." The evidence for their proposed mechanism was the observation that samples taken at early times contained a proportion of amorphous material, and that optical and electron microscopy indicated a close association of new zeolite crystals and these amorphous particles. The authors concluded that the initial nucleation period was due to a heterogeneous nucleation mechanism, and arose from the presence of macroscopic or colloidal particles in the solution. Nucleation was thought to be a surface-facilitated phenomenon. While their proposed mechanism appears to be slightly different than that of Doktor et al. [57, 58], it nonetheless involved the participation of extraneous material.

Certainly one curious factor in establishing these observations as a new proposed nucleation mechanism is that such small particulates were not observed (or at least reported) by Schoeman et al. [19–24] or by Twomey et al. [25] using

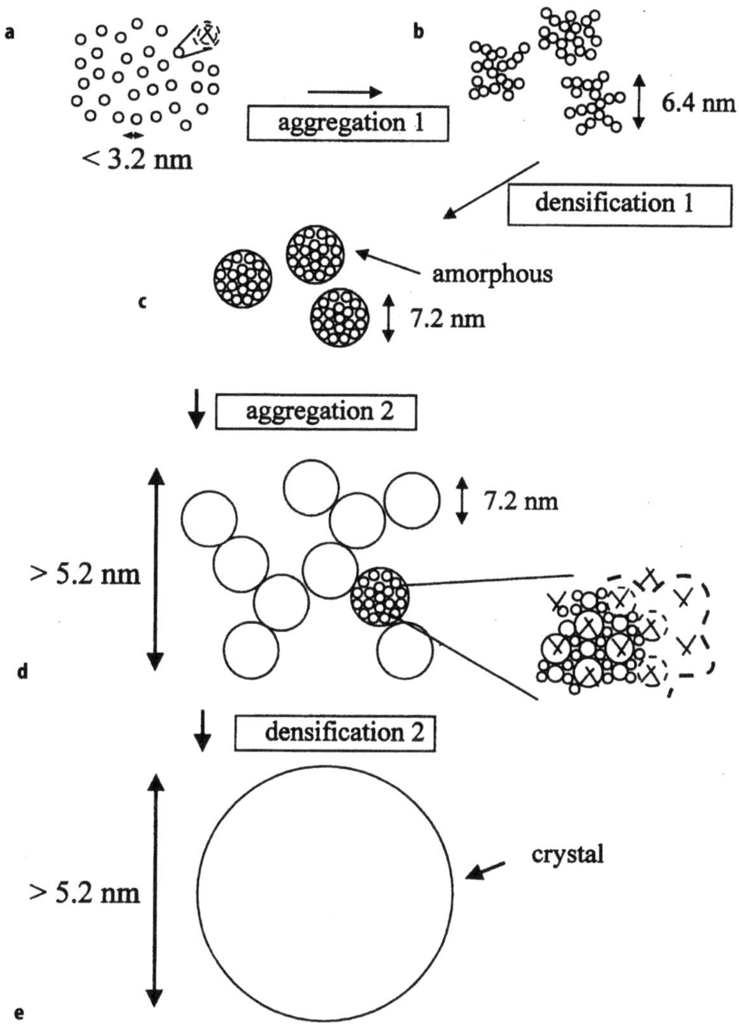

Fig. 7. Schematic illustration of the model for nucleation of silicalite from clear synthesis mixtures: **a** TPA-silicate clusters in solution, **b** primary fractal aggregates formed from the TPA-silicate clusters, **c** densification of the fractal aggregates from **b** above, **d** combination of densified aggregates into a second fractal aggregate structure, and **e** densification of the second fractal aggregates followed by crystal growth. Figure redrawn with permission from [58]

QELSS. This is especially curious in view of the detection limits of the facilities, and the fact that in at least one work [22] samples collected at early times were concentrated by centrifuging prior to analysis by light scattering. The absence of particulates either reflects the fact that there were none present in those solutions, or their size or concentration were too small to be detected.

It should be mentioned, however, that recently nanometer-sized particles have been observed in a clear solution of the zeolite NaA system by quasi-elastic laser light scattering spectroscopy [26]. It is preliminary to give much detail here, but the primary particles appeared to be approximately 1 nm in dimension, formed agglomerates of approximately 160 and 300 nm in size, and were observed in various silicate solutions prior to combining them with their corresponding aluminate solutions.

In view of these recent observations by CP-MAS-NMR, SAXS/WAXS and QELSS it is now possible to suggest that one interpretation of the hypotheses of Subotic et al. [32, 34, 35, 39–41] is that the "autocatalytic nuclei" which they have discussed previously are formed in the manner described by Doktor et al. [57, 58]. These nuclei were said to form more slowly in gel systems, due to the fact that the gel must first dissolve to form the precursor aggregates. This process could occur over a longer time period in gel systems than in clear systems, giving the appearance that "nuclei" were "popping out of the gel" as conversion of gel proceeded.

This discussion of zeolite nucleation would be incomplete without mentioning that the nucleation of zeolite crystals was suggested to occur from clear liquids via amorphous lamellae by Aiello et al. in 1970 [59], i. e., 28 years ago. The first evidence for the fact that these primary particles were amorphous was that they seemed buoyant at early times, and moved by convection, while the particles settled later in the synthesis, suggesting a change in the mass density of the particles. Electron diffraction of single particles, as well as electron microscopy, also supported the concept of zeolite crystal nucleation occurring within the amorphous lamellae.

4 Zeolite Crystal Growth

As early as the 1971 meeting of the International Zeolite Association in Worcester, Massachusetts, Zhdanov [29] reported on measurements of zeolite crystal growth in hydrothermal systems. His observations for a zeolite NaA system were that the crystal growth rate was constant for some rather long period of time, and eventually slowed down as the reagent supply became depleted. That observation was made at several temperatures, and further demonstrated that the growth rate of zeolite crystals in these systems was independent of crystal size, at least from as small sizes as could be measured by optical microscopy. In the often-cited paper by Zhdanov and Samulevich [28] they extended the technique to include a method by which the crystal growth rate and final product size distribution could be used to estimate the nucleation rate for the system. The technique was summarized by Barrer [27]. Several other research groups have used the technique since then [7, 30–32, 34], and in all cases the zeolite crystal growth rates have been reported to be constant during the early portion of the crystallization process. Crystal growth rates also have been observed to be independent of crystal size by laser light scattering techniques [19–26] for several different zeolite systems, in the nanometer size range.

Zeolite crystal growth from solution occurs by transfer of material from the solution phase, in which the solute has three dimensional mobility, to the surface of the crystal lattice being formed, and incorporation thereon in a regularly ordered framework. Thus, individual species must diffuse to the crystal surface, and then be incorporated into that crystalline structure for growth to occur, as measured by the advancement of the crystal faces, or the increase in the crystal dimensions. Consequently, it is possible that either solute diffusion or surface kinetics may be rate controlling, or they may both be of comparable magnitude. In view of the constant crystal growth rates observed throughout the literature, it might be tempting to assume that solute diffusion was the rate-limiting step, but this assumption has not been born out by experimental results, as will be detailed below.

There are two issues which are important to understand the mechanisms of zeolite crystal growth, and yet a third issue which deserves comment, those being:

1. Is either diffusion or surface kinetics the rate-limiting step to zeolite growth, or are both steps of comparable rate?
2. What is the unit, among the myriad species present in these solutions [37,38], which is responsible for growth, that is, what is the species (if it is but one) which is incorporated at the surface?
3. Why does aging the alumino-silicate solution at room temperature prior to synthesis appear to increase the inherent growth rate of zeolite crystallites?

Table 2 shows data from several sources in which the individual linear crystal growth rates for Al-free ZSM-5, or silicalite, were reported. The temperatures used in each study were very similar, except in [31], and the linear growth rates also were quite similar, except in the case of one of the systems used in [22]. The differences in that work were attributed to different synthesis conditions compared to that in [7], most notably the higher pH of ca. 12.5 in [22] compared to 10.6–11.6 in [7]. The growth rates reported for silicalite in [25] were almost identical to those found in [7], in spite of the fact that ethanol was used in one study, but not the other. Due to the similarities of the batch compositions in [22] and [25], except for ethanol, one would expect that the growth rates might be similar rather than different by an order of magnitude.

Table 3 contains values of the activation energy for linear zeolite crystal growth for several zeolite synthesis systems. The zeolite crystal growth rates

Table 3. Linear crystal growth rate activation energies

Reference	Zeolite system	Activation energy [kJ mol^{-1}]
[10]	silicalite	79 (length), 62 (width)
[22]	silicalite	45
[25]	silicalite	96
[31]	silicalite	62.5 (length), 43.7 (width)
[28]	NaX	63
[29]	NaA	44
[60]	NaA	46
	NaX	59
	NaY	63
[61]	mordenite	46

were determined by measuring the actual change in linear dimension of crystals in synthesis systems over time, rather than the slope of the "crystallization curve," and, therefore, represent true activation energies for crystal growth [27]. Several studies monitored crystal growth ex-situ, while most of the silicalite observations were made in-situ, as the growth occurred. It will be noted that the activation energies reported are all in the range of 45–80 kJ mol^{-1}, regardless of the zeolite system studied, i.e., the activation energies are all of comparable magnitude. Secondly, as noted previously [7, 10, 25, 27], the magnitude of the activation energies suggests that the resistance to crystal growth is controlled by surface kinetics rather than by diffusional transport.

The work of Schoeman et al. [22] demonstrated this conclusion more convincingly by use of a chronomal analysis of the conversion with respect to time, a technique suggested previously by Nielsen [62]. In such an analysis, the linear growth rate of the population of crystals is hypothesized to depend on certain driving forces, and the time dependence of the crystal size function is then derived for the circumstance when new crystal nucleation does not occur, as was observed in these experiments. For example, if diffusional transport from the bulk fluid phase to the surface of uniformly sized spherical particles is assumed to be rate limiting, then the change of the particle radius is given by the solution of:

$$\frac{dr}{dt} = \frac{vD(C-C^*)}{r} \tag{1}$$

where v is the molar volume, D is the diffusion coefficient, $(C-C^*)$ is the driving force for diffusion, i.e., the concentration difference above the equilibrium value, and r is the crystal diameter. The crystal size at any time relative to the final crystal size, at equilibrium is related to the change in concentration from the start by:

$$\alpha = (r/r^*)^3 = (C_0-C)/(C_0-C^*) \tag{2}$$

where r^* is the final crystal size reached at the equilibrium conversion of the solutes. By substituting Eq. (2) into Eq. (1), the following relation can be developed:

$$t = K_D\, I_D \tag{3}$$

where K_D is a grouping of constants and I_D is an integral which must be evaluated from the particle size data collected over time [22]. If the plot of I_D against time is linear, the results would suggest that diffusion is the limiting resistance to crystal growth. Similar relations were developed with first, second and third order surface kinetics hypothesized to be the rate-limiting steps, and yielding relations similar to Eq. (3), but in which the definitions of the integral term, I_i, were different. Figure 8 shows the results of the authors' analyses using the four hypothesized models for crystallization [22], as applied to experiment S100 (the batch composition and temperature are those listed in Table 2 for [22]). It can be seen from Fig. 9 that the change in size of the zeolite crystals was constant up to about 20 hours for that experiment, after which time reagent depletion caused a reduction in the growth rate. It also is noted that in Fig. 8 the

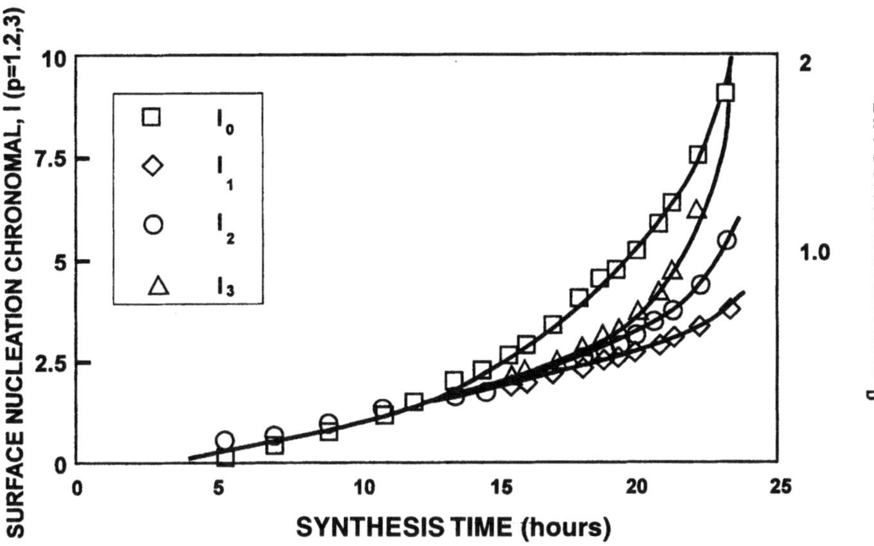

Fig. 8. Results of the chronomal analysis for silicalite crystal growth limitation by diffusion or first, second, or third order surface reaction. The limiting step is suggested by the linear relation over time, i.e., the first order surface reaction step. Reprinted by permission of the publisher from "Analysis of the crystal growth mechanism of TPA-silicalite-1" by BJ Schoeman, J Sterte, and J-E Otterstedt, Zeolites, 14, 568, copyright 1994 by Elsevier Science Inc.

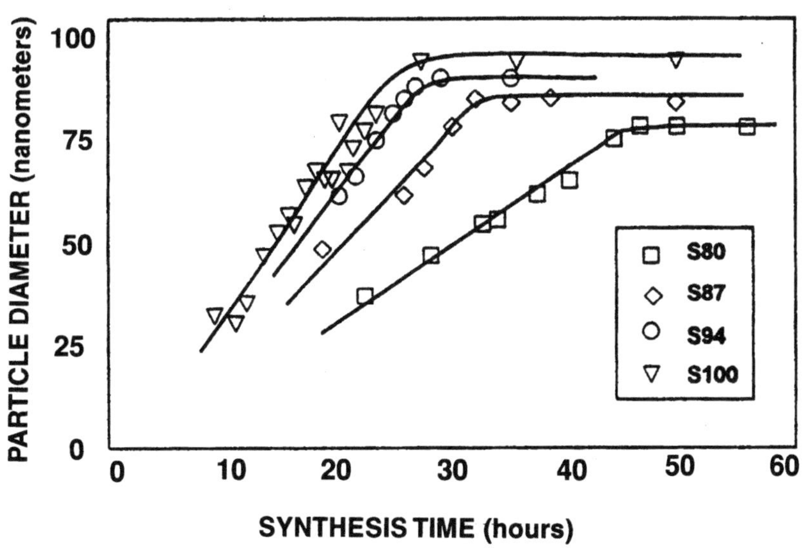

Fig. 9. Evolution of silicalite crystal size with time, showing that the growth rate was constant for experiment S100 up to about 25 hours at temperature. Reprinted by permission of the publisher from "Analysis of the crystal growth mechanism of TPA-silicalite-1," by BJ Schoeman, J Sterte, and J-E Otterstedt, Zeolites, 14, 568, copyright 1994 by Elsevier Science Inc.

chronomal for first order surface reaction is linear up to about 20 hours of synthesis. Their results showed that the best linear fit of the data to these models was for that assuming first order surface reaction kinetics to be rate-limiting, a conclusion which is at least qualitatively consistent with the rather high activation energies reported in Table 3.

Thus, it appears from the evidence cited that transport by diffusion in the liquid layer is not the rate-limiting step in zeolite crystal growth, but that the incorporation of solute by surface integration kinetics may well be. Data from [22] suggests that a first order surface reaction is the rate limiting step for silicalite crystal growth, under the conditions studied.

Lechert and Kacirek [63] presented results which appeared to explain the manner in which the Si/Al ratio of the final product is controlled in a zeolite NaX system. Their results showed that, within certain limits, the Si/Al ratio of the starting solution does not have as much influence on the final product composition as the OH^-/SiO_2 ratio in solution. They also developed an equilibrium model based on ionic reactions of aluminate monomers and silicate monomers with species on the crystal surface, the solution of which adequately described the Si/Al ratio which evolved in a variety of experiments performed by the authors using different starting Si/Al ratios and different NaOH contents in the batch. The model assumed that aluminate ions in solution could react only with silicate groups on the growing crystal surface, while silicate ions in solution could react with either silicate or aluminate surface groups. The Si/Al ratio of the product crystals was calculated based on the average of two successive layers added to the crystal surface. The surface reaction model was based on first principles and experimental observations, was consistent with Löwenstein's rule which prohibits the formation of adjacent aluminate groups in the crystal lattice, and argued against groups more complex than monomers and dimers participating in the crystal growth process. In spite of the authors' successes in describing zeolite NaX growth, they admitted that there were still some improvements which could be made, including finding an explanation for the 5–6 data points which were outliers in their Fig. 3. Nonetheless, their approach is insightful and shows promise for further advances in understanding zeolite crystal growth mechanisms.

4.1 The Tugging Chain Model

It is possible to formulate a conceptual model of the mechanism for the zeolite crystal growth process, based on the various pieces of information that have been reported in research studies to date for silicalite synthesis in clear solutions. It is speculated that this system is representative of zeolite crystal growth, that the knowledge will be applicable to other systems, and that seemingly different features of other systems stem from differences in degree rather than mechanism. Points which must be kept in mind are:

1. The growth of zeolite crystals has been demonstrated to be independent of crystal size [7, 10, 19–26, 28–32, 34, 41]

2. Rearrangement of silicate oligomers in basic media is extremely rapid [9, 37, 38]
3. Silicate oligomers in basic media have been demonstrated to form preorganized inorganic-organic structures in the presence of TPA^+ ions which have short order, and have a configuration similar to that of the final crystal structure [56]
4. In systems which nucleate on the order of 10^{12} particles cm^{-3} the particles are about 1 µm apart, on average, and are unlikely to interact, at least at early times
5. It has been suggested that the primary resistance to silicalite crystal growth in these systems is most likely to be surface reaction kinetics [10, 22, 25, 32], and perhaps a first order surface reaction mechanism [22]
6. Nucleation occurs in a single burst, and consumes a very small amount of the limiting reagent

Keeping these results in mind, one needs only focus attention on two adjacent growing zeolite crystals to hypothesize a model for zeolite growth. The implication that the primary resistance to growth is surface kinetics, and that silicate oligomers in solution attain their equilibrium distribution quickly, is that the reservoir of material between the two growing crystals can be assumed to be an equilibrium distribution of the silicate oligomers. As material is incorporated at the crystal-solution interface, the distribution of silicate oligomers between the two growing crystals, including the region closest to the crystal surfaces, re-equilibrates instantaneously, relative to the rate of crystal growth. The fact that the solution has the thermodynamic tendency to make small oligomeric structures similar to the final crystal structure also suggests that material having the same structure can be assembled at the crystal-solution interface, i.e., that the surface reaction would be expected to create similar crystalline material. Therefore, of the myriad silicate oligomer species present at the crystal-solution interface, there may be only a few which are incorporated, e.g., monomers and dimers [7–10, 25, 63]. The other larger, more complex species are continually unraveled, at a relatively fast rate, to maintain the equilibrium distribution of oligomers, or very near to it. Thus, it is quite possible that the whole process is governed by the ordering of silicates around the pertinent template species adsorbed at the crystal surface.

It also is interesting to recall that nucleation in these systems stops abruptly in spite of the fact that ample reagents appear to be present between the two existing growth centers. This observation can be interpreted to mean that the transfer of material to the crystal, by reaction at the surface, is sufficiently rapid to prohibit silicate polymerization in the bulk which would be necessary for further nucleation. In fact, a net depolymerization must be occurring in the bulk phase during crystal growth to prohibit further nucleation. However, if the growing centers are physically removed from the system, as in [25], nucleation may recommence.

With respect to the competition between nucleation and crystal growth, evidence has shown that both the nucleation "rate" and the growth rate increase with increasing temperature. However, it also has been noted that the absolute number of crystals formed actually decreases as the temperature is increased

[22, 25]. Thus, the nucleation rate increases only because the nucleation period (or the "induction time") decreases faster, i.e., the number of crystals formed divided by the induction time increases with temperature because the denominator decreases faster than the numerator. Since it has been demonstrated that nucleation will recommence if growing particles are removed from the system [25], and that the conversion increases with increasing temperature [22], then one must conclude that the reduction in the absolute number of crystals formed with increasing temperature results from the more rapid depolymerization of silicate oligomers at higher temperature, so that nothing between the hypothetical two growing particles can become an active growth center after the initial burst of nucleation has occurred.

5 Use of Seed Crystals

It is common practice to add crystals, called seed crystals, of the desired phase to a synthesis batch to increase the rate of crystallization, and in some cases to direct the outcome toward selected crystalline phases. It is worthwhile to review the literature on this subject to understand what is known of this phenomenon, however, a recent review article [5] and literature publication [64] on this subject are available, which also contain some details on the mechanism of rate enhancement stemming from the use of seed crystals.

Kerr [1] noted that the induction period, during which nuclei are formed and grow to an observable size can be eliminated by the addition of zeolite NaA to a batch designed to produce that zeolite. He also noted [65] that seeding a zeolite NaX mixture with solid material containing approximately 75% zeolite NaX reduced the crystallization time by about 20%. Adding the seeding material to the boiling sodium silicate solution and the sodium aluminate solution prior to mixing decreased the synthesis time further. Similar reduction of the crystallization time was noted when adding the seed material directly to the reaction mixture.

Mirskii and Pirozhkov [66] reported on experiments in which seed crystals were added to normal batch zeolite synthesis mixtures. In one set of experiments, different amounts of seed crystals of a desired phase (not specifically mentioned, but probably zeolite NaA) were added to the synthesis mixture, and noted to eliminate the induction time and increase the rate of crystallization. Two additional factors were investigated and reported: a) the rate of crystallization increased more with increased amounts of seed crystals added, and b) the rate of crystallization was enhanced more using the same mass of smaller seed crystals than with larger seed crystals. Both of these results were concluded to imply that the rate enhancement was due to the cumulative seed crystal surface area used to assimilate material from the solution. This point was illustrated further by adding seed crystals of one phase to a solution which nominally produced a different zeolite phase. For example, zeolite NaP seed crystals were added to a synthesis mixture, which was demonstrated to precipitate zeolite NaX, after about 30% of the amorphous reagents had already crystallized. After two additional hours of crystallization, the absolute amount of zeolite had increased, but the relative amounts of the two phases remained the same, suggesting that the crystalline phases which formed were a function of the structure of the additive crystals and the crystals already precipitated. The same effect was noted with NaX seed crystals added to a NaP mixture, and with both combinations of hydroxysodalite and zeolite NaA, and their respective synthesis mixtures.

Somewhat different results were observed in a more recent work [67] in which the sodium cation was increasingly replaced by potassium cations in a system which precipitated zeolite NaA when only sodium ions were used. At levels beyond about 20% replacement of the sodium ions by potassium ions, mixtures of zeolites A, F, and G were formed, until at 80% replacement only zeolites F and G were synthesized. Relatively low seeding levels, ca. 10–15% by weight, added to the 50% Na/K mixture resulted in formation of a new population of zeolite A crystals which did not form in the unseeded system. The same low levels of seeding with zeolite NaA crystals in the 100% K system resulted in the precipitation of zeolites F and G, but in remarkably shorter time than without seeds. That is, the presence of zeolite A seed crystals appeared to catalyze the formation of zeolites F and G. However, at a seeding level of 72.5% by weight (zeolite A crystals) in the pure K system, zeolite A was the only observable phase formed, regardless of whether seeds in the Na-form or the K-form were used. It also was obvious in those experiments that the seed crystals had grown, although not regularly, and that a new population of zeolite A crystals had been formed, unquestionably due to the presence of the seed crystals.

It also has been noted recently [5] that using silicalite seed crystals in the ammonium silicalite system results in increased levels of silicalite nuclei formed in the solution, and that a relatively large number of these new crystals seem to grow out of the seed crystals forming what has been labeled the "porcupine" morphology [5]. A typical example of that morphology is illustrated in Fig. 10, which shows several rather large silicalite seed crystals with much smaller new silicalite crystals apparently growing out of the seed crystal surfaces [68]. A unique feature of this ammonium-based system was that the gel phase became very viscous after about 30 minutes at temperature, sufficiently so that growing crystals were somewhat immobilized in the gel phase thereafter, and settling did not occur after that time. The mechanism for the formation of the porcupine morphology, described in detail in [5] and [68], involves the release of "initial-bred nuclei" from the surface of seed crystals, or from the sample. The seed crystals and the new population of crystals grow by normal means such that their growing faces advance towards one another. Ultimately, overgrowth of the two crystal types in random orientation occurs, such that the peculiar geometries are formed. Bonding between the randomly overgrown crystals is weak, as evidenced by the occurrences of their separation, noted by the "craters" in several locations in the predominant seed crystal in Fig. 10. It also was demonstrated [5] that other macroscopic-sized crystals may come into contact and grow into one another to form seemingly odd-shaped agglomerates. It also was noted that the same phenomenon occurs in other zeolite systems.

The question then arose regarding how and when the initial-bred nuclei were formed, and what control one might have over their formation. It has been demonstrated recently [64] that the initial-bred nuclei form in seed-preparation

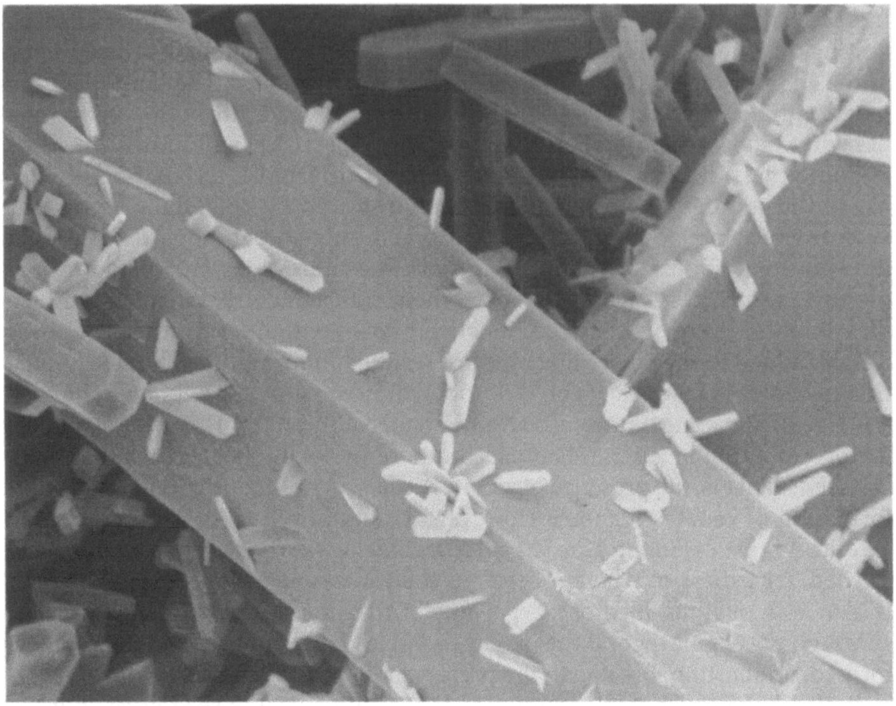

Fig. 10. SEM photomicrograph (at 1000X) of silicalite seed crystals (the largest crystals present) and the new population of silicalite crystals formed by initial breeding. The "porcupine" morphology is illustrated by the small crystals growing out of the seed crystal surfaces. Crystals grown in the NH_4^+-silicalite system by Gonthier [68], using the composition reported in [5]

systems from unconverted alumino-silicate material left in the solution. These nanometer-sized particulates may, or may not, have crystalline domains, however they were observed by light scattering to be present in filtrates from rather large seed crystals, and have the ability to form growing crystals. It is quite possible that these entities are similar to a) the particulates observed by SAXS/WAXS recently [57, 58], b) the ordered structures recently reported by Burkett and Davis [56], and c) the persistent "nuclei" reported by light scattering [25], and may form in a manner similar to the lamellae observed in 1970 [59]. Nevertheless, it has been demonstrated that initial-bred nuclei are present in several different zeolite systems [5, 26, 64], and that these particles can catalyze nucleation of a new population of crystals in a synthesis system. This concept also may explain the apparent observation that synthesis vessels have "memory," especially with inadequate cleaning between batches. That is, it is conceivable that some of these particulates cling to vessel walls, stirrer blades, heating surfaces, and other hardware parts and re-emerge in the next batch to play a role in nucleation.

Therefore, what is known about the crystallization rate enhancement resulting from the use of seed crystals at the current time is that: a) nanometer-sized (colloidal) particulates are quite likely present in most seed crystal samples, which can become viable growing zeolite crystals, b) these nanometer-sized particulates appear to be present in the mother liquor in which the seed crystals were formed, c) they are physically separate, and separable, from the seed crystals at the conclusion of the seed crystal synthesis, d) they may have been present in the seed crystal batch prior to the conclusion of the seed crystal synthesis [64], as suggested previously [25], and e) they have been used separately from the seed crystals to increase the crystallization rate of a subsequent zeolite crystallization [64, 69].

It also is known from the evidence discussed in detail in [5] that even rather large silicalite crystals can grow into and around one another, forming randomly oriented agglomerates, of perhaps as few as two crystals. In the case of the NH_4^+-silicalite system this behavior was accentuated, because the "solution" became viscous, achieving the consistency of a paste after about 30 minutes of heating. That is, after about 30 minutes the individual crystals had lost much of their mobility within the medium. However, the same phenomenon has been observed in analcime synthesis from clear unseeded solutions, which remain quite fluid, and in which the large crystals which form very quickly settle to the bottom of the vessel and continue to grow into one another, forming what might resemble a sheet of analcime crystals [70].

6 Conclusions

The synthesis of crystalline molecular sieve zeolites in hydrothermal systems involves the combination of the appropriate amounts of aluminates and silicates, usually in basic media, and usually in an aqueous medium. Syntheses generally will proceed at ambient or moderate temperatures, however, crystallization rates generally are much faster at elevated temperatures, approaching 100 °C, if pressures below one atmosphere are desired, and temperatures up to about 180 °C, if high pressure vessels are used. Most zeolites of commercial interest are metastable phases, requiring that synthesis processes be terminated at some predetermined time to avoid contamination of the solid product with denser undesirable phases.

Zeolite crystallization is a phase transformation process, since an amorphous alumino-silicate gel phase usually forms quickly after mixing the reagents in the appropriate concentrations. Clear solution syntheses have been reported, but the yield from them is typically not sufficient to generate commercial interest. It is generally agreed that the transformation from amorphous gel to crystalline zeolite occurs through the solution phase via dissolution of the amorphous gel and crystallization of the desired zeolite phase. Consequently, the normal processes of nucleation and crystal growth must occur from within the solution.

Zeolite nucleation is thought to occur via some primary mechanism, either homogeneous or heterogeneous nucleation, since neither seeding nor agitation are required for these crystallizations to proceed, which would be the case if secondary nucleation were involved. However, mathematical simulations using population balance models have suggested that the classical homogeneous nucleation mechanism probably does not apply in these syntheses. Furthermore, the works of Subotic et al. [32, 34, 35, 39–41], Hamilton et al. [44], Gonthier et al. [36], Burkett and Davis [56], Doktor et al. [57, 58], and Aiello et al. [59] have all suggested that some precursor species form in the solution, and that these species involve more than just the aluminate ions, the silicate ions, or aluminosilicate oligomers. Two prior works had suggested that the "activity" of silicates in zeolite syntheses was due to aluminate impurities in the silica source [42, 43], while Hamilton et al. [44] concluded that this activity was due to enhanced nucleation in various systems, which itself could have been due to any of several impurity levels in various silica sources. Recently, colloidal particulate matter has been observed in several synthesis solutions, and in silicate solutions prior to making the synthesis solution [26, 57, 58], which could participate in zeolite nucleation.

Two recent papers [72, 73] appeared while this manuscript was "in press", both of which discussed observations of nanometer-sized particulates in zeolite synthesis systems. Schoeman [72] reported observing particles 3 nm in size which persisted throughout the synthesis of silicalite. He also indicated that the zeolite crystal growth curve could be extrapolated back to about the same size, and that therefore, one might speculate that these particles were at least associated with nucleation, if not the nuclei themselves. Gora et al. [73] also reported observing nanometer-sized particulates which persisted throughout the synthesis of zeolite NaA in their study. However, they also noted that the same sized particles were observed to exist in the silicate solution prior to mixing with the aluminate solution, which itself did not contain any such particles. Both these reports give an indication that colloidal particles may participate in a form of heterogeneous nucleation.

The evidence seems quite clear that adding seed crystals to a new zeolite synthesis batch usually results in enhanced crystallization rates, and occasionally the presence of seed crystals determines the phases precipitated in the process. Results also seem to point to the fact that seed crystal samples contain in them sub-micron sized particulate matter, either crystalline or amorphous, that has the ability to catalyze the nucleation of new zeolite crystals in the system [5, 26, 64, 67–69, 71].

Growth of zeolite crystals in hydrothermal systems has repeatedly been observed to be independent of crystal size, and constant over time until the reagent concentration begins to decrease. Results with aluminum-free ZSM-5 (silicalite-1) suggested that zeolite crystal growth rates increased with a concentration driving force, as determined by changing solubilities of amorphous gel and zeolite crystal phases with pH [13]. Analysis of the synthesis of aluminum-free ZSM-5 from clear solutions indicated that the growth rate was limited by the incorporation of material at the solution-crystal interface by a first order surface kinetics reaction [22]. Diffusional transport rates in the solution appeared to be sufficiently rapid to keep the surface supplied with material. The same conclusion could be inferred by true activation energies for zeolite crystal growth reported by numerous research groups (see Table 3).

7
References

1. Kerr GT (1966) J Phys Chem 70:1047
2. den Ouden CJJ, Thompson RW (1992) I&EC Res 31:369
3. Randolph AD, Larson MA (1988) Introduction to the theory of crystallization processes, 2nd ed. Academic, London
4. Thompson RW (1992) Population balance analysis of zeolite crystallization. In: Catlow CRA (ed) Modelling of structure and reactivity in zeolites. Academic, London, p 231
5. Gonthier S, Thompson RW (1994) Effects of seeding on zeolite crystallisation, and the growth behaviour of seeds. In: Jansen JC, Stöcker M, Karge HG, Weitkamp J (eds) Advanced Zeolite Science and Applications. Elsevier, Amsterdam, p 43
6. Sung CY, Estrin J, Youngquist GR (1973) AIChE J 19:957
7. Cundy CS, Lowe BM, Sinclair DM (1990) J Crstl Gr 100:189
8. Knight CTG (1990) Zeolites 10:140
9. Gilson J-P (1992) Organic and inorganic agents in the synthesis of molecular sieves. In: Derouane EG et al. (eds) Zeolite microporous solids: synthesis, structure, and reactivity. Kluwer Academic, Netherlands, p 19
10. Cundy CS, Lowe BM, Sinclair DM (1993) Fara Discuss 95:235
11. Casci JL, Lowe BM (1983) Zeolites 3:186
12. Lowe BM (1983) Zeolites 3:300
13. Fegan SG, Lowe BM (1986) J Chem Soc, Faraday Trans 82:785
14. Ueda S, Koizumi M (1979) Amer Miner 64:172
15. Ueda S, Sera T, Tsuzuki Y, Koizumi M, Takahashi S (1983) J Clay Sci 23:60
16. Ueda S, Kageyama N, Koizumi M (1984) Crystallization of zeolite Y from solution phase. In: Olson D, Bisio A (eds) Proceedings of the 6th international zeolite conference. Butterworths, Guildford UK, p 925
17. Ueda S, Kageyama N, Koizumi M (1983) Crystallization of zeolite Y from aqueous solution. Proceedings of the 1st international symposium on hydrothermal reactions, p 695
18. Wenqin P, Ueda S, Koizumi M (1986) Synthesis of zeolite NaA from homogeneous solutions and studies of its properties. In: Murakami Y, Iijima A, Ward JW (eds) Proceedings of the 7th international zeolite conference. Elsevier, Amsterdam, p 177
19. Schoeman BJ, Sterte J, Otterstedt J-E (1993) J Chem Soc, Chem Comm 13:994
20. Schoeman BJ, Sterte J, Otterstedt J-E (1994) Zeolites 14:110
21. Schoeman BJ, Sterte J, Otterstedt J-E (1994) Zeolites 14:208
22. Persson AE, Schoeman BJ, Sterte J, Otterstedt J-E (1994) Zeolites 14:557; and Schoeman BJ, Sterte J, Otterstedt J-E (1994) Zeolites 14:568
23. Schoeman BJ, Sterte J, Otterstedt J-E (1994) The synthesis of discrete colloidal zeolite particles. In: Hattori T, Yashima Y (eds) Zeolites and microporous crystals. Kodansha/Elsevier, Tokyo, p 49
24. Schoeman BJ (1994) Ph D dissertation, Dept of Eng Chem, Univ of Goteborg, Sweden
25. Twomey TAM, Mackay M, Kuipers HPCE, Thompson RW (1994) Zeolites 14:162
26. Gora L (1995) Ph D dissertation, Dept of Chem Eng, WPI, Worcester MA (USA)
27. Barrer RM (1982) The hydrothermal chemistry of zeolites, 1st edn. Academic, London
28. Zhdanov SP, Samulevich NN (1981) Nucleation and crystal growth of zeolites in crystallizing alumino-silicate gels. In: Rees LVC (ed) Proceedings of the 5th international conference on zeolites. Heyden, London, p 75
29. Zhdanov SP (1971) Some problems of zeolite crystallization. In: Flanigen EM, Sand LB (eds) Adv Chem Ser 101:20
30. Sand LB, Sacco A, Thompson RW, Dixon AG (1987) Zeolites 7:387
31. Feoktistova NN, Zhdanov SP, Lutz W, Bulow M (1989) Zeolites 9:136
32. Golemme G, Nastro A, B.Nagy J, Subotic B, Crea F, Aiello R (1991) Zeolites 11:776
33. Ciric J (1968) J Coll Int Sci 28:315

34. Bronic J, Subotic B, Smit I, Despotovic LA (1988) Influence of gel ageing on zeolite nucleation processes. In: Grobet PJ, Mortier WJ, Vansant EF, Schulz-Ekloff G (eds) Innovation in zeolite materials science. Elsevier, Amsterdam, p 107
35. Subotic B, Bronic J (1993) Modelling and simulation of zeolite crystallization. In: von Ballmoos R, Higgins JB, Treacy MMJ (eds) Proceedings of the 9th international zeolite conference. Butterworth-Heineman, Stoneham, MA, p 321
36. Gonthier S, Gora L, Guray I, Thompson RW (1993) Zeolites 13:414
37. McCormick AV, Bell AT, Radke CJ (1986) Application of 29Si and 27Al NMR to determine the distribution of anions in sodium silicate and sodium alumino-silicate solutions. In: Murakami Y, Iijima A, Ward JW (eds) Proceedings of the 7th international zeolite conference. Elsevier, Amsterdam, p 247
38. Keijsper JJ, Post MFM (1989) Precursors in zeolite synthesis: a critical review. In: Occelli ML, Robson HE (eds) Zeolite synthesis. American Chemical Society, Washington DC, p 28
39. Subotic B, Graovac A (1980) On kinetic equations of zeolite crystallization. In: Sersale R, Colella C, Aiello R (eds) Recent Progress Reports and Discussion of 5th International Conference on Zeolites. Giannini, Naples Italy, p 54
40. Subotic B, Graovac A (1985) Kinetic analysis of autocatalytic nucleation during crystallization of zeolites. In: Drzaj B, Hocevar S, Pejovnik S (eds) Zeolites: synthesis, structure, technology and application. Elsevier, Amsterdam, p 199
41. Subotic B (1989) Influence of autocatalytic nucleation on zeolite crystallization processes. In: Occelli ML, Robson HE (eds) Zeolite synthesis. American Chemical Society, Washington DC, p 110, and Katovic A, Subotic B, Smit I, Despotovic LjA, Curic M (1989) Role of gel aging in zeolite crystallization. Ibid, p 124
42. Freund EF (1976) J Crystl Gr 34:11
43. Lowe BM, MacGilp NA, Whittam TV (1980) Active silicates and their role in zeolite synthesis. In: Rees LVC (ed) Proceedings of the 5th international conference on zeolites. Heyden, London, p 85
44. Hamilton KE, Coker EN, Sacco, A, Dixon AG, Thompson RW (1993) Zeolites 13:645
45. den Ouden CJJ, Thompson RW (1991) J Coll Int Sci 143:77
46. Charnell JF (1971) J Cryst Gr 8:291
47. Neels H, Schmitz W, Berger E-M, Lutz D (1978) Krist Tech 13:1345
48. Gutsze A, Kornatowski J, Neels H, Schnitz W, Finger G (1985) Cryst Res Technol 20:151
49. Kornatowski J, Finger G, Schmitz W (1987) Polish J Chem 61:155
50. Schmitz W, Kornatowski J, Finger G (1988) Cryst Res Technol 23:K25
51. Kornatowski J, Finger G, Schmitz W (1990) Cryst Res Technol 25:17
52. Scott G, Dixon AG, Sacco A, Thompson RW (1989) Synthesis of zeolite Na-A in the presence of triethanolamine. In: Jacobs PA, van Santen RA (eds) Zeolites: facts, figures, future. Elsevier, Amsterdam, p 363
53. Scott G, Thompson RW, Dixon AG, Sacco A (1990) Zeolites 10:44
54. Morris M, Sacco A, Dixon AG, Thompson RW (1991) Zeolites 11:178
55. Coker EN, Thompson RW, Dixon AG, Sacco A, Nam SS, Suib SL (1993) J Phys Chem 97:6465
56. Burkett SL, Davis ME (1994) J Phys Chem 98:4647
57. Doktor WH (1994) Ph D dissertation, Chemical Technology, TU Eindhoven, The Netherlands
58. Doktor WH, van Garderen HF, Beelen TPM, van Santen RA, Bras W (1995) Angew Chem Int Ed Engl 34:73
59. Aiello R, Barrer RM, Kerr IS (1971) Stages of zeolite growth from alkaline media. In: Sand LB, Flanigen EM (eds) Molecular Sieve Zeolites-I. Adv Chem Ser No 101, American Chemical Society, New York, p 44
60. Breck DW, Flanigen EM (1968) Synthesis and properties of Union Carbide zeolites L, X, and Y. In: Barrer RM (ed) Molecular sieves. Society of Chemical Industry, London, p 47
61. Domine D, Quobex J (1968) Synthesis of mordenite. In: Barrer RM (ed) Molecular sieves. Society of Chemical Industry, London, p 78
62. Nielsen AE (1964) Kinetics of precipitation, Pergamon, Oxford, UK

63. Lechert H, Kacirek H (1991) Zeolites 11:720
64. Gora L, Thompson RW (1995) Zeolites 15:526
65. Kerr GT (1968) J Phys Chem 72:1385
66. Mirskii YaV, Pirozhkov VV (1970) Russ J Phys Chem 44:1508
67. Warzywoda J, Thompson RW (1991) Zeolites 11:577
68. Gonthier S (1993) MS Thesis, Dept of Chem Eng, WPI, Worcester, MA USA
69. Tsokanis EA, Thompson RW (1992) Zeolites 12:369
70. Brock AB, Link GN, Poitras PS, Thompson RW (1993) J Mater Chem 3:907
71. Dutta PK, Bronic J (1994) Zeolites 14:250
72. Schoeman BJ (1997) Zeolites 18:119
73. Gora L, Streletzky K, Thompson RW, Phillies GDJ (1997) Zeolites 18:119

Synthesis of Porosils: Crystalline Nanoporous Silicas with Cage- and Channel-Like Void Structures

H. Gies · B. Marler · U. Werthmann

Institut für Mineralogie, Ruhr-Universität Bochum, D-44780 Bochum, Germany.
E-mail: hermann.gies@ruhr-uni-bochum.de

1 Porosils: An Ever-Increasing Class of Materials

The many different silicate zeolite structure types [1] are summarized in a general formula as solid solution series: $A^{x+}_{(4-w)y/x}[Si_{1-y}T^{w+}_{y}O_2]*zH_2O*nM$, where A denotes mono- or divalent cations, T tetra-, tri-, di-, or monovalent cations tetrahedrally coordinated by oxygen, and M neutral atomic or molecular guest species. The general formula highlights the specific compositional breadth typical of zeolites and gives valuable information on composition-dependent properties. All-silica end members, the porosils [2], are included in this description; however, there is a fundamental difference in properties between conventional silicate zeolites and all-silica nanoporous materials [3]. Since the tetrahedral SiO_2 frameworks are neutral, no framework-charge balancing cations are contained in the material and all materials are hydrophobic. In the different structure types, the void space is generated by neutral structure-directing agents, SDA, also called template molecules M, which determine by their size and geometry in advance the size and geometry of the pore. Since the bonding energy of the silica framework resembles that of natural silica polymorphs, e.g. quartz, the thermal stability of this class of material is generally higher than for structurally related zeolites, e.g. in dry atmosphere up to about 1300 K for MTN (MTN is a mnemonic used for a particular structure type; for this and similar structure-type codes as well as for details, refer to [1]) [4].

So far, only some zeolite structure types cover a wide range of Si/Al ratios as framework constituents, e.g. SOD [1]. Most of the typical silicate zeolites with low Si/Al ratios have not yet been obtained as silica end members, neither in direct synthesis nor in post-synthesis treatment. Similarly, all-silica and high-silica materials are difficult to synthesize in low Si/Al ratios. The reason is the difference in synthesis concepts for all- and high-silica zeolites and low-silica zeolites. The simplified all-silica synthesis system reduces to $SiO_2{*}nM{*}wH_2O$, with SiO_2 forming the three-dimensional, four-connected host framework and M as structure-directing template. The template M is the synthesis variable allowing for the choice of the porosil structure type crystallized during the synthesis. In addition, there is a marked influence of the intensive synthesis variables p and T on the structure type formed. Several nanoporous silicas have been obtained through post-synthesis treatment of the reaction product such as steaming and $SiCl_4$-treatment. Although the process is very important for particular applications, no reference is made in this chapter to those materials. For detailed information the interested reader is referred to the literature [5].

The porosils are further subdivided into clathrasils and zeosils depending on the pore geometry which is cage-like and channel-like, respectively. In Figs. 1 and 2, the cages found in porosils and representative channels for zeosils are shown. The skeletal drawings show the Si atoms as knots of the three-dimensional host framework. Oxygen atoms are omitted for clarity and are close to the midpoint between two silicon sites. The cages in clathrasils are bound by at most 6MR of $[SiO_4]$ units, suppressing properties such as sorption/desorption of organic molecules. In contrast, zeosils have at least 8-MR windows with 4.0 Å pore width. For small molecules this is large enough to penetrate. Since the porosity of the different porosil structure types is created only after calcination of the as-synthesized material, the specific properties of the two subclasses of porous silicas become only obvious after the removal of the organic template. The synthesis of clathrasils and zeosils, however, should be subject to the same general rules and will be treated together. The synthesis of porosils has been most successful under mild hydrothermal conditions. There are also reports on syntheses in nonaqueous polar solvents such as alcohols and amines [6, 7, 8]. A compilation of all porosils known to date is given in Table 1, which also includes crystallographic and structural details of the porous host silica frameworks.

In the temperature range 130–250 °C, solutions of silicic acid and amphiphilic structure-directing agents have been subject to autogenous pressure in auto-

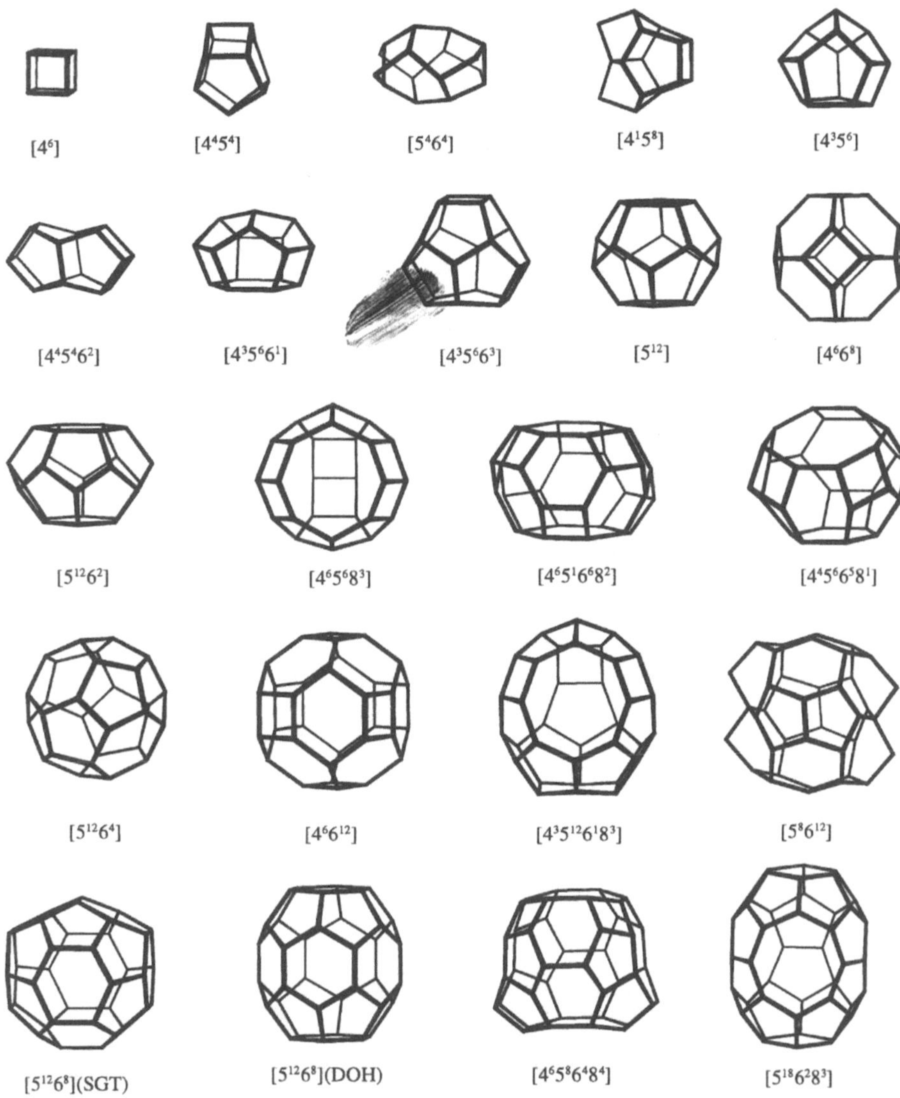

Fig. 1. Compilation of the cage-like voids observed in clathrasils

claves for a few days up to several months. The most successful class of amphiphilic SDA is the group of aliphatic and cyclic amines with the hydrophilic amino headgroup and the hydrophobic tail. More than 100 SDAs have been used so far for the synthesis of 26 different porosil structure types. In Table 2 a compilation of representative templates used as SDA for various porosil structure types is given. Because of the vast number of successful SDAs for porosil synthesis, only typical template molecules are listed here. A more exhaustive survey of SDAs and porosil structures can be found in the literature [3, 20, 36, 37].

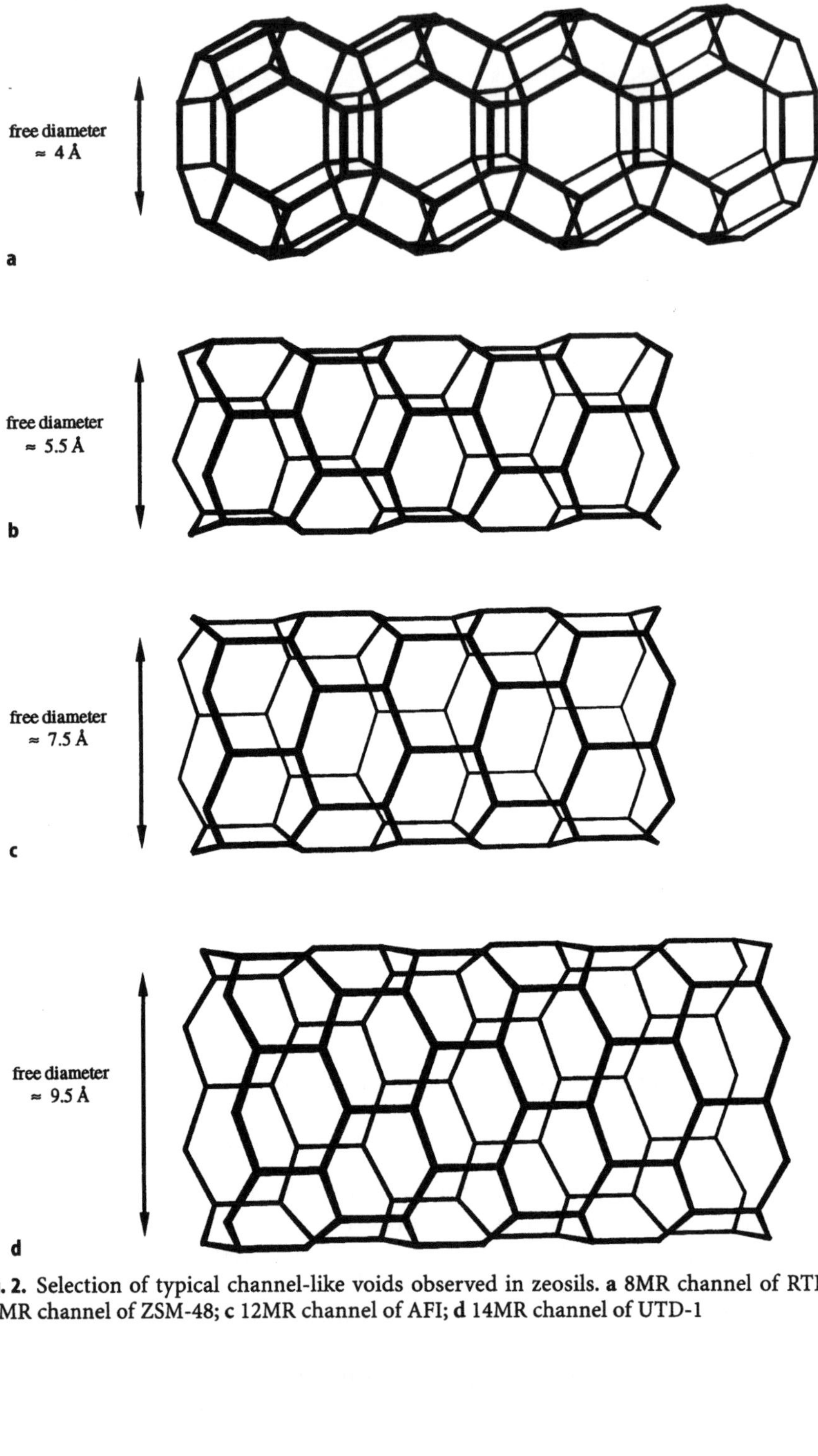

Fig. 2. Selection of typical channel-like voids observed in zeosils. **a** 8MR channel of RTE; **b** 10MR channel of ZSM-48; **c** 12MR channel of AFI; **d** 14MR channel of UTD-1

Table 1a. Structural properties of clathrasils

Structure type	Structure type code	Composition per unit cell[a]	Lattice parameters [Å]	Maximum topological space group	Number of cages per unit cell	Cage type	Cage symmetry	Cage volume [$Å^3$]	Ref.: Synthesis, structure
RUB-3	RTH	$24\,[SiO_2]$ $\times 2M^{10}2M^{16}$	$a_o = 14.0$, $b_o = 13.7$, $c_o = 7.4$ $ß = 102.5°$	C2/m	2 2	$[4^45^46^2]$ $[4^65^46^68^2]$	2/m 2/m	35 300	9,1
Silica-Sodalite	SOD	$12\,[SiO_2] \times 2M^{14}$	$a_o = 8.8$	Im3m	2	$[4^66^8]$	m3m	130	6, 1
Decado-decasil 3H	–	$120\,[SiO_2] \times 6M^{10}$ 9 $M^{12}1M^{15}4M^{19}1M^{23}$	$a_o = 13.9$, $c_o = 40.9$	P3m	6 9 1 4 1	$[4^35^66^1]$ $[5^{12}]$ $[4^65^68^3]$ $[4^35^{12}6^18^3]$ $[5^{18}6^28^3]$	3m m3 62m 3 m 6/mmm	35 80 230 350 540	10, 10
Decado-decasil 3R	DDR	$120\,[SiO_2]$ $\times 6M^{10}9M^{12}6M^{19}$	$a_o = 13.9$, $c_o = 40.9$	$R\bar{3}m$	6 9 6	$[4^35^66^1]$ $[5^{12}]$ $[4^35^{12}6^18^3]$	3m m3 3m	35 80 350	11, 1
Octa-decasil	AST	$20\,[SiO_2] \times 2M^62M^{18}$	$a_o = 9.2$, $c_o = 13.4$	I4/m	2 2	$[4^6]$ $[4^66^{12}]$	4/mmm 4/mmm	5 280	12, 1
Sigma-2	SGT	$64\,[SiO_2] \times 8M^94M^{20}$	$a_o = 10.2$, $c_o = 34.4$	$I4_1/amd$	8 4	$[4^35^6]$ $[5^{12}6^8]$(SGT)	2 mm $\bar{4}$	25 390	13, 1
Dode-casil 1H	DOH	$34\,[SiO_2]$ $\times 3M^{12}2M^{12'}1M^{20}$	$a_o = 13.8$, $c_o = 11.2$	P6/mmm	3 2 1	$[5^{12}]$ $[4^35^66^3]$ $[5^{12}6^8]$(DOH)	m3 62 m 6/mmm	80 70 430	14, 1
Dode-casil 3C	MTN	$132\,[SiO_2]$ $\times 16M^{12}8M^{16}$	$a_o = 19.4$	Fd3m	16 8	$[5^{12}]$ $[5^{12}6^4]$	m3 m3	80 240	14,1
Melano-phlogite	MEP	$46[SiO_2]$ $\times 2M^{12}6M^{14}$	$a_o = 13.4$	Pm3n	2 6	$[5^{12}]$ $[5^{12}6^2]$	m3 $\bar{4}$ m2	80 160	15,1

Table 1a (continued)

Structure type	Structure type code	Composition per unit cell[a]	Lattice parameters [Å]	Maximum topological space group	Number of cages per unit cell	Cage type	Cage symmetry	Cage volume [Å³]	Ref.: Synthesis, structure
Nonasil	NON	$88[SiO_2] \times 8M^8 8M^9 4M^{20}$	$a_o = 22.2$, $b_o = 15.1$, $c_o = 13.6$	Fmmm	8	$[5^4 6^4]$	222	25	16, 1
					8	$[4^1 5^8]$	mm2	30	
					4	$[5^8 6^{12}]$	mmm	290	
ITQ-3	ITE	$64[SiO_2] \times 4M^{22}$	$a_o = 20.6$, $b_o = 9.7$, $c_o = 19.6$	Cmcm	4	$[5^4 5^4]$		20	17, 17
						$[4^6 5^8 6^4 8^4]$		500	
RUB-10	RUT	$36[SiO_2] \times 4M^{10} 4M^{16}$	$a_o = 13.1$, $b_o = 12.9$, $c_o = 12.4$, $ß = 102.5°$	C2/m	4	$[4^4 5^4 6^2]$	2/m	35	18, 1
					4	$[4^4 5^6 6^5 8^1]$		230	

[a] M^n: M = Template molecule, n = the number of faces of the occluding cage.

Table 1b. Structural properties of zeosils

Structure type	Structure type code	Composition per unit cell	Lattice parameters [Å^3]	Maximum topological space group	Ring size	Cross section of the limiting pore [Å]	Direction of the channel	Dimensionality of the pore system	Form of the channel	Ref.: Synthesis, structure
RUB-3	RTH	$24[SiO_2] \times M^{8MR}$	$a_o = 14.0$, $b_o = 13.7$, $c_o = 7.4$ $ß = 102.5°$	C2/m	8-MR	4.2×3.9		1-D	straight	9, 1
ITQ-3	ITE	$64[SiO_2] \times M^{8MR} \times M^{8MR}$	$a_o = 20.6$, $b_o = 9.7$, $c_o = 19.6$	Cmcm	8-MR 8-MR	3.8×4.3 5.8×2.3	[010] [001]	2-D	straight straight	17, 17
Decadodecasil 3H	–	$120[SIO_2] \times M^{8MR} \times M^{8MR}$	$a_o = 13.9$, $c_o = 40.9$	P3m	8-MR 8-MR	4.4×3.8 4.4×3.8	[100] [010]	2-D	zigzag zigzag	10, 10
Decadodecasil 3R	DDR	$120[SiO_2] \times M^{8MR} \times M^{8MR}$	$a_o = 13.9$, $c_o = 40.9$	R$\bar{3}$m	8-MR 8-MR	4.4×3.6 4.4×3.6	[100] [010]	2-D	zigzag	11, 1
Silica-ferrierite	FER	$36[SiO_2] \times M^{8MR} \times M^{10MR}$	$a_o = 18.6$, $b_o = 13.9$, $c_o = 7.2$	Immm	8-MR 10-MR	4.8×3.5 5.4×4.2	[010] [001]	2-D	straight straight	8, 1
Silica-theta-1	TON	$24[SiO_2] \times M^{10MR}$	$a_o = 13.8$, $b_o = 17.4$, $c_o = 5.0$,	Cmcm	10-MR	5.5×4.4	[001]	1-D	zigzag	19, 1
Silica-ZSM-23	MTT	$24[SiO_2] \times M^{10MR}$	$a_o = 21.5$, $b_o = 11.1$ $c_o = 5.0$	Pmmn	10-MR	5.2×4.5	[001]	1-D	zigzag	19, 1

Table 1b (continued)

Structure type	Structure type code	Composition per unit cell	Lattice parameters [Å³]	Maximum topological space group	Ring size	Cross section of the limiting pore [Å]	Direction of the channel	Dimension-ality of the pore system	Form of the channel	Ref.: Synthesis, structure
Silica-ZSM-48	–	48 $[SiO_2]$ × M^{10MR}	$a_o = 14.2$, $b_o = 20.1$, $c_o = 8.4$,	Cmma	10-MR	5.6 × 5.3	[001]	1-D	straight	20, 21
Silica-MCM-22	MWW	72 $[SiO_2]$ × M^{10MR} × M^{10MR}	$a_o = 14.2$, $c_o = 24.9$	P6/mmm	10-MR 10-MR	5.5 × 4.0 5.5 × 4.0	[100] [010]	2-D	sinusoidal sinusoidal	22, 23
Silica-ZSM-5	MFI	96 $[SiO_2]$ × M^{10MR} × $M^{10'MR}$	$a_o = 20.1$, $b_o = 19.9$, $c_o = 13.4$	Pnma	10-MR	5.6 × 5.3 5.5 × 5.1	[100] [010]	3-D	sinusidal straight	24, 1
Silica-ZSM-11	MEL	96 $[SiO_2]$ × M^{10MR} × M^{10MR}	$a_o = 20.1$, $c_o = 13,4$	$I\bar{4}m2$	10-MR 10-MR	5.4 × 5.3 5.4 × 5.3	[100] [010]	2-D	straight straight	25, 1
Silica-ZSM-12	MTW	56 $[SiO_2]$ × M^{12MR}	$a_o = 24.9$, $b_o = 5.0$, $c_o = 24.3$, ß = 107.7°	C2/c	12-MR	5.9 × 5.5	[010]	1-D	straight	26, 1
VPI-8	VET	17 $[SiO_2]$ × M^{12MR}	$a_o = 13.0$, $c_o = 5.0$,	$P\bar{4}$	12-MR	6.0 × 5.9	[001]	1-D	straight	27, 1
Silica-SSZ-24	AFI	24 $[SiO_2]$ × M^{12MR}	$a_o = 13.7$, $c_o = 8.4$	P6/mcc	12-MR	7.3 × 7.3	[001]	1-D	straight	28, 1
Silica-NCL-1/SSZ-31	–	28 $[SiO_2]$ × M^{12MR}	$a_o = 14.9$, $b_o = 8.4$, $c_o = 12.4$, ß = 106.0°	P2/m (Poly-morph A)	12-MR	8.6 × 5.7	[010]	1-D	straight	29, 30

Table 1b (continued)

Structure type	Structure type code	Composition per unit cell	Lattice parameters [Å³]	Maximum topological space group	Ring size	Cross section of the limiting pore [Å]	Direction of the channel	Dimensionality of the pore system	Form of the channel	Ref.: Synthesis, structure
ITQ-4	IFR	$32\,[SiO_2] \times M^{12MR}$	$a_o = 18.7$, $b_o = 13.5$, $c_o = 7.6$, $ß = 102.0°$	I2/m	12-MR	7.2 × 6.2	[001]	1-D	sinusoidal	31, 32
Silica-beta	BEA	$64\,[SiO_2] \times M^{12MR} \times M^{12MR} \times M^{12'MR}$	$a_o = 12.7$, $c_o = 26.6$ Å	$P4_122$ (Polymorph A)	12-MR 12-MR 12-MR	5.5 × 5.5 5.5 × 5.5 7.6 × 6.4	[100] [010] [001]	3-D	straight straight spiral	33, 1
UTD-1	–	$64\,[SiO_2] \times M^{14MR}$	$a_o = 19.0$, $b_o = 8.4$, $c_o = 23.0$	Imma	14-MR	10.0 × 7.5	[010]	1-D	straight	34, 34
CIT-5	CFI	$32\,[SiO_2] \times M^{14MR}$	$a_o = 13.7$, $b_o = 5.0$, $c_o = 25.5$	Imma	14-MR	10.0 × 9.9	[010]	1-D	straight	35, 35

Table 2a. Selection of structure-directing agents (SDA) that have been used successfully for the synthesis of different clathrasil structure types

Structure type	Structure type code	Typical structure-directing agents			
		acyclic	monocyclic	bicyclic	tricyclic
RUB-3	RTH			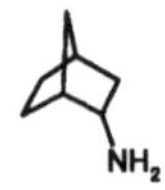	
Silica-Sodalite	SOD	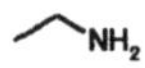	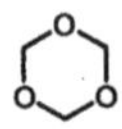 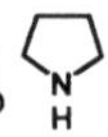		
Decadodecasil 3H	–				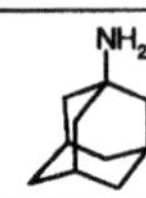
Decadodecasil 3R	DDR			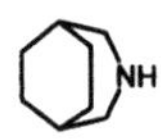	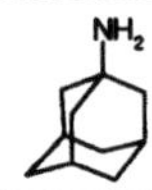
Octadecasil	AST	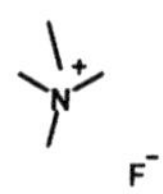		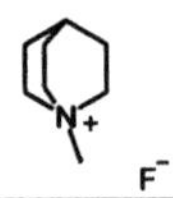	
Sigma-2	SGT			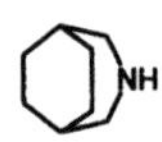	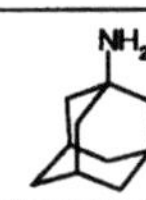
Dodecasil 1H	DOH	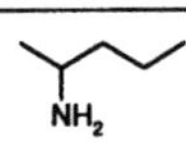	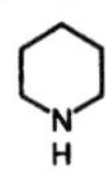 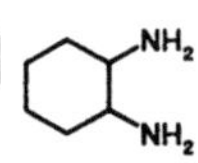	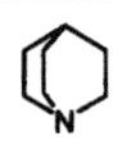	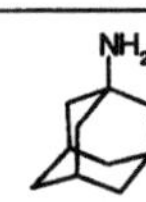
Dodecasil 3C	MTN		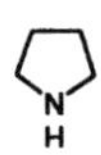 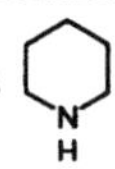	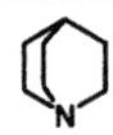	
Melanophlogite	MEP	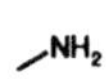			
Nonasil	NON	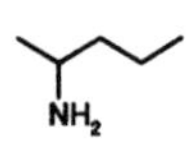	 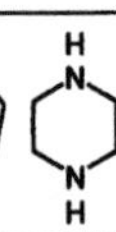		
ITQ-3	ITE			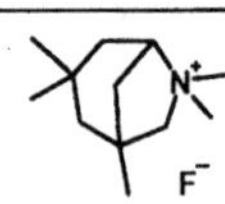	
RUB-10	RUT		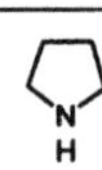		

Table 2b. Selection of structure-directing agents (SDA) that have been used successfully for the synthesis of different zeosil structure types

Structure type	Structure type code	Typical structure-directing agents		
		acyclic	monocyclic	bicyclic/tricyclic
Silica-ferrierite	FER			
Silica-theta-1	TON	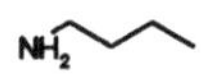		
Silica-ZSM-23	MTT	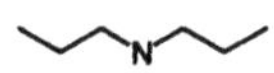	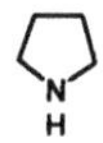	
Silica-ZSM-48	-	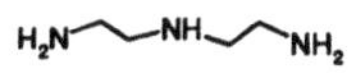	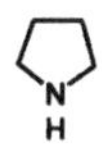	
Silica-MCM-22	MWW		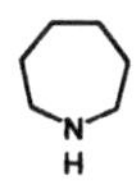	
Silica-ZSM-5	MFI	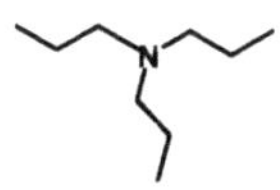		
Silica-ZSM-11	MEL	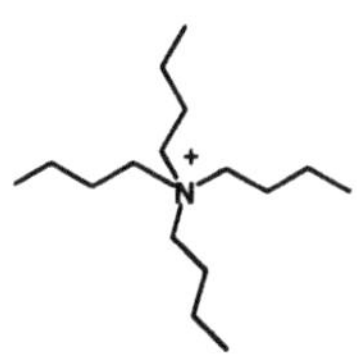		
Silica-ZSM-12	MTW		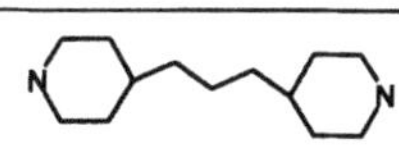	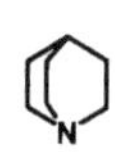
VPI-8	VET	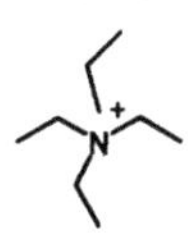		
Silica-SSZ-24	AFI			

Table 2b (continued)

Structure type	Structure type code	Typical structure-directing agents		
		acyclic	monocyclic	bicyclic/tricyclic
ITQ-4	IFR			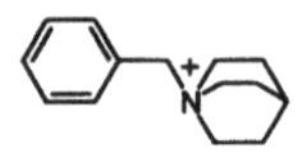
Silica-beta	BEA		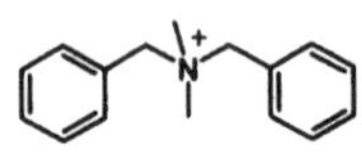	
UTD-1	-		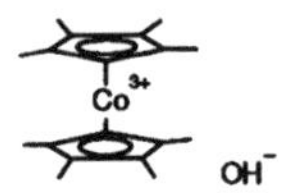	
CIT-5	CFI			

2 Synthesis Strategies and Synthesis Mechanism

2.1 Interaction Between Structure-Directing Agent, Solvent, and Silica Framework

Since every SDA is distinct in its properties, the influence of the SDA on the structure type formed during the porosil synthesis is obvious. The vast amount of experimental data on the synthesis of porosils available in the literature allows the deduction of a number of empirical synthesis rules [20, 36, 37, 38, 39, 40]. Very little is known of atomistic details of the synthesis mechanism [41, 42].

It is clear from the nature of the successful amphiphilic templates that the interaction with the solvent as well as with the silica surface is most important. During the crystallization process, the basic functional group, most frequently the hydrophilic amino groups in organic molecules, participates in the hydrogen bonding system with the silanol groups of the silica surface and water as solvent. The hydrophobic part of the template fills the voids of the half-completed cavities which are exposed at the growth face and interacts with the silicate wall of these cavities. As a model, the SDA 1-aminoadamantane docked in the open $[5^{12}6^{8}]$-cage of the growth face of dodecasil 1H (DOH) is visualized (Fig. 3).

Although the nucleation and growth to a stable crystallite are still mysterious, the host-guest interactions between template and silicate monomers and oligomers are most important. From investigations of the SDA mechanism in the synthesis of silica-MFI, it is indicated that preorganized organic-inorganic composites exist in the synthesis mixture [43]. They condense to stable crystallites which might be of the size of only a few unit cells (< 80 Å) [44]. Under the influence of the template the periodic silica framework of MFI grows in a typical

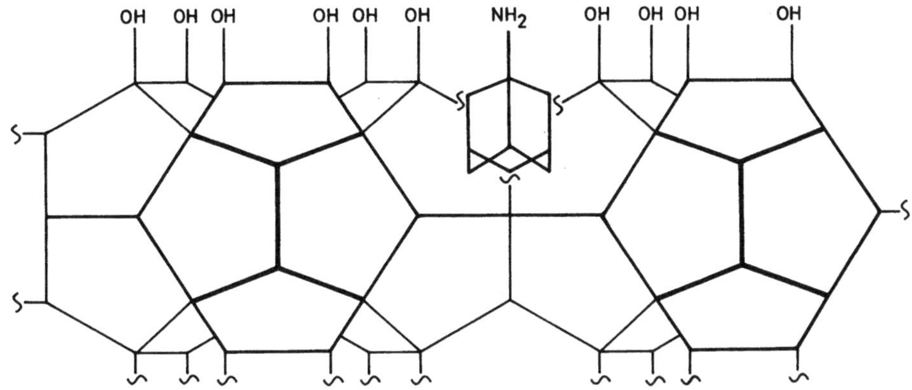

Fig. 3. Schematic drawing of the surface of a DOH crystal showing the SDA 1-aminoadamantane docked in the half-cavity of the [$5^{12}6^8$] cage

morphology with regular, smooth growth faces. The growth of the crystal is controlled by the sorption of SDA on the surface and the following condensation of silica entities [3, 38, 40, 45]. Since there are only weak van der Waals interactions, the growth rate-determining step would be the establishment of the sorption/desorption equilibrium of the SDA in half-cavities on the surface of the porosil. In cases where a pronounced anisotropy in crystal morphology is observed, the fast growing faces are dominated by the condensation of silica. For example, the porosils containing one-dimensional channels grow fast along the channel axis. Here, the continuous growth of the silicate wall through the condensation of silica dominates. Perpendicular to the channel axis, the creation of new channel-like pores requires the sorption of SDAs.

Crystal faces which are largest represent the slowest growing faces and are representative for the rate-determining step in crystal growth. In porosils, those faces which contain the highest density of sorption sites for SDAs in the structure-dominating pores are always the largest [3]. Therefore, the morphology of every structure type developed by the porosil crystal is typical and depicts a macroscopic protocol of the processes occurring on a microscopic scale. Theoretical calculations on the efficacy of the template during synthesis of porosils support this view, although the calculations only consider the fit of the template in the unperturbed silica framework and neglect solvent interaction with the template and the framework [46, 47].

2.2
Properties of Structure-Directing Agents Decisive for the Pore Geometry

2.2.1
Size and Shape of the Templating Molecule

From the nucleation and growth mechanism, it is obvious that the size and the shape of the SDA determine the size and the shape of the pore of the porosil.

Spherical molecules favour cage-like voids, whereas chain-like molecules lead to 1-D channels. SDAs with branched chains lead to intersecting channel systems. The correspondence of the size and shape of the SDA and the pore of the porosil at the molecular level has consequences for the choice of successful SDAs. As can be seen from Table 2, SDAs of very different size, shape, and conformation have been used for the templating of cages and channels in the various porosil structure types. For almost every structure type there are several SDAs indicating the flexibility of the silica framework. Those forming a particular structure type have been classified according to the size and geometry of the pore they occupy [3]. As examples, all SDAs which are to date known to crystallize DOH are listed in Table 3. At the same time, there are several SDAs templating different structure types (Tables 2 and 4), but under different synthesis conditions. This shows that the intensive synthesis parameters (p, T) and also the extensive parameters, e.g., the concentration of SDA, play a crucial role in the process of formation, however, of secondary importance compared to the size and shape of the SDAs. The shape of the SDA for the clathrasils is spherical. Larger cages, therefore, require larger spherical and rigid molecules. Effective templates for clathrasils with large cages, therefore, are polycyclic compounds which occur only in a limited number of conformations. The most successful SDAs are the tricyclic amines 1-amino-adamantane, quinuclidine, and azabicyclononane which all act as SDA for four different structure types. For example, 1-amino-adamantane as SDA is occluded in four different cages of different size [42]. Whereas DDR with the 19-hedral cage-[$4^3 5^{12} 6^1 8^3$] and SGT with the 20-hedral cage-[$5^{12} 6^8$](SGT) crystallize only at lower synthesis temperatures ($T < 160\,°C$) and high concentrations of SDA (> 50 mg SDA/ml silicate solution), DOH with the 20-hedral cage-[$5^{12} 6^8$](DOH) is the most flexible structure for the SDA 1-aminoadamantane. It crystallizes in the temperature range from 150–250 °C and at low to medium concentrations of SDA. DD3H is a difficult to obtain variant of the deca-dodecasil series of polytypes. Whereas in DDR the stacking sequence of the deca-dodecahedral layer is ABCABC, in the polytype DD3H the stacking is AABAAB, yielding the largest clathrasil cage observed so far, the [$5^{18} 6^2 8^3$] cage.

From sorption experiments the efficacy of a sorbate has been measured as heat of adsorption and described as "nest effect", relating size and shape of the sorbate with the surface curvature of the pore [48]. Recently, host-guest complexes have been formulated quantitatively in terms of van der Waals interactions. Lewis et al. [47] calculated the "nonbonded interactions energy" of the SDA within the cavities of different silica zeolites, which was in good agreement with the experimental synthesis experience. The computational strategy developed in this study should stimulate the systematic search for new effective SDAs for the synthesis of new porosil structures with tailored pore geometry [49].

2.2.2
Flexibility, Basicity, Amphiphily, and Charge of Templating Molecules

The size of the SDA is usually estimated by the number of atoms in the molecule, including hydrogen atoms. With increasing size of the aliphatic SDA, the mani-

Table 3. Summary of aliphatic and cyclic molecules that have be used successfully as structure-directing agents (SDA) for the synthesis of the clathrasil dodecasil 1H (DOH)

Number of skeletal atoms	Typical structure-directing agents			
	acyclic	monocyclic	bicyclic	tricyclic
5	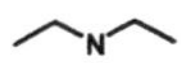			
6	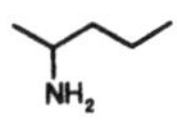	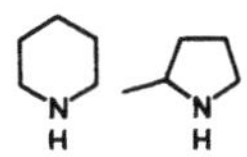		
7		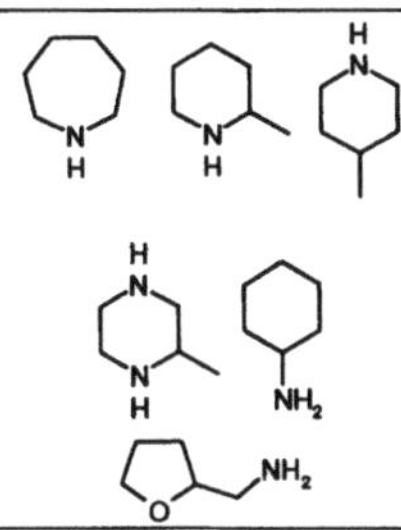		
8		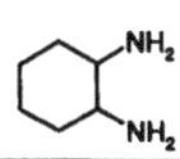	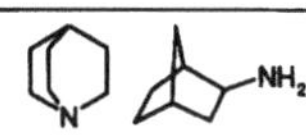	
9		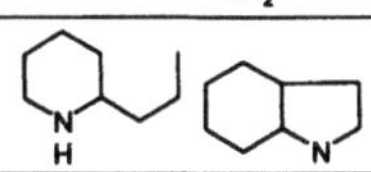	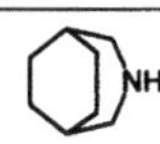	
10		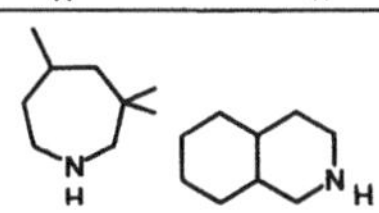		
11		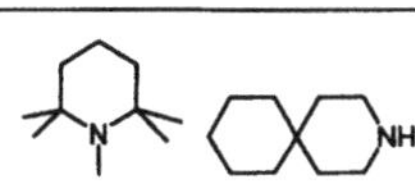	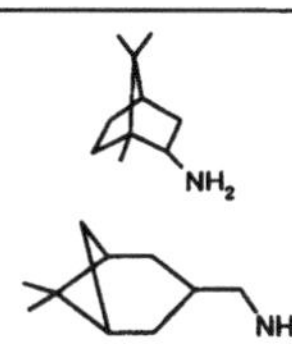	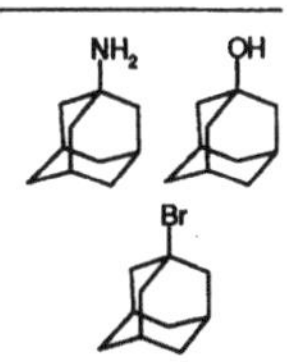
12				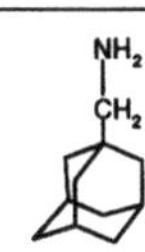

fold of conformations of the molecule generally increases and, therefore, weakens its potential as template. For the same number of atoms in a template molecule, more rigid, branched or cyclic molecules are the better SDAs. In a synthesis series different templates with five and eight skeletal atoms (C + N atoms) were used (Table 5). Clearly, the linear alkylamine reacted only after prolonged reaction times, whereas the branched 2-aminobutane and the cyclic pyrrolidine

Table 4. Tricyclic amines used as structure-directing agents for the formation of different porosil structure types. Reaction mixture: x SiO_2 : y SDA : 20–80 H_2O (: additives)

Structure-directing agent	Synthesis temperature	x SiO_2 : y SDA	Additives	Obtained structure type
Quinuclidine	low (preferentially 160 °C)	1 : 1	–	MTW
Quinuclidine	high (preferentially 220 °C)	1 : 1	–	DOH
Quinuclidine	intermediate (preferentially 180 °C)	1 : 4	–	MTN
Quinuclidine	intermediate (preferentially 170 °C)	1 : 1	Hydrofluoric acid	AST
3-Azabicyclo [3.2.2]-nonane	very low (preferentially 150 °C)	1 : 1	–	RUB-4 [a]
3-Azabicyclo [3.2.2]-nonane	low (preferentially 170 °C)	1 : 1	–	DDR
3-Azabicyclo [3.2.2]-nonane	intermediate (preferentially 180 °C)	1 : 1	–	SGT
3-Azabicyclo [3.2.2]-nonane	high (preferentially 220 °C)	1 : 1	–	DOH
1-Aminoadamantane	very low (preferentially 150 °C)	1 : 1/3	–	DDR
1-Aminoadamantane	very low (preferentially 150 °C)	?	–	SGT
1-Aminoadamantane	low (preferentially 160 °C)	1 : 1/3	–	DD3H
1-Aminoadamantane	high (preferentially 220 °C)	1 : 1/20	–	DOH

[a] RUB-4 has a disordered structure which is closely related to the RTE structure type.

Table 5. Influence of the flexibility of the structure-directing agent on the rate of clathrasil formation

Structure type	Number of skeletal atoms	Structure-directing agent (SDA), time ← increasing flexibility			
MTN	5	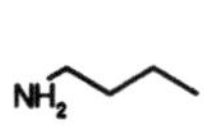 > 150 days	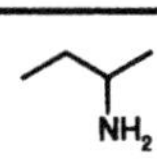 78 days	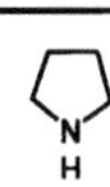 3 days	
DOH	8	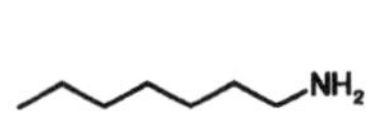 > 150 days	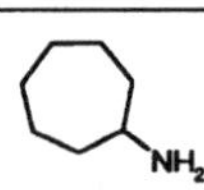 ≈ 30 days	 11 days	 16 days

Table 6. Summary of help bases used to enhance the crystallization of porosils

Help base	Ethanol-amine	Ethylen-diamine	Triethylamine	Lithium-hydroxide	Sodium-hydroxide
pKA value	9.5	10.7/7.6	10.8	13.8	14
Typical concentration in synthesis mixtures	1 M	1 M	1 M	0.2 M	0.1 M

acted more effectively as templates. With the exception of polyaza compounds, aliphatic amines with a carbon chain length longer than six have not been used successfully as templates so far, whereas polycyclic amines with up to 11 C-atoms acted as SDA in porosil synthesis.

For the mobilization of silica the basicity of the synthesis solution is important. In the absence of mineral bases the basicity of the synthesis solution depends only on the Brønsted basicity of the SDA. The higher the basicity of the SDA in solution, the higher is the concentration of polysilicate anions of different molecular weight and, hence, the higher is the reaction rate of porosil formation. In order to enhance the reaction rate, basic help molecules (Table 6) have been used extensively together with the SDAs. The help base promotes the mobilization of silica, but does not act as SDA in the formation of porosils. In the presence of only ethylenediamine (*en*), for example, no porosils crystallize even after 9 months of reaction time. When *en* is used in combination with an SDA, porosils crystallize faster, but only the SDA is occluded in the structure-determining porosil void. As minor component the help bases might also be included in other voids of the silica frameworks [50].

The influence of the amphiphily of the SDA shows up clearly in the synthesis of porosils. Since the framework carries no charges and, therefore, the SDA is neutral, the interaction at the surface of the crystalline solid and the synthesis solution plays a crucial role in the reaction progress. It is known from large amphiphilic molecules such as n-alkyl-trimethylammonium cations with an alkyl chain length greater than 8 that they self-organize and form micellar aggregates, tubular- or layer-like units, or liquid crystalline phases of different structure. Self-organization of amphiphilic molecules prevents the structure-directing activity of the individual molecule and suppresses the formation of porosils. On the other side, the stronger the amphiphily of the neutral SDA, the better the potential of the molecule to act as SDA. Organic amines have proven to be the most effective class of amphiphilic molecules for the synthesis of porosils. The amino-functional group in organic molecules ideally serves as hydrophilic head group and, at the same time, provides Brønsted basicity for the mobilization of silica. The hydrocarbon tail is the hydrophobic part of the molecule. In contrast to tensides, the hydrophobic tail should be bulky and rigid in order to prevent self-organization of the amphiphilic agent to separate entities. The overview on SDAs given in Table 2 clearly supports the explanation, but also

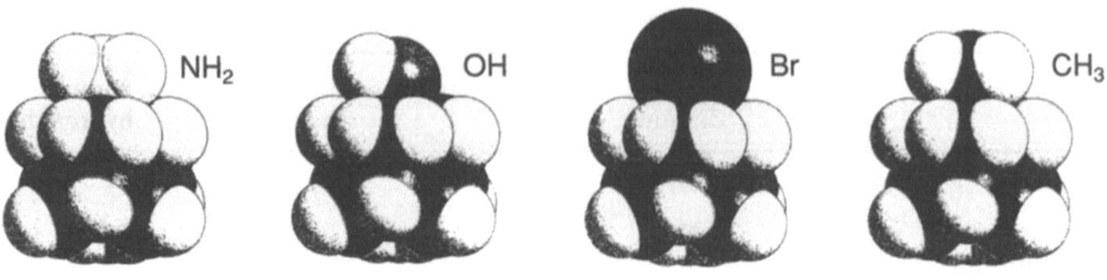

A = NH_2 > OH > Br > CH_3

Fig. 4. Adamantane derivatives as structure-directing agents of similar size and shape but different functional groups. The molecules are arranged according to decreasing activity A

shows the restriction of the concept to molecules of limited size only. Larger molecules either do not dissolve enough in aqueous solution, thus preventing the crystallization of porosils, or self-aggregate, leading in the presence of silica, under certain circumstances, to new, noncrystalline mesoporous phases of silica.

As an example of the influence of basicity and amphiphily of SDAs, the isosteric series of four adamantane derivatives has been investigated (Fig. 4). Whereas the most basic 1-aminoadamantane is one of the most active and versatile structure-directing template molecules, the apolar 1-methyl-adamantane has not been a successful SDA. 1-Hydroxyadamantane is a poor SDA and 1-bromo-adamantane is even less active.

Charge-carrying templates have also been used successfully as SDAs in porosil synthesis. The template is incorporated in the structure as ion pair, usually with the ammonium cation in the large structure-determining void and the charge-balancing anion in small, separate cavities. The synthesis technique successfully employed is the so-called fluoride route [51]. Under acidic conditions, the flouride anions serve as mobilizing agent for silica and, at the same time, as co-guest molecules in small cavities of the porosil structure. The enclathration of the F^- anion is indicated by the significant low-field shift of the ^{19}F-NMR signal as compared to the free fluoride anion (Table 7) [53]. The synthesis route allows for mild synthesis conditions giving the most open porosil frameworks [17]. Calcination of the more open porosils in their as-synthesized form removes the ion pair. In the case of typical clathrasils such as octadecasil or nonasil, however, the calcination leads to the collapse of the silica framework, since the removal of the templates requires higher calcination temperatures, leading to the thermal decomposition of the organic cation but also to the formation of hydrofluoric acid. This enhances the transformation to thermodynamically stable SiO_2 phases. So far, five porosil structures have been obtained via the fluoride route only (AST, UTD-1, ITQ-3, ITQ-4, silica FER) (Table 7); many others known from neutral SDA synthesis have also been obtained in this way. Table 7 also lists some materials which have been carefully analyzed in their as-synthesized form including the detailed location of the F^- ion elucidating the crystallization mechanism. Figure 5 shows the results of

Table 7. Selection of porosils synthesized by the fluoride method from ion pairs of the form $SDA^+ F^-$

Porosil	Structure type code	SDA^+	Location of the SDA^+	Location of the fluoride ion	Chemical shift ^{19}F MAS NMR spectroscopy
Nonasil	NON	Cobalticinium	$[5^8 6^{12}]$ cage (V = 290 Å^3)	$[4^1 5^8]$ cage (V = 30 Å^3)	–76.4 ppm [52]
Octadecasil[a]	AST	Quinuclidinium	$[4^6 6^{12}]$ cage (V = 280 Å^3)	$[4^6]$ cage (V = 5 Å^3)	–38.2 ppm [52]
Dodecasil 3C	MTN	Tetramethylammonium	$[5^{12} 6^4]$ cage (V = 240 Å^3)	$[5^{12}]$ cage (V = 30 Å^3)	–81 ppm [52]
Dodecasil 1H	DOH	1,1-Dimethylcobalticinium	$[5^{12} 6^8]$ cage (V = 430 Å^3)	$[4^3 5^6 6^3]$ cage (V = 70 Å^3) $[5^{12}]$ cage (V = 80 Å^3)	–78.9 ppm [52] –82.3 ppm [52]
ITQ-3[a]	ITE	1,3,3,6,6-Pentamethyl-6-azonium[3.2.1]bicyclooctane	$[4^6 5^8 6^4 8^4]$ cage (V = 500 Å^3)	probably in the $[4^4 5^4]$ cage (V = 20 Å^3)	unknown
Silica-Theta-1	TON	Dipropylammonium	10-MR channel	unknown	–74.8 ppm [53]
Silica-ZSM-23	MTT	Dipropylammonium	10-MR channel	unknown	–74.2 ppm [53]
ITQ-4[a]	IFR	N-benzyl-1-azonium[2.2.2]bicyclooctane	12-MR channel	probably in the $[4^3 5^2 6^1]$ cage (V = 30 Å^3)	unknown
Silica-ZSM-5	MFI	Tetrapropylammonium	intersection of the 10-MR channels	probably in the $[5^8]$ cage (V = 20 Å^3)	–64.0 ppm [53]

[a] Structure type which could so far only be synthesized by the fluoride method.

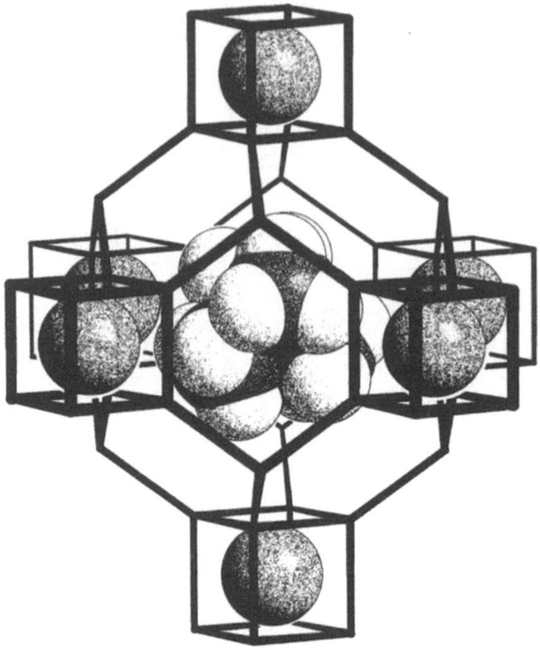

Fig. 5. The structure of the octadecasil (AST) synthesized with the tetramethylammonium fluoride ion pair. The tetramethylammonium ion occupies the large $[4^6 6^{12}]$ cage, whereas the F^- ion occupies the small $[4^6]$ cage

the single crystal structure analysis of octadecasil (AST) [12] which unambiguously proved that the small $[4^6]$-cage contains the F^- anion as a second SDA. Theoretical calculations supported this result in confirming the contribution of the F^- anion to the overall lattice energy of the structure of the clathrasil [54].

2.2.3
Help Guest Species in Porosil Synthesis

The dominating influence of the SDA on the crystallization of porosils is supported and sometimes enforced by additional guest molecules which are occluded in smaller voids of the silica framework [55]. The help guest species are often atmospheric gases present during synthesis or organic bases used to mobilize silica and to increase the reaction rate. They are most helpful for the synthesis of clathrasils since all structure types except sodalite contain more than one type of cage. The SDA usually occupies the largest cage, therefore, determining the structure type formed. The help guest species fill the smaller cages and contribute to the overall lattice energy. In studies of the clathrasils melanophlogite (MEP) [56] and dodecasil 3C (MTN) [57], the atmospheric gases, in particular Ar and N_2, were found with almost 100% occupancy in the smaller pentagondodecahedra, i.e. the $[5^{12}]$-cages, whereas the SDAs were housed in the larger $[5^{12}6^{2}]$- and $[5^{12}6^{4}]$-cages. On several occasions, the presence or the absence of help guest species was crucial for the formation of a particular structure type. With pyrrolidine as SDA the zeosil ZSM-48 crystallized in the absence of help guest species, whereas the clathrasils MTN and DOH crystallized when those were present [58].

A similar role is played by the fluoride anion in the fluoride synthesis route when the ion pair is actively templating. In the examples given above, the presence of the fluoride anion is decisive for the formation of the particular structure type, and, as has been shown in structure analyses and NMR experiments, the fluoride anion occupies specific small cavities in clathrasil phases (see Table 7).

2.3
Influence of Synthesis Parameters on the Porosil Formation

Besides the choice of the templating SDA, the synthesis parameters concentration of silica, SDA, ratio of concentration [silica]/[SDA], pressure, and temperature are of major importance in the planning of the synthesis. Since the formation of the porosil is a kinetically controlled process [60], the interaction energy between silica host and templating SDA at the growth surface plays a crucial role in the formation of the crystalline solid. No porosil formation will occur for too high temperatures with too short residence times for the SDA at the growth surface to become physically trapped by the condensing silica. The highest temperatures at which porosils crystallize are below 300 °C and reflect the magnitude of the host guest interaction energy which is in the order of RT. For example, pyrrolidine acts as SDA even at 250 °C for DOH and is one of the most effective SDAs in porosil synthesis. Porosil synthesis is usually carried out in the temperature range between 150 and 200 °C. Since the synthesis conditions in classical porosil synthesis are only weakly basic, i.e. pH ≈ 9 to pH ≈ 11, lower synthesis temperatures result in low crystallization rates and limit the potential of the method. Nevertheless, within the chosen temperature range, the actual synthesis conditions still have a decisive influence on the crystallization product.

A general observation is the fact that the framework density of the porosil phases increases with increasing synthesis temperature. Although there is considerable overlap between the synthesis regimes of different structure types, the upper limit of the synthesis temperature increases with increasing framework density. This is thought to be indicative of the increasing importance of the contribution of the Si-O bonding energy of the silica framework to the overall lattice energy of the material. Up to now, however, the influence of the synthesis pressure on the product formed has not been investigated in detail.

2.3.1
Influence of Synthesis Temperature and Pressure

As can be seen from Tables 2 and 4, there are several guest molecules which serve as SDA for different porosil structure types. The most sensitive synthesis parameter for the template efficacy is the temperature of synthesis. The space re-

Table 8. Influence of the synthesis temperature on the type of clathrasil formed with a given structure-directing agent

Structure type	Volume of the cage [Å³] housing the SDA	Structure-directing agents (SDA): Number of skeletal atoms									
		2	5	6	6	7	7	7	8	9	11
DOH	430		250 °C 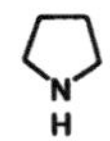	250 °C 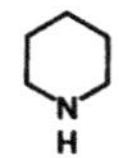	220 °C 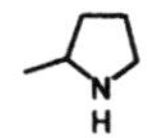	220 °C 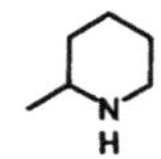	160 °C 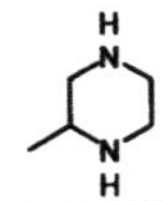	200 °C 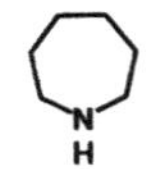	220 °C 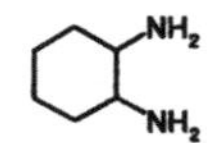	200 °C 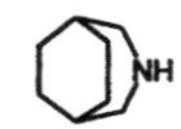	200 °C 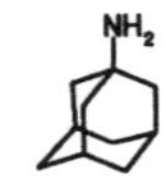
SGT	390									180 °C 	
DDR	350										160 °C 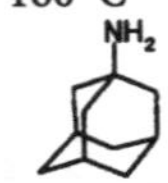
NON	290				180 °C 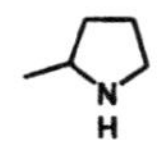	190 °C 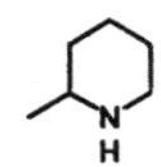	150 °C 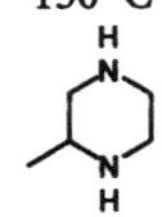	180 °C 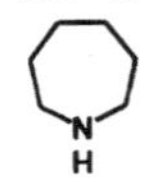	180 °C 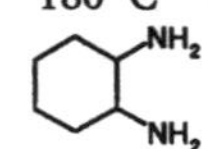		
MTN	240	250 °C 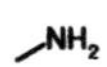	200 °C 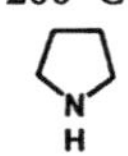	200 °C 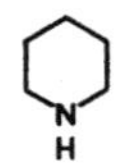							
MEP	160	200 °C 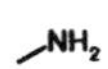									

quirement of the SDA and the free space in the cavity housing the SDA change with temperature in opposite directions. The volume of the SDA increases with temperature and those SDAs which fit tightly in the cavities at lower synthesis temperatures require larger cavities at higher synthesis temperatures. At the same time, the thermal motion of the framework atoms at higher temperatures increases and reduces the accessible free volume of the cavity. The result is that a new porosil structure type with larger void space is formed. This is demonstrated by the tricyclic amines (see examples in Table 4). More detailed information on specific porosil systems, $SiO_2 * M$, is compiled in Table 8 with a selection of SDA stabilizing different clathrasil structure types as a function of the synthesis temperature. Here, in the low-temperature porosil structure, the SDA just fits the cavity and interacts tightly with the silica framework. The cavity of the high-temperature form has in each case a much larger volume.

The influence of the synthesis pressure has not yet been investigated in detail. It is clear that the filling factor of the autoclave and the concentration of the organic SDA have considerable influence on the overall pressure. For those SDAs which dissolve poorly in aqueous solutions, an increased pressure would increase the active concentration of the SDA in solution and, thus, has a considerable influence on the type of reaction product formed. Qualitatively, this phenomenon has been observed for the system $SiO_2/CH_3NH_2/H_2O$. The gaseous SDA is highly soluble in water and an excellent SDA for MTN. For aqueous solutions saturated with CH_3NH_2 (~40% CH_3NH_2), much higher pressure builds up at the synthesis temperature of 160 °C, leading to exclusive crystallization of melanophlogite. A more detailed description of the concentration dependence is given in the following section.

2.3.2
Influence of SDA Concentration

The concentration of SDA in the synthesis solution has not been discussed so far. In most porosil syntheses, the synthesis solution is saturated with SiO_2 at the temperature of synthesis. The SDA concentration is important first for the nucleation process and, secondly, it is needed to maintain crystal growth. Whereas the nucleation of the porosil depends on the presence of an SDA, crystal growth of a stable crystallite might continue with only a fraction of cages occupied by SDA or help guest species. In this way, DOH membranes have been synthesized for the separation of hydrogen gas in gas mixtures [59].

A number of SDAs lead to different structure types depending on their concentration in the synthesis solution. Lower concentrations of these SDAs give porosils with lower cage density for the SDA housing cage, whereas higher concentrations of SDA increase the cage density, leading to different structure types. In all examples found so far, the density of cages per unit volume, CD, increases with the concentration of SDA in the synthesis solution. In Table 9 an overview of the various porosil systems with detailed information on the concentration dependence is given. As an example we consider 1-aminoadamantane which at 160 °C leads to the crystallization of DOH. For SDA concentrations above 50 mg/ml, DDR is formed. In DOH, the cage density, CD, of the

Table 9. Influence of the SDA concentration on the type of porosil formed at a given temperature

Structure type	Cage housing the SDA	Cage density [cages/1000 Å^3]	Structure-directing agent (SDA), SiO_2/SDA ratio					
			160 °C	160 °C	150 °C	180 °C	160 °C	160 °C
SOD	[$4^6 6^8$]	2.9	 1 :10.3	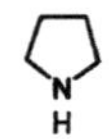 1 : 9.9				
SGT	[$5^{12} 6^8$]	1.2					 1 : 2	
MTN	[$5^{12} 6^4$]	1.1	 1 : 0.4	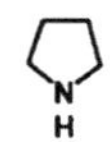 1 : 0.8	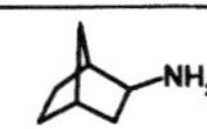 1 : 5.3	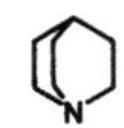 1 : 4		
NON	[$5^8 6^{12}$]	0.9			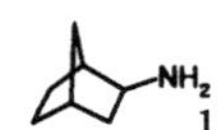 1 : 15			
DDR	[$4^3 5^{12} 6^1 8^3$]	0.9					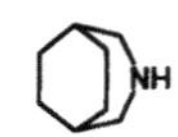 1 : 0.5	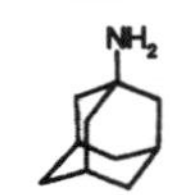 1 : 0.33
DOH	[$5^{12} 6^8$]	0.5				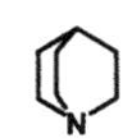 1 : 1	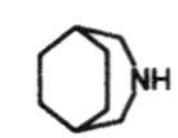 1 : 0.1	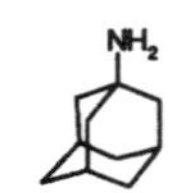 1 : 0.05

[$5^{12}6^{8}$]-cages housing the SDA is 0.5 cages per 1000 Å^3, whereas in DDR the CD for the [$4^{3}5^{12}6^{1}8^{3}$]-cages is 0.9. More examples are summarized in Table 9. A quantitative explanation for this observation has not yet been provided. However, since nucleation determines which porosil phase will crystallize, there should be a concentration dependence of the pre-nucleation complexes, i.e. the SDA/polysilicate anions, formed in solution.

In all cases observed so far, a higher concentration of the SDA leads to a higher CD for the cavity housing the SDA with the highest CD of 2.9 for the [$4^{6}6^{8}$]-cages in silica-sodalite. Therefore, it should be a general strategy to increase the active concentration of guest species in solution during synthesis for structures with higher porosity.

3 Alternative Synthesis Routes Used for Porosil Synthesis

Although hydrothermal synthesis has been very successful and has produced all porosil structure types, there is a continuous search for alternative synthesis routes. The most widely investigated ones are syntheses in nonaqueous solution and dry-gel synthesis. In both cases, the reaction is still of the solvo-thermal type and, therefore, comparable to hydrothermal synthesis. However, the conditions employed are extreme in that either the solvent water is almost completely replaced by an organic solvent or the solvent water is reduced to very low amounts.

3.1 Synthesis in Nonaqueous Solvents

Since the first synthesis of silica-sodalite from ethylene glycol solution [6] considerable efforts have been made to exploit the particular advantages of nonaqueous solvents for the synthesis of porosils [60]. Although a number of porosils were obtained, no new structure type could be synthesized so far [7, 8, 61]. Polar protic solvents usually containing water as traces or amounts up to ~5% were most successful. In Table 10 a summary of solvents used and reaction products obtained is listed. It is worthwhile mentioning that single crystals several mm in length could be grown in nonaqueous solvents, which is important for the use as advanced materials in optical, magnetic, and electronic processes [61]. Most likely, the nucleation rate for porosil crystals is very low in organic solvents, whereas crystal growth proceeds similar to aqueous conditions keeping the crystallites within the Ostwalt-Miers region.

3.2 Dry Gel Synthesis

The synthesis of porosils from dry gels takes advantage of the reactivity of the silica gel which usually contains considerable amounts of silanol groups as reactive sites. An intimate mixture of silica gel and SDA, together with very small amounts of water sufficient to activate the condensation process, is reacted in

Table 10. Porosil synthesis in non-aqueous systems

Organic solvent	Obtained porosil
Glycol + NaOH	SOD
Glycerol + NaOH	SOD
Sulfolane + NaOH	SOD
Dimethylsulfoxide + NaOH	SOD
Ethanol + NaOH	SOD
Pyridine + NaOH	SOD
C6 and C7 alcohols + NaOH	SOD
Butanol + glycol	SOD + ZSM-48
Butanol + glycol + propylamine + NaOH	MTN
Glycol + tetrapropylammonium bromide + NaOH	SOD + MFI
Pyridine/HF	MTN
Pyridine/HF + propylamine	FER
Pyridine/HF + propylamine + Tetrapropylammonium bromide	MFI

autoclaves at temperatures between 150 and 200 °C and autogenous pressures [62]. Two routes have been developed using basic [62, 63, 64] and acidic (65) conditions in vapour phase. This synthesis procedure allows the use of very high concentrations of SDA/SiO_2, so that high cage densities in the porosil structure can be obtained. In Table 11 a compilation of dry gels and SDA together with the reaction products of the different synthesis routes is given. The process has been used mainly for the reproduction of the synthesis of known nanoporous structures under modified conditions [62–68]. The finding of new templates for the synthesis of silica sodalite and the first synthesis of silica RUB-10, however, shows the potential of the method [18].

3.3
Condensation of Preorganized Layer Silicates

Recently, a new hydrous layer silicate has been synthesized which contained an intercalated organic template [69]. Careful annealing at temperatures of ~550 °C transformed the layer silicate into pure silica ferrierite (FER). An X-ray structure analysis of the starting material confirmed that the silicate layer already represents a two-dimensional slap of ferrierite which condenses in a topotactic reaction to a three-dimensional network [70]. Similar findings are reported for MCM-22 (ERB-1) [23, 71]. There are several crystalline hydrous layer-silicates which show remarkable resemblance to building blocks of porosil frameworks [72, 73]. They also might be used as precursors in order to create the zeolite framework in a secondary condensation step. In addition, intercalation of organics and grafting of the silicate surface open a multitude of possible modifications of the precursor leading to modified framework structures.

Table 11. Porosil synthesis from dry gels

Dry gel	Additional reactants	Reaction conditions pH, Temperature	Obtained porosil
$Na_2O + SiO_2 + Al_2O_3$ + (little) H_2O	Vapor of triethylamine + ethylenediamine + H_2O	basic, 453 K	MFI + FER [64]
$K_2O + Na_2O + SiO_2 + Al_2O_3$ + tetraethylammonium oxide + (little) H_2O	Vapor of H_2O	basic, 453 K	BEA [67]
$Na_2O + SiO_2 + Al_2O_3$	Vapor of triethylamine + ethylenediamine + H_2O	basic, 453 K	MFI [62]
$SiO_2 + Al_2O_3$	Tetramethylammonium cloride + NH_4F	acidic, 483 K	MTN [68]
$SiO_2 + B_2O_3$	Tetrapropylammonium bromide + NH_4F	acidic, 473 K	MFI [68]
$SiO_2 + Al_2O_3 + B_2O_3$	1,4-Diazabicyclo[2.2.2] octane + NH_4F	acidic, ?	MTW [68]
SiO_2	Cobalticinium fluoride + NH_4F	acidic, 413–433 K	NON [52]
SiO_2	Cobalticinium fluoride + NH_4F	acidic, 433–463 K	AST [52]
SiO_2	1,1-Dimethylcobalticinium fluoride + NH_4F	acidic, 413–453 K	DOH [52]
SiO_2 + (little) H_2O	Pyrrolidine + NH_4F	basic, 433 K	RUT + MTN + SOD [18]

4
Perspectives for the Future

Since its discovery almost 20 years ago [24], the number of synthetic clathrasils and zeosils have steadily increased. The variety of template molecules, which were successfully employed as structure-directing agents, has exceeded 100 showing experimentally the validity of the synthesis concept. Although there is no thorough theoretical description of the nucleation process and the synthesis mechanism, first results show that the synthesis of new materials might benefit from computer-assisted template design, increasing the efficacy of the more intuitive or combinatorial synthesis approach. The demand for new materials with even larger pores of high thermal and chemical stability is still strong. In view of the increasing importance of zeolites as catalysts for fine chemical synthesis, fine tuning of structural properties such as pore width and dimensionality, acidity and purity are important goals for the future.

In the development of new ideas, the porosils will always play an important role as model compounds of the corresponding aluminosilicates (zeolites). Porosils as simplified end-member materials allow for a detailed analytical and mechanistic study leading to a better understanding of the fundamental processes involved. The recent renaissance of the interest in all-silica species with the invention of a series of new structure types, UTD-1, ITQ-3, ITQ-4, SSZ-31, NCL-1, CIT-5, clearly supports this.

5
References

1. Meier WM, Olson DH, Baerlocher C (1996) Atlas of Zeolite Structure types, 4th edn. Elsevier, London
2. Liebau F, Gies H, Gunawardane RP, Marler B (1986) Zeolites 6:373
3. Gies H, Marler B (1996) Crystalline microporous silicas as host-guest systems. In: MacNicol DD, Toda F, Bishop R (eds) Supramolecular chemistry, vol 6. Elsevier Science, Oxford, Chap 26, 851 pp
4. Park SH, Große-Kunstleve R-W, Graetsch H, Gies H (1997) The thermal expansion of the zeolites MFI, AFI, DOH, DDR, and MTN in their calcined and as synthesized forms. In: Chon H, Ihm S-K, Uh YS (eds) Studies in Surface Science and Catalysis, vol 105 °C. Elsevier, London, p 1989
5. Szostak R (1991) Modified zeolites. In: Van Bekkum H, Flanigen EM, Janssen JC (eds) Introduction in Zeolite Science and Practice. Elsevier, Amsterdam, p 153
6. Bibby DM, Dale MP (1985) Nature 317:157
7. Huo Q, Feng S, Xu R (1989) Syntheses of pentasil silica-zeolites from nonaqueous systems. In: Jacobs P, van Santen RA (eds) Studies in Surface Science and Catalysis, vol 49. Elsevier, London, p 291
8. Kuperman A, Nadimi S, Oliver S, Ozin GA, Garces JM, Olken MM (1993) Nature 365:239
9. Marler B, Grünewald-Lüke A, Gies H (1995) Zeolites 12:388
10. Gies H, Czank M (1986) Z Kristallogr 174:64
11. Gies H (1986) Z Kristallogr 175:93
12. Caullet P, Guth JL, Hazm J, Lamblin JM, Gies H (1991) Eur J Solid State Inorg Chem 28:345
13. Stewart A (1989) Zeolites 9:140
14. Gies H, Gerke H, Liebau F (1982) Angew Chem 94(3):215
15. Gies H, Gerke H, Liebau F (1982) Neues Jahrb Mineral Monatsh 3:119

16. Marler B, Dehnbostel N, Eulert HH, Gies H, Liebau F (1986) J Inclusion Phenomena 4:339
17. Camblor MA, Corma A, Lightfoot P, Villaescusa LA, Wright PA (1997) Angew Chem 109(23):2774
18. Werthmann U, Gies H, Marler, B (in preparation) Microporous Materials
19. Parker LM, Bibby DM (1993) Zeolites 3:8
20. Franklin KR, Lowe BM (1989) Hydrothermal crystallisation of silica molecular sieves and clathrasils from amine containing reaction mixtures. In: Jacobs P, van Santen RA (eds) Studies in Surface Science and Catalysis, vol 49. Elsevier, Amsterdam, p 179
21. Schlenker JL, Rohrbaugh WJ, Chu P, Valyocsik EW, Kokotailo GT (1985) Zeolites 5:355
22. Camblor MA, Corell C, Corma A, Diaz-Cabanas MJ, Nicolopoulos S, Gonzalez-Calbet JM, Vallet-Regi M (1996) Chemistry of Materials 8:2415
23. Leonowicz ME, Lawton JA, Lawton SL, Rubin MK (1994) Science 264:1910
24. Flanigen EM, Bennett JM, Grose RW, Cohen JP, Patton RL, Kirchner RM, Smith JV (1978) Nature 271:512
25. Bibby DM, Milestone NB, Aldridge LP (1979) Nature 280:664
26. Marler B (1986) Z Kristallogr 174:141
27. Annen MJ, Davis ME (1993) Microporous Materials 1:57
28. Lobo RF, Davis ME (1994) Microporous Materials 3:61
29. Kumar R, Reddy KR, Raj A, Ratnasamy P (1993) In: von Ballmoos R, Higgins JB, Treacy MMJ (eds) Proc 9th Int Zeolite Conference. Butterworth-Heinemann, Boston, p 189
30. Lobo RF, Tsapatsis M, Freyhardt CC, Chan I, Chen C-Y, Zones SI, Davis ME (1997) J Am Chem Soc 119:3732
31. Camblor MA, Corma A, Villaescusa LA (1997) Chem Commun:749
32. Barrett PA, Camblor MA, Corma A, Jones RH, Villaescusa LA (1997) Chem Mater 9:1713
33. van der Waal JC, Rigutto MS, van Bekkum H (1994) J Chem Soc Chem Commun (1994):1241
34. Freyhardt CC, Tsapatsis M, Lobo RF, Balkus KJ, Davis ME (1996) Nature 381:295
35. Wagner P, Yoshikawa M, Lovallo M, Tsuji K, Tsapatsis M, Davis ME (1997) Chem Commun:2179
36. Gies H (1985) Nachr Chem Tech Lab 33:387
37. Gies H, Marler B (1992) Zeolites 12:42
38. Davis ME, Lobo RF (1992) Chem Mater 4:756
39. Kubota Y, Helmkamp MM, Zones SI, Davis ME (1992) Microporous Materials 6:213
40. Davis ME, Zones SI (1995) A perspective on zeolite synthesis: How do you know what you'll get? In: Occelli ML, Kessler H (eds) Synthesis of porous materials. Marcel Dekker, New York, p 1
41. Barrer RM (1982) Hydrothermal chemistry of zeolites. Academic Press, London, chaps 3 and 4
42. Szostak R (1989) Molecular sieves. Principles of synthesis and identification. Nostrand Reinhold, New York, chaps 2 and 3
43. Burkett SL, Davis ME (1995) Chem Mater 7:920
44. Jacobs PA, Derouane EG, Weitkamp J (1981) Chem Commun:59
45. Jansen JC (1991) The preparation of molecular sieves. In: van Bekkum H, Flanigen EM, Jansen JC (eds) Introduction in Zeolite Science and Practice. Elsevier, Amsterdam, p 77
46. Santilli DS, Harris TV, Zones SI (1993) Microporous Materials 1:329
47. Lewis DW, Freeman CM, Catlow CRA (1995) J Phys Chem 99:11194
48. Derouane EG, Andre J-M, Lucas AA (1988) J Catal 110:58
49. Lewis DW, Willock DJ, Catlow CRA, Thomas JM, Hutchings GJ (1996) Nature 382:604
50. Ripmeester JA, Desando MA, Handa YP, Tse JS (1988) J Chem Soc Chem Commun:608
51. Guth JL, Kessler H, Wey R (1986) In: Murakami Y, Jijima A, Ward JW (eds) New Developments in Zeolite Science and Technology. Elsevier, Amsterdam, p 121
52. van de Goor G (1995) Mikrostrukturierte SiO_2-Komposite mit organischen, metall-organischen und anorganischen Gastsubstanzen. Hartung-Gorre Verlag, Konstanz
53. Delmotte L, Soulard M, Guth F, Seive A, Lopez A, Guth JL (1990) Zeolites 10:778
54. George AR, Catlow CRA (1997) Zeolites 18:67

55. Gunawardane RP, Gies H, Liebau F. (1987) Z Anorg Allg Chem 546:189
56. Gies H (1983) Z Kristallogr 164:247
57. Gies H (1984) Z Kristallogr 167:73
58. Davis ME, Katz A, Ahmad WR (1996) Chem Mater 8:1820
59. Grebner MD, Reich A, Schüth F, Unger KK, Franz KD (1993) Zeolites 13:139
60. Braunbarth CM, Behrens P, Felsche J, van de Goor G, Wildermuth G, Engelhardt G (1995) Zeolites 16:1996
61. van Erp WA, Kouvenhoven HW, Nanne JM (1987) Zeolites 7:286
62. Xu W, Dong J, Li J, Li J, Wu F (1990) J Chem Soc Chem Commun:755
63. Kim MH, Li HX, Davis ME (1993) Microporous Materials 1:191
64. Matsukata M, Nishiyama N; Uyama (1996) Microporous Materials 7:109
65. Althoff R, Unger K, Schüth F (1994) Microporous Materials 2:557
66. van de Goor G, Lindlar B, Felsche J, Behrens P (1995) J Chem Soc Chem Commun:2559
67. Hari Prasad Rao PR, Matsukata M (1996) J Chem Soc Chem Commun:1414
68. Deforth U, Unger KK, Schüth F (1997) Microporous Materials 9:287
69. Schreyeck L, Caullet P, Mougenel JC, Guth JL, Marler B (1996) Microporous Materials 6:259
70. Schreyeck L, Caullet P, Mougenel JC, Guth JL, Marler B (1997) A new layered (alumino) silicate and its transformation into FER-type material by calcination. In: Chon H, Ihm S-K, Uh YS (eds) Studies in Surface Science and Catalysis, vol 105 C. Elsevier, London, p 1949
71. Millini R, Perego G, Parker WO, Bellussi G, Carluccio L (1995) Microporous Materials 4:221
72. Oberhagemann U, Bayat P, Marler B, Gies H, Rius J (1996) Angew Chem 108(23/24):3041
73. Gies H, Marler B, Vortmann S, Oberhagemann, Bayat P, Krink K, Rius J, Wolf I, Fyfe CA (1998) Microporous Materials (in press)

Synthesis of more Recent Aluminosilicates with a Potential in Catalysis and Adsorption

Stefan Ernst

Department of Chemistry, Chemical Technology, University of Kaiserslautern, Erwin-Schrödinger-Strasse 54, D-67663 Kaiserslautern, Germany

1 Introduction and Scope

Improvements in chemical processes are very often based on the discovery or development of new catalysts or adsorbents. One particularly exciting example in the field of zeolite catalysis is the replacement of the formerly used amorphous silica-alumina catalysts in fluid catalytic cracking (FCC) of vacuum gas oil by rare earth exchanged X-type zeolites [1]. This resulted in considerably improved yields of the desired gasoline and, hence, a much more efficient utilization of the crude oil. Further examples are the introduction of zeolite HZSM-5 as catalyst in the synthesis of ethylbenzene from benzene and ethylene [2], for xylene isomerization [3] and for the conversion of methanol to high-

Molecular Sieves, Vol. 1

octane gasoline [4]. With the advent of an increasing number of synthetic zeolites and zeolite-like materials the opportunities for improving existing and even developing new processes (and products) have strongly expanded.

It is the aim of the present paper to review the synthesis of a selected number of microporous materials which have more recently attracted the attention of zeolite scientists. The following selection criteria were applied: (i) the review is restricted to aluminosilicates, hence, isomorphously substituted zeolites, SAPOs, MeAPOs and the like are not considered, (ii) crystalline microporous aluminosilicates are exclusively dealt with, the synthesis of ordered mesoporous molecular sieves (i.e. MCM-41) is not discussed and, (iii) especially those zeolites have been selected which, based on their particular structural or physicochemical properties, have a certain potential for applications in catalysis or adsorption. Some readers might miss such important and interesting zeolites like Beta, EMC-2, ZSM-12, etc. However, although there is still intensive research going on in the synthesis of these zeolites, their main synthesis principles have been summarized earlier [5, 6] and will therefore not be discussed here.

In the following, the discussion of the syntheses of the selected zeolites is ordered based on their largest pore apertures in the sequence 8-, 10-, 12- and 14-membered ring openings.

2 Eight-Membered Ring Zeolites

Among the small pore zeolites, erionite and zeolite A have been or are currently in use commercially. A bifunctional erionite-type catalyst (i.e. in its acidic form and loaded with small amounts of nickel as hydrogenation/dehydrogenation component) was applied in Mobil's Selectoforming process, which was designed to increase the octane number of the (gasoline) product from catalytic reforming by shape selective hydrocracking of linear alkanes [7]. This process was later replaced by the so-called M-Forming, which uses a ZSM-5-type catalyst [8].

Zeolite A is extensively used as selective adsorbent in several purification and separation processes [9]. The most prominent examples are the enrichment of oxygen from air [e.g. 9, 10] and the separation of linear and branched alkanes [e.g. 9, 11]. Among others, zeolites Rho and ZK-5 have been studied in detail as catalysts for the selective amination of methanol with ammonia to dimethylamine [e.g. 12, 13]. Also, the potential of small pore zeolites for the conversion of methanol to lower olefins has been explored [e.g. 14, 15].

More recently, zeolite ZSM-58 [16–21] has been identified by a couple of groups as a promising adsorbent and catalyst or catalyst support. ZSM-58 is a high-silica zeolite [16, 17] isotypic with the clathrasil deca-dodecasil 3R [22, 23] and zeolite Sigma-1 [18]. Its structure consists of larger intracrystalline cavities which are accessible through windows formed from 8-membered oxygen rings [22, 23].

Zeolite ZSM-58 (and its counterpart Sigma-1) have been found useful in the conversion of methanol to mainly lower olefins [16, 18] and for the oligomerization of propylene [17]. Moreover, it has been shown that shape selective hydrogenation catalysts can be prepared by solid-state ion exchange of zeolite HZSM-58 with noble metal salts (e.g. $PtCl_2$ or $PdCl_2$). The resulting catalysts selectively hydrogenate linear olefins in a mixture with trimethyl-branched olefins [21].

As one would have expected based on its structure, aluminum-free deca-dodecasil 3R readily adsorbs small molecules like CO_2, N_2O and NH_3 [20]. On the other hand, alkanes and alkenes with three or four carbon atoms show only retarded adsorption at 296 K, while 1,3-butadiene is readily adsorbed under these conditions. This peculiar behavior has been attributed to the more elliptical cross section of the diolefin due to the presence of sp^2-hybridized carbon atoms. The former obviously has a better "fit" to the elliptical windows in the DDR structure than the circular cross section of saturated hydrocarbons [20]. In the following, the synthesis of materials with the DDR framework topology will be described in more detail.

2.1
Synthesis of Materials with the DDR Framework Topology

So far, essentially two templates have been reported to promote the crystallization of materials with the DDR framework topology, viz. 1-aminoadamantane (ADA) and methyltropiniumiodide. Their structures are depicted in Fig. 1. Both are relatively bulky molecules which, after hydrothermal synthesis, are encapsulated inside the large cavities of the DDR structure and have to be removed by calcination in an oxygen-containing atmosphere at temperatures of at least 813 K. Materials which have been prepared using 1-aminoadamantane are denoted as deca-dodecasil 3R [20, 22] or Sigma-1 [18], while zeolite ZSM-58 is prepared using methyltropiniumiodide [16, 17, 19].

The synthesis of the pure silica form of deca-dodecasil 3R was first described by Gies [22, 24] and later adapted by den Exter et al. [20]. A typical molar composition of the synthesis gel is 47 ADA : 100 SiO_2 : 404 EN : 11240 H_2O. Tetramethylorthosilicate is used as silica source and hydrolyzed through the addition of the base ethylenediamine (EN). The crystallization occurs in glass tubes [22,

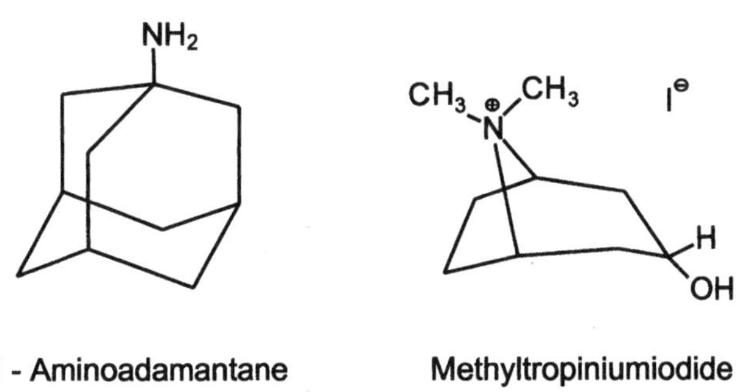

Fig. 1. Templates reported in the literature for the synthesis of zeolites with the DDR framework topology

Table 1. Influence of temperature, dilution and degree of agitation on the crystallization of a gel with the molar composition 47 ADA : 100 SiO_2 : 404 EN : x H_2O (DOH denotes the clathrasil dodecasil 1H); after [20]

T, °C	x	t, days	rotation	product(s)
190	5600	11	–	DOH
160	5600	25	–	DOH/DDR
160	7450	25	–	DOH/DDR
160	11240	25	–	DOH/DDR
160	11240	25	+	DDR

24] or in Teflon-lined stainless steel autoclaves at temperatures around 433 K. Recently, den Exter et al. [20] investigated the influence of the crystallization temperature, the dilution of the gel and the degree of agitation during synthesis on the nature of the products. Some pertinent results are summarized in Table 1. It can be seen that the formation of the hexagonal six-membered ring clathrasil dodecasil 1H (DOH) is favored by high temperatures and concentrations, while (still impure) deca-dodecasil 3R could only be obtained at lower temperatures (433 K) and from diluted mixtures. In all cases, relatively large crystallites with average sizes around 125 µm to 150 µm were obtained. Pure DDR was finally crystallized when a low temperature and a high dilution were combined with agitation (rotating the autoclave) during synthesis (cf. Table 1, [20]). The crystallites had now an average size of 5 to 10 µm. The authors interpreted this observation with the occurrence of a faster nucleation upon rotating the autoclave. The results presented in Table 1 were obtained using autoclaves with an internal volume of 45 cm^3 which had been filled to two thirds with the synthesis mixture. It is reported, however, that an up-scaling to a 900 cm^3 autoclave has also been succesful, which yielded 20 g of deca-dodecasil 3R [20].

Stewart et al. [18] describe the preparation of aluminum-containing deca-dodecasil 3R from gels containing 1-aminoadamantane as structure directing agent and using silica sol as silica source and sodium aluminate as source of alumina (in addition, NaOH was used occasionally to adjust the pH or sodium content). A typical gel composition was as follows: 3 to 5 Na_2O : 20 ADA : 1 Al_2O_3 : 60 SiO_2 : 2400 H_2O. After reacting this mixture for five to six days at 453 K in a stirred autoclave (stirring rate: 500 min^{-1}) pure Sigma–1 was obtained. As compared to the gel composition for the pure silica form (vide supra), a lower template concentration and a more concentrated gel can be applied for the synthesis of an aluminosilicate with the DDR structure. In addition, a higher synthesis temperature and concomitantly shorter synthesis times are possible. Hence, the presence of aluminum in the synthesis gel and a relatively low pH ($pH \approx 11.7$) seem to favor the crystallization of the DDR phase. However, this is only true for a very narrow range of SiO_2/Al_2O_3. Increasing this ratio from 60 to ca. 100 and keeping all other conditions constant results in the co-crystallization of zeolite Sigma–2, which is a dense material [25]. Sigma–2 (aluminum-free) can be obtained in the pure form from the corresponding Al-free gel. If, on the other hand, the SiO_2/Al_2O_3 ratio of the gel is decreased to 40 and further to 30, Sigma-1 first contains zeolite NU-3 as an impurity and finally NU-3 is obtained as a pure phase [18]. Zeolite NU-3 has the framework topology of levyne (LEV, [26]), viz. it is an eight-membered ring zeolite.

Valyocsik et al. [16] and Rodewald and Valyocsik [17] describe the synthesis of materials with the DDR topology from gels containing methyltropiniumiodide (MTI) as template. This organic is normally not commercially available, but it can be easily prepared by refluxing tropine with iodomethane in ethanolic solution for three days. Typically, a gel with the molar composition 11.5 Na_2O : 17.5 MTI : 1 Al_2O_3 : 70 SiO_2 : 2800 H_2O is prepared from silica sol, sodium aluminate, sodium hydroxide, methyltropiniumiodide and water. The synthesis is performed in stirred stainless steel autoclaves at 433 K for four days. It is claimed that SiO_2/Al_2O_3 can be varied from 50 to 1000.

Based on the information from the patent literature [16, 17], Ernst and Weitkamp systematically varied the template content and the SiO_2/Al_2O_3 ratio of the gel [19]. The X-ray powder pattern of a ZSM-58 sample, synthesized from a gel with the above molar composition and calcined in air for 16 hours at 813 K to remove the template, is shown in Fig. 2. It reveals good crystallinity and phase purity. The scanning electron micrograph of this sample (cf. Fig. 3) shows the presence of ball-shaped crystallites with diameters from 1 to 2 µm. This is comparable to the morphology reported for Sigma-1 crystallites, viz. aluminum-containing DDR synthesized in the presence of 1-aminoadamantane.

The influence of the aluminum content of the gel on the synthesis products is summarized in Table 2. It can be seen that ZSM-58 can be obtained from gels with $SiO_2/Al_2O_3 = 40$ up to virtually aluminum-free gels. With increasing alu-

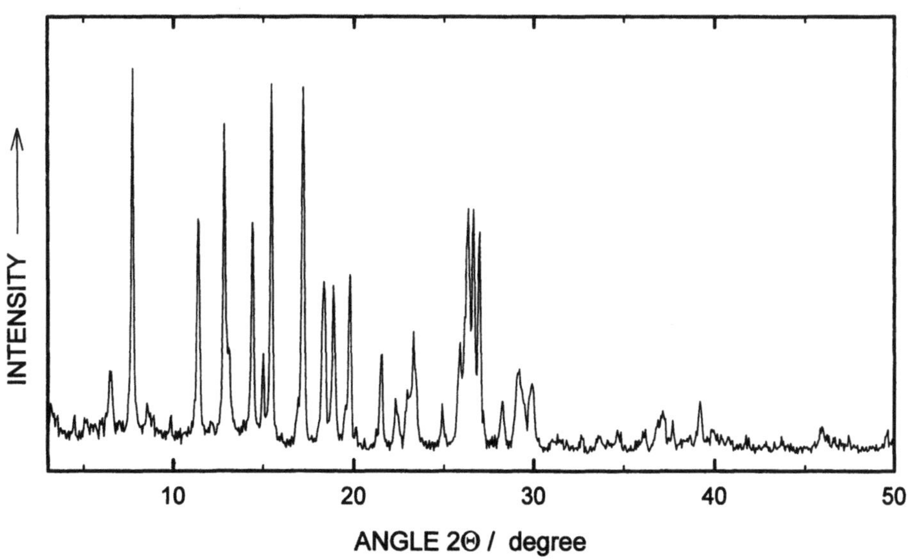

Fig. 2. Typical X-ray powder pattern for zeolite ZSM-58 after calcination in air for 16 hours at 813 K (CuKα-radiation)

Fig. 3. Typical crystallite size and morphology of a "standard" ZSM-58 sample

Table 2. Variation of the aluminum content in the gel for the synthesis of zeolite ZSM-58. Molar gel composition: 11.5 Na_2O : 17.5 MTI : x Al_2O_3 : 70 SiO_2 : 2800 H_2O ($x = 0$ to 7; $T = 433$ K); after [19]

$n_{SiO_2}/n_{Al_2O_3}$	t, days	product
10	4	Phillipsite
20	10	amorphous
40	6	ZSM-58
70	4	ZSM-58
100	4	ZSM-58
250	4	ZSM-58
∞	3	ZSM-58

minum content, the time required for complete crystallization increases. For SiO_2/Al_2O_3 below 40, either an amorphous phase or a crystalline phase with the phillipsite topology are obtained. With increasing SiO_2/Al_2O_3 ratio of the gel, the crystallite size of ZSM-58 increases, and the shape of the crystallites becomes more regular. Figure 4 shows SiO_2 with the structure of ZSM-58. This material crystallizes in the shape of prismatic particles with a diameter up to ca. 30 µm and a thickness of ca. 15 µm.

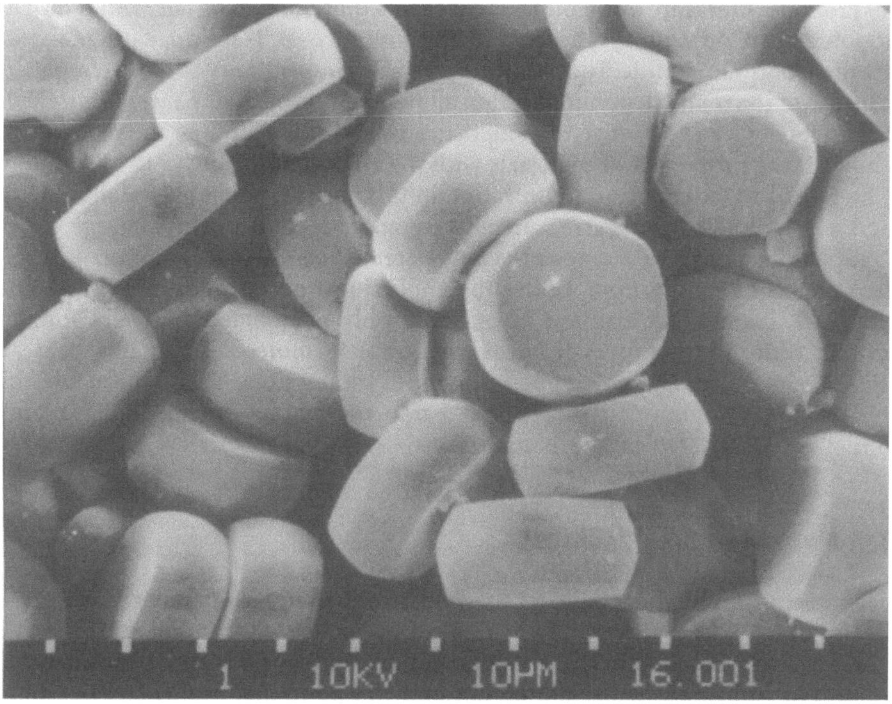

Fig. 4. Scanning electron micrograph of zeolite ZSM-58 obtained from an aluminum-free gel

The template concentration in the gel composition for the synthesis of zeolite ZSM-58 initially derived from the patent literature (vide supra) was relatively high (MTI/SiO_2= 0.25). Therefore, Ernst and Weitkamp also systematically varied the template content of the gel between 0.30 and zero [19] keeping all other factors constant. These authors found that MTI/SiO_2 can be reduced down to 0.15 (i.e. nearly one half of the initial value) without significantly affecting the quality of the crystalline product. Upon further reduction of the template content, and even in the absence of methyltropiniumiodide and in new autoclaves, pure zeolite ZSM-5 was obtained. Hence, by chance a template-free synthesis for zeolite ZSM-5 was discovered.

From the variation of the aluminum content in zeolite ZSM-58 it is possible to tailor the density of acid sites for a given catalytic application. Frequently, however, it is also desirable to systematically vary the strength of the acid sites. It has been demonstrated for, e.g. zeolite ZSM-5 that this can be achieved by substituting the aluminum source in the synthesis gel with the corresponding gallium, iron or boron compounds (e.g. [27, 28]). This so-called isomorphous substitution has been found to be also possible for the framework of zeolite ZSM-58 if the following molar gel composition is applied: 11.5 Na_2O : 17.5 MTI : 1 Me_2O_3 : 70 SiO_2 : 2800 H_2O and $Ga(NO_3)_3 \cdot 9\,H_2O$, $Fe_2(SO_4)_3 \cdot 7\,H_2O$ and H_3BO_3 were used as sources for the respective Me^{3+} cations [29].

From the literature available so far, it seems that pure materials with the DDR topology can only be synthesized in the presence of 1-aminoadamantane or methyltropinium iodide. This points to a very high specificity of these templates. The use of the methyltropinium cation allows the synthesis of ZSM-58-type zeolites with a very wide range of compositions, and it is therefore the template of choice if the properties of materials with the DDR topology have to be tailored. The role of the methyltropinium cation in this case is probably twofold: First it acts as a template for the structure and second, due to its positive charge, it can compensate a negative framework charge created by the incorporation of trivalent metals like aluminum. Under the applied synthesis conditions for zeolite Sigma-1, 1-aminoadamantane is certainly only protonated to a very limited extent, while the major portion of the template molecules remains neutral. Hence, almost no charge compensation will be possible. This is probably the reason why aluminum is incorporated into the framework of Sigma-1 only under very restricted synthesis conditions.

3 Ten-Membered Ring Zeolites

The most prominent example for an industrially used ten-membered ring zeolite is ZSM-5. The discovery of this zeolite by Argauer and Landolt more than 25 years ago [30] resulted in the development of a number of new or considerably improved processes in the petroleum and the petrochemical industries (e.g. [3]). Until very recently, however, there was virtually no other member of the group of medium-pore zeolites in use commercially. However, this situation has changed completely in the past years: a bifunctional form of the microporous silicoaluminophosphate SAPO-11 is used as catalyst for the dewaxing of lubricating oils by isomerization [e.g. 31]; a zeolite type from the ferrierite family is probably being commercialized as catalyst for the isomerization of n-butene to iso-butenes [e.g. 32], and finally zeolite HMCM-22, which is reported to be superior to zeolite HZSM-5 as a catalyst for benzene alkylation with ethylene [33]. On the basis of these examples it has to be expected that additional processes using medium pore zeolites as catalysts will emerge in the future.

Given their particular structures and on the basis of our own experience with these materials in catalysis and adsorption, zeolites MCM-22 [34] and NU-87 [35] seem to have a certain potential as catalysts and/or adsorbents. Therefore, the present section will focus entirely on the synthesis of these two materials.

The structure of zeolite MCM-22 (structure code MWW) was determined by Leonowicz et al. [34]. In brief, it contains large supercages (ca. 0.71 nm in diameter and ca. 1.8 nm in height) which are accessible via ten-membered ring windows, and an independent second channel system of two-dimensional sinusoidal channels with ten-membered ring openings. In agreement with this model it has been concluded from a Spaciousness Index of 8 that the effective pore size of MCM-22 is in the intermediate region between those of ten-membered ring zeolites and the very open twelve-membered ring structures [36]. Further materials with the MWW framework topology are zeolites PSH-3, which was in fact synthesized before the disclosure of MCM-22 [37], zeolite SSZ-25 [38] and the borosilicate ERB-1 [39].

Zeolite MCM-22 in its Brønsted-acid form has been described in the literature as a useful catalyst for a variety of acid-catalyzed reactions, such as iso-alkane/olefin alkylation [e.g. 40, 41], skeletal and double-bond isomerization of olefins [42] and ethylbenzene synthesis via alkylation of benzene with ethylene [43], to name merely a few. Moreover, due to its very large intracrystalline cavities, zeolite MCM-22 has also been demonstrated to be a suitable host material for a variety of catalytically active guests, e.g. transition metal complexes which are useful in selective oxidation [44] or hydrogenation [45] reactions. Due to these interesting properties it seems worthwhile to focus on the synthesis features of MCM-22 (see below).

The principal structural features of zeolite NU-87 (structure code NES), a high-silica zeolite, are quite similar to those of materials with the MWW topology, viz. larger intracrystalline cavities which are accessible via ten-membered rings only [35]. In more detail, the NES topology consists of parallel ten-membered ring channels with a crystallographic diameter of 0.47×0.60 nm. These channels are interconnected by larger twelve-membered ring "bridges" [35, 46]. Currently, there is only one further synthetic material known to possess the NES topology, viz. zeolite SSZ-37 [47, 48], which can be synthesized using an organic template derived from a Diels-Alder reaction. Most interestingly, a natural counterpart of zeolite NU-87, designated Gottardiite, has recently been discovered in the Antarctica [49, 50]. Its SiO_2/Al_2O_3 ratio of 12.4 is the highest found so far in nature. From its further chemical composition (containing Na^+, K^+, Mg^{2+} and Ca^{2+} cations) and the geologic conditions at the place where Gottardiite was found, it could probably be possible to deduce synthesis conditions and gel compositions which allow the synthesis of NU-87 with high aluminum content and without the use of an organic.

Zeolite NU-87, if containing Brønsted-acid sites, is an active catalyst for a large variety of acid catalyzed reactions like toluene disproportionation, alkylation of benzene with ethylene, amination of methanol to methylamines etc. [51]. Moreover, it was found to possess interesting shape selective properties in the conversion of m-xylene [52] and of polynuclear aromatics, e.g. methylnaphthalenes [53]. On non-acidic (i.e. Cs^+-exchanged) zeolite NU-87, loaded with small amounts of platinum, n-alkanes like n-hexane or n-octane can be dehydrocyclized in high yields to the corresponding aromatics [54].

3.1
Synthesis of Zeolite MCM-22

The synthesis of zeolite PSH-3, the first material having the MWW topology, was disclosed by Puppe and Weisser as early as 1982 using hexamethyleneimine (cf. Fig. 5) as structure directing agent [37]. The same organic was later disclosed as a template for the synthesis of zeolite MCM-22 [55]. For the synthesis of the isostructural zeolite SSZ-25, an adamantane-derived template (*N,N,N*-trimethyl-1-adamantammonium-hydroxide, cf. Fig. 5) is used [47, 48]. Due to the fact that hexamethyleneimine (HMI) is a commercially readily available chemi-

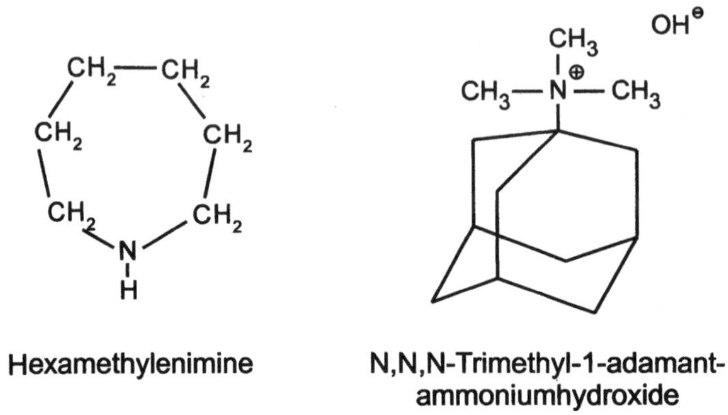

Fig. 5. Organic templates used for the synthesis of zeolites with the MWW framework topology

cal and the use of this template molecule for the synthesis of materials with the MWW topology predominates in the scientific literature, the present review will focus on the the use of HMI as structure directing agent.

Beside those disclosed in the patent literature, several additional recipes for the synthesis of zeolite MCM-22 have recently been published [36, 56–59]. In the following, two typical recipes which were succesfully used several times in our own group will be presented [36]. The first one is based on Mobil's patents [40, 42, 55]. In this case, fumed silica (Cab-*o*-sil M5, Fluka) and sodium aluminate (Riedel-de Haën, 54 wt.-% Al_2O_3, 41 wt.-% Na_2O) are used as silica and alumina sources, respectively. The synthesis gel is prepared as follows: A solution of 0.1 g concentrated sulfuric acid (98 wt.-%) in 5 g distilled water is slowly added under stirring to a solution of 0.53 g NaOH and 0.86 g sodium aluminate in 20 g distilled water. The resulting solution is stirred into a suspension of 8.8 g fumed silica in 85 g H_2O. Finally, 4.8 g hexamethyleneimine (HMI) are blended thoroughly into the mixture. The molar composition of the resulting gel is 32.2 $SiO_2 : Al_2O_3 : 2.7\ Na_2O : 10.6\ HMI : 1333\ H_2O : 0.22\ H_2SO_4$. From this gel, MCM-22 with $SiO_2/Al_2O_3 = 22$ is obtained after 12 days at 423 K under continuous agitation. In the second synthesis route silica sol (30 wt.-% SiO_2 in water, Levasil VP 4038, Bayer AG) and $Al_2(SO_4)_3 \cdot 18\ H_2O$ (Riedel-de Haën) are used as silica and alumina sources, respectively. The mode of gel preparation resembles the one employed for the former synthesis recipe, however, the molar gel composition is now closer to the one typically used in the initial patent on the synthesis of zeolite PSH-3 [37], viz. 49.3 $SiO_2 : Al_2O_3 : 14.5\ Na_2O : 23.7\ HMI : 1980\ H_2O : 8.8\ H_2SO_4$. Crystallization of MCM-22 is achieved after ten days at 423 K under continuous agitation and results in a material with $SiO_2/Al_2O_3 = 42$. Note that this latter synthesis procedure has also been successfully scaled up to a stirred autoclave (stirring rate = 150 min^{-1}) with a volume of 5 l. The yield of calcined product then amounted to ca. 250 g as compared to less than 10 g for the small-scale preparation [36].

A typical X-ray powder pattern obtained for uncalcined zeolite MCM-22 is depicted in Fig. 6. The pattern for this sample contains both sharp and broad

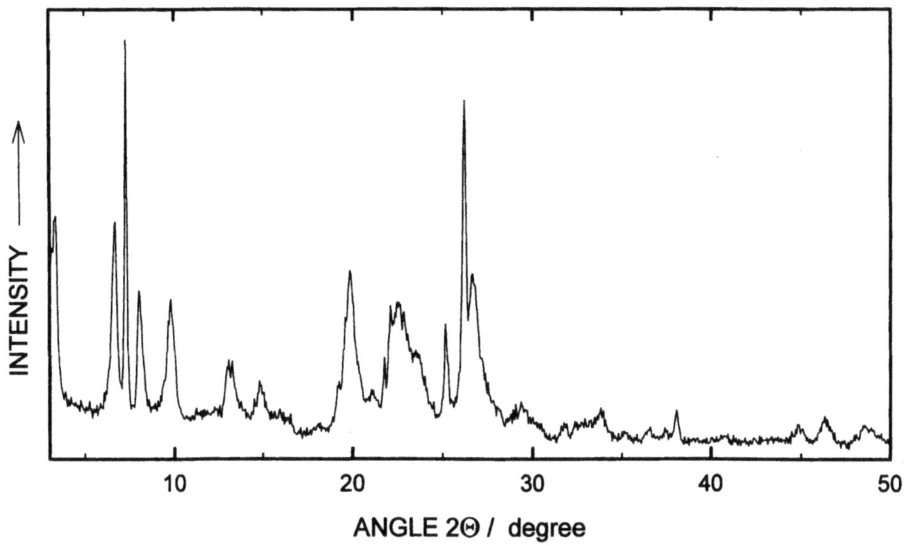

Fig. 6. Typical X-ray powder pattern of the as-synthesized (uncalcined) zeolite MCM-22 precursor

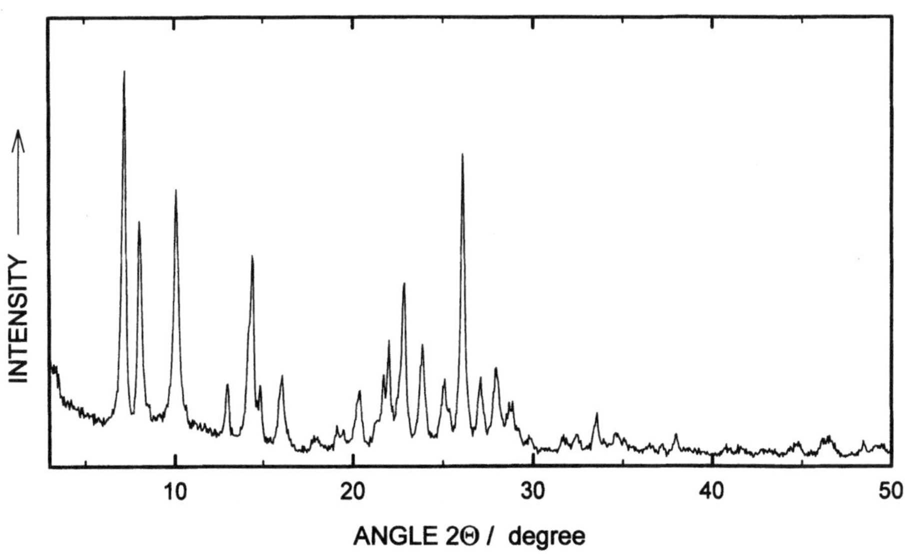

Fig. 7. X-ray powder pattern of zeolite MCM-22 after calcination in air at 813 K

reflections indicative of structural disorder. After calcination at 813 K significant changes in the X-ray powder pattern are observed (cf. Fig. 7), which cannot only be accounted for by the template removal. This phenomenon was investigated in much detail by Millini et al. for the case of the borosilicate ERB-1, which is isostructural with MCM-22 [39]. They concluded that as-synthesized ERB-1 has a bi-dimensional layered structure which, during calcination at 543 K, form the three-dimensional microporous structure known for MCM-22. From X-ray powder diffraction measurements as well as temperature programmed desorption and ^{29}Si-MAS-NMR studies they were able to show that the formation of the three-dimensionally ordered structure occurs through the condensation of silanol groups present on the surface of the layers of the "precursor" material [39].

The typical crystallite morphology of zeolite MCM-22 is shown in Fig. 8. Usually the crystallites possess a disk-like shape with a diameter below 1 µm and a thickness of ca. 0.1 µm only. Frequently, these disks are grown together.

Recently, the synthesis of zeolite MCM-22 (or of its precursor) in the presence of hexamethyleneimine has been studied by several groups in some detail [36, 56–59]. Corma et al. reported that for SiO_2/Al_2O_3 ratios around 30, zeolite MCM-22 can be obtained as a pure phase and with good yields in a relatively broad range of the OH/SiO_2 ratios (0.18 to 0.30), however, it was found that, with increasing pH, the efficiency of silicon incorporation into the framework decreases [57]. The water content of the synthesis gel does not seem to have a significant influence on the crystallization process. One major factor which

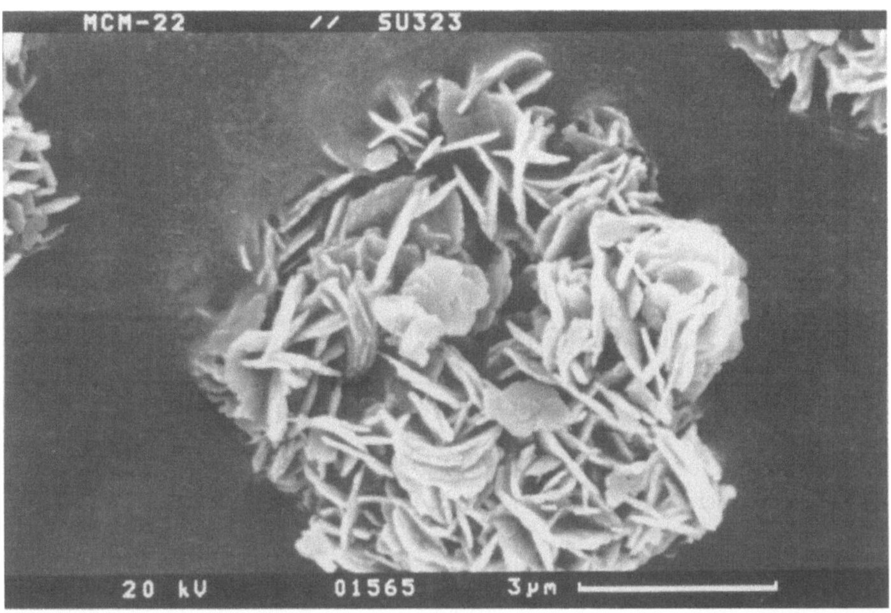

Fig. 8. Scanning electron micrograph of zeolite MCM-22

influences the synthesis of zeolite MCM-22 seems to be the SiO_2/Al_2O_3 ratio of the gel. Although MCM-22 is still formed if this ratio is increased from 30 to 70 or higher, the risk of contamination of the product with zeolite ZSM-5 increases. To combat the ZSM-5 formation, a reduction of the routinely applied synthesis temperature of 423 K to 408 K was found useful. Further impurity phases which may crystallize are zeolite ZSM-12 (for $SiO_2/Al_2O_3 = 200$, $OH/SiO_2 = 0.12$ and a synthesis temperature of 408 K) and a ferrierite-type material which is formed if the crystallizing gel is not agitated. The occurrence of a relatively narrow range of SiO_2/Al_2O_3 ratios from which zeolite MCM-22 can be crystallized has been attributed to difficulties in the nucleation step for low aluminum concentrations. The authors therefore speculate that it should be possible to synthesize MCM-22 with very low aluminum content if its nucleation rate could be enhanced [57].

In a systematic study, Mochida et al. investigated the influence of the aluminum content, the concentration of hexamethyleneimine and the addition of seeds on the synthesis time and products [59]. In agreement with Corma et al. [57] they found that MCM-22 can only be synthesized as a pure phase from a relatively narrow range of aluminum concentrations in the gel. Moreover, they were able to reduce the synthesis time required at 443 K for complete crystallization of a gel with $SiO_2/Al_2O_3 = 25.1$ from ca. 90 hours down to 40 hours or less by seeding (cf. Fig. 9). The seeds used in this study were obtained by crystallizing a similar synthesis mixture for ca. 22 hours at 443 K. This yielded a seeding material which exhibited only a very low crystallinity as determined by X-ray powder

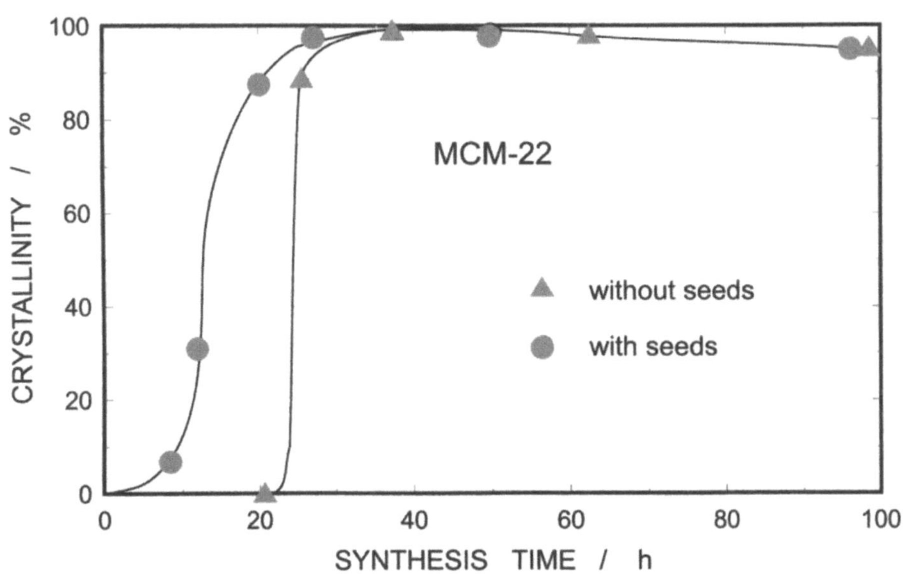

Fig. 9. Crystallization kinetics of zeolite MCM-22 from an unseeded and a seeded synthesis mixture under otherwise identical conditions (ca. 0.7 wt.-% seeds based on the silica content of the gel have been used) [59]

diffraction. Roughly 0.7 wt.-% of such seeds were used, based on the amount of silica in the synthesis gel. Appropriate seeding could also be a versatile tool to overcome the problems of MCM-22 nucleation in the case of low aluminum concentrations in the synthesis gel (vide supra). In this way it should in principle be possible to synthesize MCM-22 with higher SiO_2/Al_2O_3 ratios as reported so far.

For quite a number of crystalline aluminosilicates the gallium- and iron-silicate analogues have been synthesized in the past as well. Therefore, it is not too surprising that the isomorphous substitution of gallium [60] or iron [61] in the MWW framework has recently been reported. Again, as in the case of zeolite ZSM-58, this offers the opportunity of fine-tuning the strength of the Brønsted-acid sites in such materials.

In the typical synthesis recipes for the preparation of the MCM-22 precursor, the molar ratio of the organic template to inorganic cations in the gel (HMI/Na) is always higher than 2.0. If this ratio is decreased to a value below 2.0, a material is formed which, already in its uncalcined state, has an X-ray powder pattern very similar to that of calcined MCM-22. This material has been designated MCM-49 by Mobil researchers [62, 63].

In as-synthesized MCM-22- or ERB-1-type materials, the majority of the organic template used for the synthesis is intercalated between the layers and can be replaced by other molecules [39]. Based on these intercalation properties of the MCM-22 precursor the preparation of a new material (MCM-36) has been claimed which is based on the swelling of the MCM-22 precursor with cetyltrimethylammonium cations followed by pillaring the swollen material with tetraethylorthosilicate, hydrolysis with water and calcination at 813 K [64-66]. This procedure leads to materials with distinct adsorption properties as compared to zeolite MCM-22. From the measured pore size distributions and the adsorption equilibrium and kinetics of a variety of hydrocarbons it has been concluded that pillaring of the MCM-22 precursor which leads to MCM-36 most probably increases the accessibility of one kind of pore, while others (most likely the sinusoidal ten-membered ring channels) remain unaffected [66]. Hence, by expanding the distance between the layers in the MCM-22 structure through pillaring, in a sense a hybrid material is obtained which possesses micro- and mesoporous regions at the same time. It has to be seen in the future, whether this interesting feature turns out to be of advantage in adsorptive and/or catalytic applications.

3.2 Synthesis of Zeolite NU-87

The synthesis of zeolite NU-87 was first reported by Casci and Stewart in 1990 [67]. As structure directing agent, bis-quaternary organic compounds (cf. Fig. 10) were claimed, with decane-1,10-bis-(trimethylammonium)dibromide (decamethonium-bromide, "diquat"; $DeBr_2$) as the preferred template. Typically, SiO_2/Al_2O_3 ratios from 25 to 60 and $NaOH/SiO_2$ ratios around 0.3 are applied. A recommended synthesis mixture is prepared from silica sol, sodium aluminate, sodium hydroxide, water and the template. In our own experiments for the

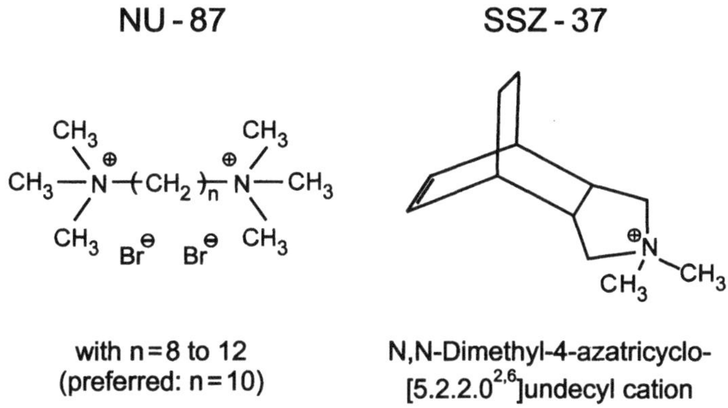

Fig. 10. Structure directing agents used for the synthesis of zeolites NU-87 and SSZ-37 (both possessing the NES framework topology)

optimization of the synthesis of zeolite NU-87 we found the following procedure useful and reproducible [53]: A solution of 0.38 g sodium aluminate (Riedel-de Haën, 54 wt.-% Al_2O_3, 41 wt.-% Na_2O) and 1.59 g sodium hydroxide in 47 g water is added under continuous stirring to 22.2 g colloidal silica sol (30 wt.-% SiO_2 in water, Levasil VP 4038, Bayer AG). To the resulting mixture a second solution is stirred consisting of 5.8 g decamethoniumbromide (Fluka) dissolved in 37.5 g water. The resulting gel had the following molar composition: 55 SiO_2 : Al_2O_3 : 11.5 Na_2O : 6.8 $DeBr_2$: 2800 H_2O : 2.4 NaBr, which is very close to the one recommended by Casci et al. [46, 67]. For crystallization the gel is transferred into a stainless steel autoclave with a volume of ca. 300 cm^3 and rotated with a frequency of 30 min^{-1} at 453 K for ten days. The X-ray powder pattern of zeolite NU-87 synthesized in this way is depicted in Fig. 11. It agrees very well with data reported in the literature [46, 67]. The crystallite size and morphology of the synthesized sample can be seen from the scanning electron micrograph depicted in Fig. 12: The crystallites have the typical cuboid shape with a length of around 2 μm and a thickness of ca. 0.5 μm.

Beside the publications of Casci et al. [46, 67] there are apparently only the reports from Moini et al. which deal (in part) with the synthesis of zeolite NU-87 [68, 69]. These authors investigated the influence of the length of the alkane chain in the diquaternary ammonium compounds on the crystallizing zeolite structure. Under their conditions it was found that the decamethonium cation is the only diquaternary compound which promotes the formation of the NES topology. Moreover, slight variations of pH, the nature of the alkali cation used or the water content resulted in the formation of other crystalline phases such as ZSM-12 and ZSM-5. From additional studies using computer modelling they proposed as location of the diquaternary cation the twelve-membered ring channels of NU-87 which interconnect the parallel ten-membered ring pores. This very specific geometrical orientation of the organic species suggests the existence of a strong templating effect during synthesis and could explain the strong influence of the size of the diquats on the structures formed [68, 69].

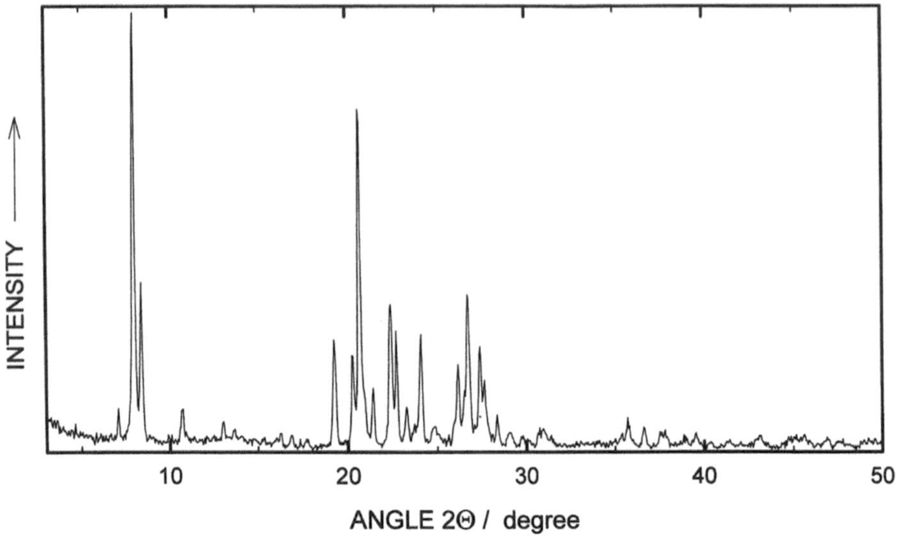

Fig. 11. X-ray powder pattern of as-synthesized zeolite NU-87

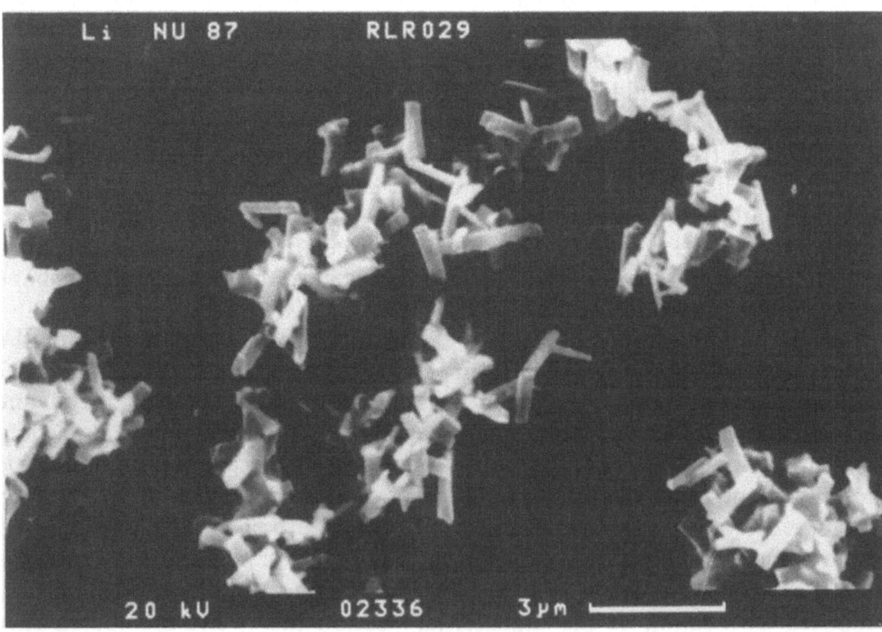

Fig. 12. Scanning electron micrograph of zeolite NU-87

Zeolite SSZ-37 is believed to possess the NES framework topology as well. It was first synthesized by Nakagawa [47] using a Diels-Alder derived template, viz. the N,N-dimethyl-4-azatricyclo[5.2.2.0^{2,6}]undecyl cation (cf. Fig. 10). By contrast to the preferred range of SiO_2/Al_2O_3 ratios from 25 to 60 in the synthesis of zeolite NU-87, SSZ-37 is preferentially synthesized from gels having lower aluminum contents, i.e. with SiO_2/Al_2O_3 starting from ca. 60. Below this value, a material with the chabazite structure begins to co-crystallize with SSZ-37. From all-silica synthesis gels, zeolite SSZ-31 crystallizes [70]. A structural model for this zeolite was recently presented [71]. In essence, SSZ-31 is a one-dimensional large pore zeolite. Its pore apertures are elliptical with major and minor axes of ca. 0.86 × 0.57 nm. It is a very complex structure which can be described as an intergrowth of four structurally related but different polymorphs. Due to its very large pore diameter and the very small crystallite sizes usually obtained, zeolite SSZ-31 is expected to be a good acid catalyst with unique selectivities [71]. From a comparison of the X-ray powder pattern of zeolite SSZ-31 with that of zeolite NCL-1 [72, 73] one can suspect that both materials are structurally related. Zeolite NCL-1 is synthesized from aluminosilicate gels containing hexane-bis-(triethylammonium) cations [72, 73].

4 Twelve-Membered Ring Zeolites

There is an ongoing trend towards shape selective catalysis involving larger reactants, transition states and products [74–76]. One major reason for this development is the increasing number of investigations related to the zeolite-catalyzed preparation of organic intermediates, fine chemicals and even pharmaceuticals. The molecules in such reactions usually are much bulkier than those encountered in typical processes of the petroleum or petrochemical industries (e.g. linear and branched alkanes and olefins; mononuclear or at best binuclear aromatics). Hence, there is an increasing need for new large-pore zeolite structures in order to enable the selection of tailored catalysts for a given application. Fortunately, in recent years quite a number of (sometimes somewhat "older") zeolites have been recognized to be large-pore materials, and their synthesis has been explored. In addition, the proper choice of structure directing agents has brought us new large-pore materials. The synthesis of selected large-pore aluminosilicates will be reviewed in the present chapter.

4.1 Synthesis of Zeolite ZSM-10

Zeolite ZSM-10 was first synthesized by Ciric more than 25 years ago from aluminosilicate gels containing potassium cations and a diquaternary compound, viz. the 1,4-dimethyl-diazonia(2.2.2)bicyclooctane cation (cf. Fig. 13; [77]). Since its first description, only two further reports on the synthesis appeared, one Japanese patent [78] and a very recent publication by Higgins and Schmitt [79], the latter also describing the framework topology of zeolite

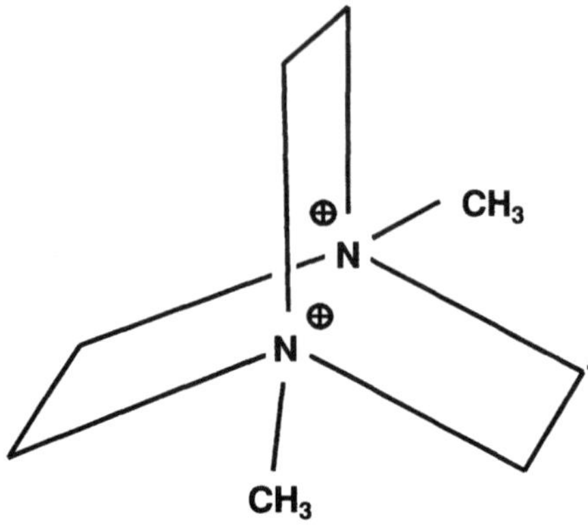

1,4-Dimethyldiazonia[2.2.2]bicyclooctane

Fig. 13. Organic cation used in the synthesis of zeolite ZSM–10

ZSM–10. Its structure is made up from columns of alternating cancrinite cages and double six-rings. The resulting pore system contains two different one-dimensional twelve-membered ring channels which are interconnected by distorted eight-membered ring channels. These are probably too small to permit the diffusion of organic molecules. One of the two twelve-membered ring channels is identical to the one found in zeolite L (LTL), the other one is the channel occurring in zeolite offretite (OFF). There are twice as many offretite-type channels as L-type channels in the structure of zeolite ZSM–10.

The synthesis of zeolite ZSM–10 and the role of the organic cation during its synthesis were studied in detail by Higgins and Schmitt [79]. They found that the result of the synthesis mainly depends on the presence of the organic cation during the nucleation/aging step: if there is no aging step at room temperature and direct heating of the synthesis mixture to crystallization temperatures of 373 or 413 K is applied, no ZSM–10 is obtained. The authors were able to show that this is due to the decomposition of the organic cation which starts already under mild conditions. If, on the other hand, an aging step at room temperature is involved (ca. 3 days) or the synthesis mixture is heated with a slow heating rate (i.e. over two days) to the final crystallization temperature, zeolite ZSM–10 could be synthesized. This suggests that the organic is indeed required to promote nucleation of zeolite ZSM–10. Once nucleated, the "template" seems to be no longer necessary for the growth of the crystallites. This is in-line with the observation, that the directing agent is not incorporated in the final product [79]. It will be very interesting to explore the shape selective properties of this peculiar zeolite structure.

4.2 Synthesis of Zeolite ZSM–18

As in the case of zeolite ZSM–10, the first synthesis of zeolite ZSM–18 was first reported decades ago [80]. Ciric used a quite unusually large and relatively

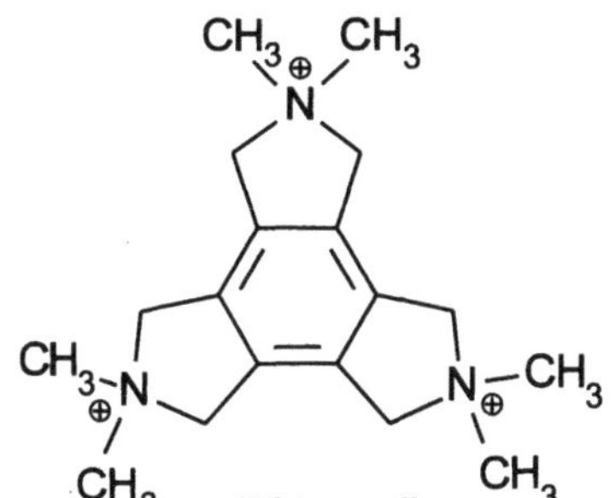

Fig. 14. Trisquaternary organic cation used as structure directing agent in the synthesis of zeolite ZSM-18

rigid trisquaternary ammonium cation (2,3,4,5,6,7,8,9-octahydro-2,2,5,5,8,8-hexamethyl-1*H*-benzo[1,2-c:3,4-c′:5,6-c″]tripyrrolium) which is shown in Fig. 14. It is anticipated here that the multistep synthesis required for the preparation of this organic cation [80, 81] has prevented zeolite ZSM-18 from being synthesized and investigated by zeolite researchers for so many years. This situation has changed with the solution of the framework topology of zeolite ZSM-18 [82]. In essence, its void structure is characterized by a linear one-dimensional twelve-membered ring channel with an approximate pore opening of 0.7 nm. The twelve-membered ring channels are lined with pockets that are capped by seven-membered ring "windows" with dimensions of 0.28×0.35 nm. The tight fit between the size and shape of the intracrystalline voids and the organic trisquaternary cation suggests that a strong templating effect might be responsible for the formation of this unusual zeolite structure. After its solution, the peculiar framework topology of zeolite ZSM-18 and the host/guest interactions occurring between the zeolite structure and the encapsulated templating agent were studied theoretically by a number of groups using molecular modelling techniques and quantum and molecular mechanics calculations [83–87].

The first syntheses of zeolite ZSM-18 reported by Ciric used the hydroxide form of the trisquaternary cation ("triquat"), tetramethylorthosilicate as silica source and sodium aluminate as alumina source. A typical molar gel composition was 2 triquat : 1.14 Na_2O : Al_2O_3 : 15 SiO_2 : 500 H_2O. From this synthesis mixture, ZSM-18 with SiO_2/Al_2O_3 ratios around 11 crystallizes after ca. 5 days at 398 to 403 K [80]. A typical X-ray powder pattern of as-synthesized zeolite ZSM-18 is shown in Fig. 15. It seems that, provided the complex template is available, the synthesis of zeolite ZSM-18 is straightforward, and it would be very interesting to explore its catalytic and adsorptive properties. Unfortunately, however, it turned out that the removal of the trisquaternary ammonium cations by calcination in air leads to a destruction of the zeolite framework. This is probably due to the relatively high aluminum content of this zeolite. As a consequence, attempts were made to develop methods for careful removal of the template without destroying the structure of the zeolite. So far, two different procedures have been disclosed: In the first method, a three-step procedure is

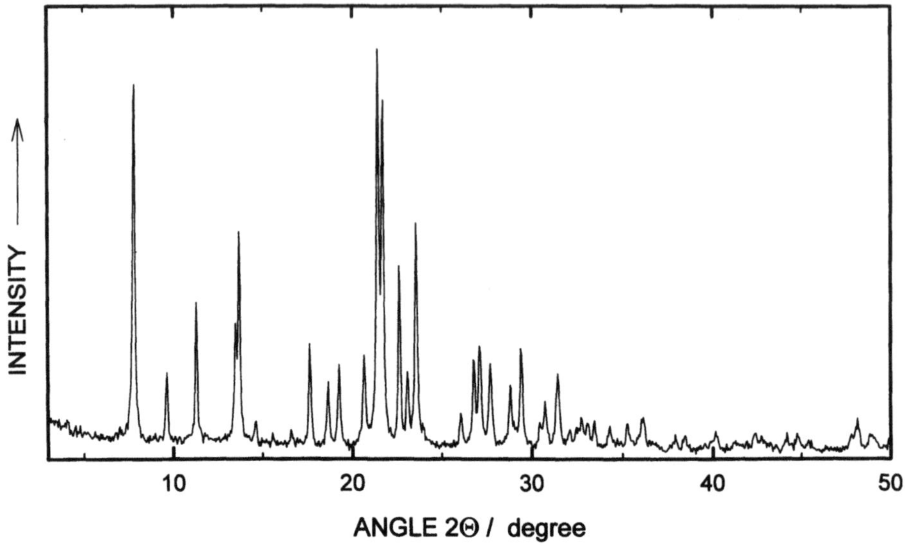

Fig. 15. X-ray powder pattern of as-synthesized zeolite ZSM-18

applied which involves (i) contacting the as-synthesized zeolite with an aqueous solution of ammonium hexafluorosilicate, (ii) ion-exchanging the zeolite with an alkali or alkaline earth metal salt (e.g. KCl) and (iii) calcining the ion-exchanged zeolite [88]. Obviously, a mild dealumination is applied in this procedure to increase the thermal stability of the zeolite framwork during calcination. In the second approach, zeolite ZSM-18 is loaded with a small amount of platinum, either via the addition of the noble metal component (e.g. a platinum tetraamine complex) to the synthesis gel of the zeolite or by post-synthetic impregnation of the template-containing zeolite with a platinum-containing compound [89]. Presumably, the role of the platinum introduced is to catalyze the oxidation of the encapsulated template at lower temperatures than usually required, thereby preventing the zeolite framework from being damaged.

Recently, Schmitt and Kennedy reported a novel method for the synthesis of zeolite ZSM-18 [90, 91]. Their rationale was to keep the same total charge and symmetry as in the originally used triquat, but to select those compounds, which can be easily synthesized and are acyclic in order to ensure a relatively easy removal of the template. With the help of computer modelling they finally ended up with the two trisquaternary compounds depicted in Fig. 16. Using triquat I it was possible to synthesize pure zeolite ZSM-18 in a high yield, while triquat II requires the addition of ZSM-18 seed crystallites in order to promote crystallization. Moreover, both new templates can be easily removed from the as-synthesized zeolites by calcination without significantly affecting the framework integrity [91].

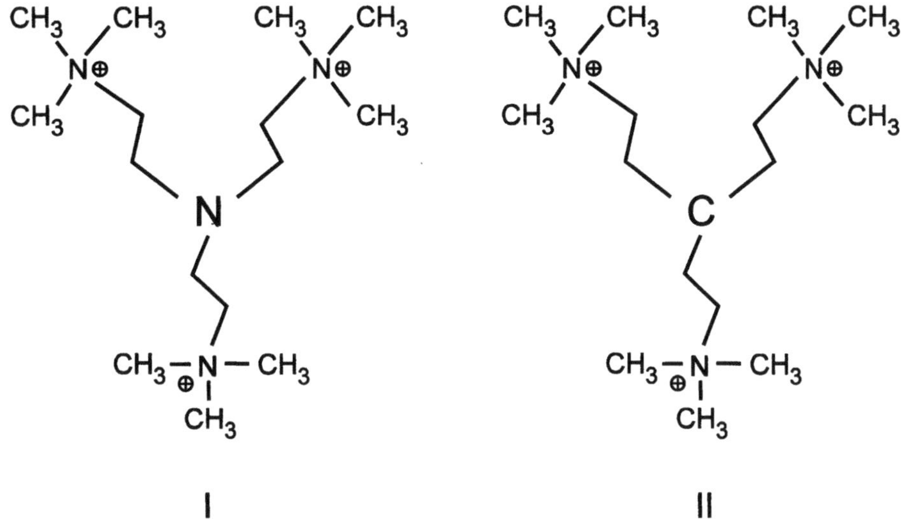

Fig. 16. Novel trisquaternary organic cations found useful for the synthesis of zeolite ZSM-18

4.3
Synthesis of Zeolite MCM-58

Zeolite MCM-58 was recently described for the first time in the patent literature [92, 93]. It belongs to the group of zeolites which can be synthesized with high-silica contents, viz. SiO_2/Al_2O_3 ratios from ca. 30 to 100 have been claimed [92]. From a comparison with the X-ray powder pattern of zeolite SSZ-42, which also was recently claimed by Zones and Rainis [94], it could be concluded that zeolites MCM-58 and SSZ-42 probably possess the same framework topology. The structure of zeolite SSZ-42 was recently determined [95]. It is characterized by an undulating, one-dimensional channel system with twelve-membered ring pore openings. The minimum and maximum pore diameters are 0.64 and 1.0 nm, respectively [95], suggesting that zeolites MCM-58 and SSZ-42 should be interesting catalysts for the conversion of relatively bulky molecules.

So far, a couple of organic compounds have been claimed as structure directing agents for zeolites MCM-58/SSZ-42, viz. benzyltropanium salts [93], benzylquinuclidinium (or *N*-benzyl-1-azabicyclo[2.2.2]octane) cations [92, 94] and N-benzyl-1,4-diaza-bicyclo[2.2.2] cations [94]. In our own work [96], we started from Valyocsik's patent on the synthesis of zeolite MCM-58 using benzylquinuclidiniumbromide (BC) as templating agent [92]. The latter can be conveniently prepared by refluxing an equimolar mixture of benzylbromide and quinuclidine in ethanolic solution for 16 hours and recovering the solid after cooling the reaction mixture in an ice bath. In a typical synthesis [96], 1.99 g $Al_2(SO_4)_3 \cdot 18\,H_2O$ are dissolved in 88 g of distilled water. To this solution 3.37 g KOH and 8.46 g benzylquinuclidiniumbromide are added under continuous stirring. Subsequently, 30 g colloidal silica sol (30 wt.-% SiO_2 in water; Levasil VP 4038, Bayer AG) are slowly added under stirring. A low-viscosity liquid results

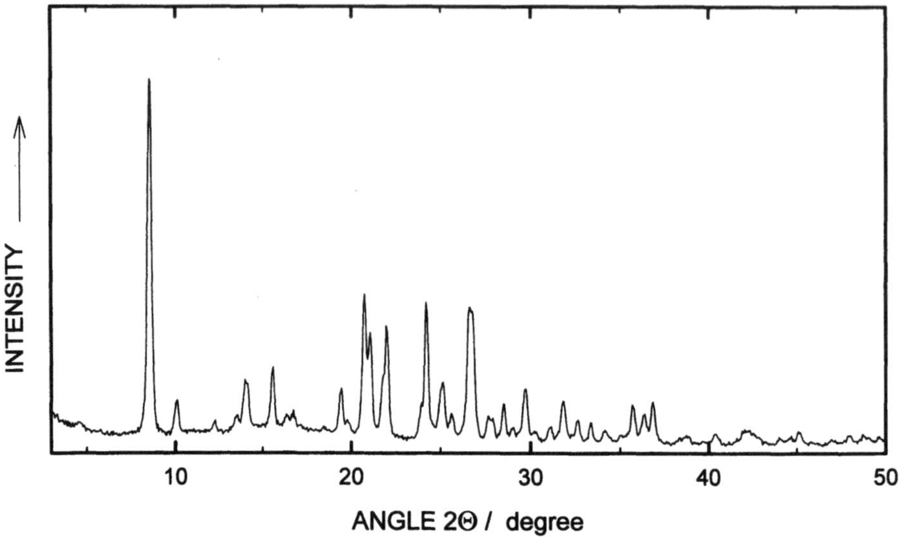

Fig. 17. X-ray powder pattern of zeolite MCM-58 after calcination in air at 813 K

which is further homogenized for two minutes and then transferred to a stainless steel autoclave with a volume of ca. 250 cm^3. The synthesis mixture prepared in this manner has the following molar composition: 12.5 K_2O : 10 BC : 1 Al_2O_3 : 50 SiO_2 : 2000 H_2O. The autoclave is rotated for eight days at 443 K. The X-ray powder pattern of the material obtained is depicted in Fig. 17. It is in very good agreement with literature data [92]. From Fig. 18 it can be seen that the crystallites are elliptical in shape with maximum dimensions below 1 μm. Chemical analysis reveals a SiO_2/Al_2O_3 ratio of 38, hence aluminum is slightly enriched in the zeolite as compared to the composition of the synthesis mixture. In further experiments, the influence of a reduction of the template concentration and of the aluminum content in the synthesis mixture was investigated [96]. It was found that the initial ratio of $BC/SiO_2 = 0.2$ could be reduced by 25% to 0.15 without significantly affecting the quality of the synthesized materials. Upon further reducing this ratio, however, α-quartz begins to co-crystallize with zeolite MCM-58 and is finally obtained as pure phase when the template is completely omitted. A similar effect is observed when BC/SiO_2 is kept constant at 0.15 and the SiO_2/Al_2O_3 ratio is gradually increased: already at a value of ca. 100, α-quartz appears as impurity (ca. 20%) under the conditions applied in this study [96]. Valyocsik reported the occurrence of mordenite as an impurity phase if either the aluminum content of the gel is too high (i.e. $SiO_2/Al_2O_3 = 15$) or sodium cations are used instead of potassium cations in the synthesis gel [92].

In a very recent publication by Camblor et al. [97], the synthesis of an aluminum-free material has been claimed which is believed to also possess the framework topology of zeolites MCM-58/SSZ-42. These authors used benzylquinuclidiniumhydroxide as templating agent and tetraethylorthosilicate as

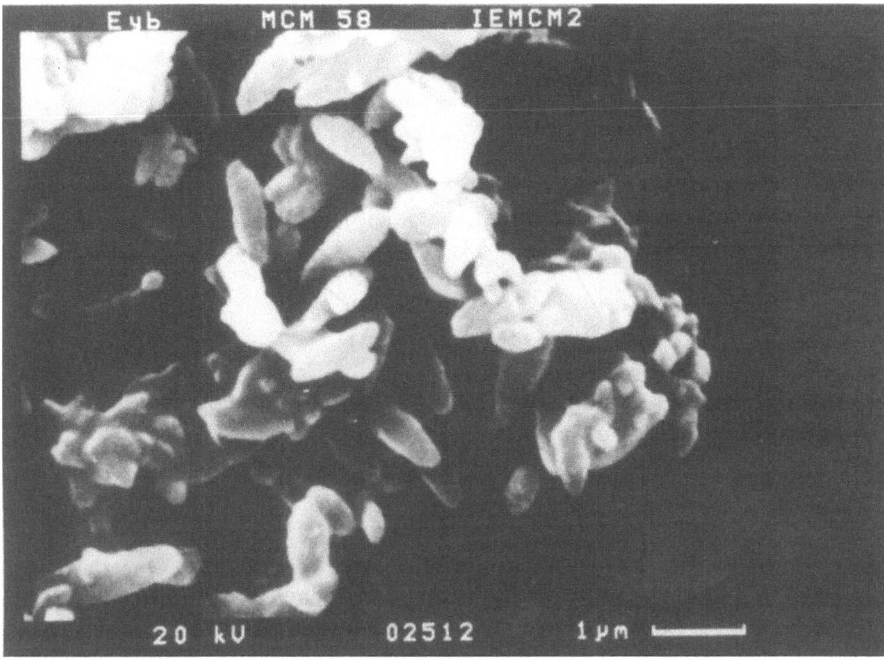

Fig. 18. Scanning electron micrograph of zeolite MCM-58

silica source in a fluoride medium to synthesize this new microporous silica polymorph, which they designated ITQ-4.

4.4
Synthesis of SSZ-24

SSZ-24 is a high-silica zeolite which was first synthesized by Zones and co-workers [98, 99]. Its pore structure consists of linear, non-interconnected twelve-membered ring channels with a circular diameter of 0.73 nm [99]. Its attractiveness from a catalytic and/or adsorption point of view is at least twofold: First, SSZ-24 possesses the AFI framework topology, viz. it is isostructural with the aluminophosphate $AlPO_4$-5 and its silicon- or metal-substituted derivatives, e.g. SAPO-5, MeAPO-5, etc. [99, 100]. Hence this offers the unique opportunity for studying the catalytic and adsorptive properties having one and the same framework topology but with a very broad range of chemical compositions. Secondly, molecular sieves possessing the AFI topology show an unusual behavior in adsorption and catalysis involving C_6 alkanes [101, 102]. In essence, SSZ-24 and molecular sieves of the $AlPO_4$-5 familiy exhibit a preference for the adsorption (from a mixture of C_6 alkanes) or formation (from hydrocracking of n-hexadecane) of the bulkier isomers (especially dimethylbutanes and methylpentanes). This preference for the sterically more demanding species has neither been found for molecular sieves with smaller pore diameters (e.g. zeolite

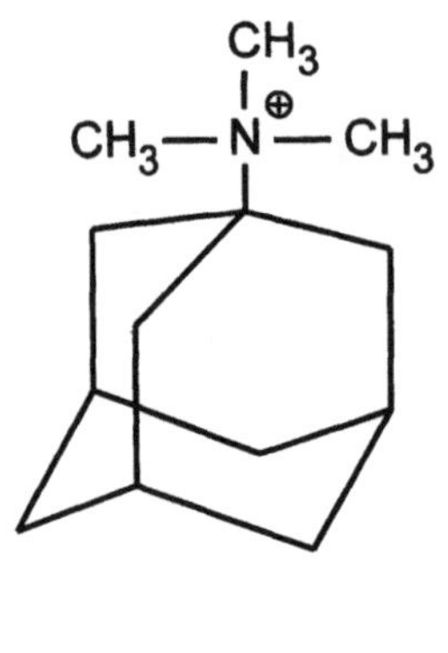

N,N,N-Trimethyl-1-adamantyl-ammonium cation

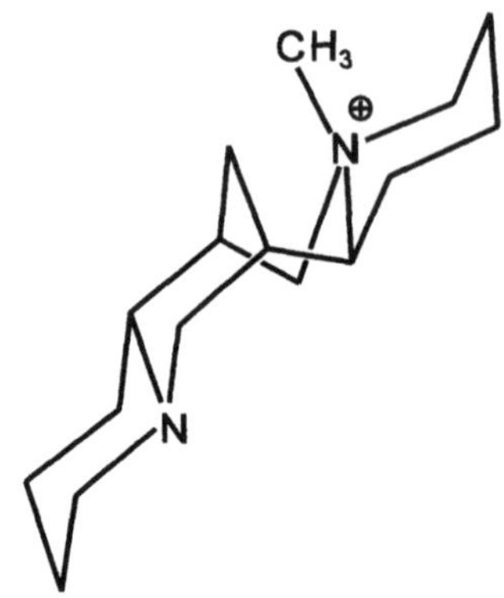

N-(16)-Methylsparteinium

Fig. 19. Structure directing agents for the synthesis of SSZ-24

ZSM-5), nor in the absence of shape selectivity (e.g. zeolite Y). It has been suggested that this selectivity for the adsorption/formation of branched alkanes over linear ones is caused by attractive forces between the molecular sieve walls and the hydrocarbons. Interesting applications in petroleum refining could be thought of, e.g. the selective formation of multibranched hydrocarbons in the C_5/C_6 range in hydrocracking of long chain hydrocarbons on AFI-containing hydrocracking catalysts [e.g. 103].

The first synthesis of SSZ-24 was accomplished using N,N,N-trimethyl-1-adamantylammonium cations as structure directing agent [98, 99] (cf. Fig. 19, left-hand side). It has been reported that the synthesis of SSZ-24 is critical with respect to the starting materials and the gel composition used [99]. A typical synthesis of an all-silica SSZ-24 material starts from a gel with the following molar composition (T=template): K_2O (or Na_2O)/SiO_2=0.10; T/SiO_2=0.15; H_2O/SiO_2= 15 to 44; OH-/SiO_2=0.25. Typically, Cab-*o*-sil (fumed silica) is used as the silica source. Crystallization usually occurred within six days at 423 to 433 K with or without mild agitation [99]. If the crystallization conditions are not properly controlled, then other crystalline phases have been reported to form [101], viz. SSZ-23 (structure unknown), SSZ-25 (MWW topology, see above) and SSZ-13 (a high-silica zeolite with chabazite topology, [104]). Hence, even if the proper template is used, the compositional window from which SSZ-24 can be crystallized is very narrow.

The synthesis of SSZ-24 is usually described to occur from aluminum-free gels. On the other hand, the preparation of an SSZ-24-type material which could be converted to its Brönsted-acid form suitable for catalysis requires the presence of at least a minor amount of aluminum in the framework. This however, as well as the isomorphous substitution of other atoms for framework Si by direct synthesis, turned out to be difficult. This has changed with the recently reported technique of substituing boron for silicon using a calcined form of boron-substituted zeolite Beta as the boron source or even as a combined source for boron and silicon [101, 105]. This resulted not only in the formation of boron-containing SSZ-24 but also in a considerable reduction of the required crystallization time from ca. six days to only one day at temperatures around 423 K [101, 105]. The concept of using boron-containing zeolites, in particular boron-substituted zeolite Beta, as silicon and boron supplying reagent was recently extended by Zones and Nakagawa to the synthesis of other zeolites and its consequences with respect to the crystallizing structures was discussed in detail [106, 107].

Once boron-containing SSZ-24 is obtained, it can be easily converted to a form having tetrahedrally coordinated framework aluminum species by first calcining the boron-containing material to remove boron from the lattice and then incorporating Al into the created defect sites by refluxing the material in an aluminum nitrate solution [101]. True aluminum incorporation into the framework was proven by ^{27}Al MAS NMR spectroscopy and catalytic activity in Brønsted-acid catalyzed reactions [101].

Gittleman et al. studied in more detail the synthesis of all-silica SSZ-24 [108]. In agreement with earlier studies they found that the use of the trimethyl-1-aminoadamantyl cation and fumed silica as SiO_2 source is essential for the formation of pure SSZ-24. Moreover, they also confirmed that the molar ratios template/SiO_2 and OH-/SiO_2 should exceed values of 0.15 and 0.25, respectively [108]. Especially for lower pH or alkali content, the formation of non-zeolitic silicate phases such as quartz, cristobalite and layered silicates are reported to be favored.

Recently, Lobo and Davis [109] reported the synthesis of pure-silica as well as boron-substituted SSZ-24 using a new structure directing agent, viz. *N*-(16)-methyl-sparteinium bromide (cf. Fig. 19, right-hand side). This new template allows for the direct incorporation of boron into the zeolite framework using sodium borate instead of zeolite boron-Beta as the source of boron in the synthesis mixture. From their investigations, the careful purification of the template (by recrystallization) seems to be essential for synthesizing a pure zeolite product. Moreover, seeding of the gel with as-synthesized SSZ-24 crystallites or the addition of as-synthesized boron-substituted zeolite Beta was found to suppress the formation of Kenyaite, a layered sodium silicate which was frequently found as by-product in SSZ-24 syntheses using *N*-(16)-methyl-sparteinium bromide as structure-directing agent. After synthesis, the organic molecule is found intact inside the molecular sieve pores. In addition, the amount of isomorphously incorporated boron is related to the number per unit cell of template molecules occluded inside the zeolite pores. Based on earlier work of de Ruiter et al. [110] conducted for the case of boron-substitution in the MFI framework, it has been speculated [109] that the incorporation of boron into the framework of SSZ-24 is closely related to the hydrophobicity of the structure directing agent: while the new template has a positive charge surrounded by a large hydrophobic and relatively rigid organic environment, the trimethylated form of 1-aminoadamantane contains the positive charge on a nitrogen atom which is separated from surrounding water by three methyl groups only. If this speculation were true, it would offer interesting new opportunities for the design of new templates for the synthesis of boron-silicates.

4.5
Synthesis of Zeolites with Intersecting Ten- and Twelve-Membered Ring Pores

Only very few molecular sieves are known which possess large and intersecting pores. The potential advantages of such a pore system would be a better resistance against deactivation by coke formation and an improved diffusivity of reactants and products to and from the intracrystalline catalytically active sites. Two of these multi-dimensional large-pore structures are of considerable commercial interest, viz. zeolites Y and Beta. While the former zeolite has been used in commercial practice for decades, the latter has a high potential as catalyst in quite a number of emerging commercial processes, e.g. dewaxing by isomerization [111], cumene synthesis via alkylation of benzene with propene [112], etc. These few examples suggest, that the availability of more zeolites with relatively large and three-dimensionally interconnected pore systems is highly desirable. While the channel systems of zeolites Y and Beta are made-up from twelve-membered ring windows and pores, respectively, a natural mineral was recently discovered, the pore system of which consists of intersecting ten- and twelve-membered ring channels. This mineral was designated Boggsite [113, 114] and discovered in trace amounts in close proximity to the natural counterpart of zeolite Beta, called Tschernichite [115].

Very recently, molecular sieves have been synthesized (based, in part, on a rational design of the organic structure directing agent as well as a lot of experience collected in zeolite synthesis), which possess interconnected ten- and twelve-membered ring pores, viz. the aluminosilicate SSZ-26 [116–118] and the borosilicates SSZ-33 [117, 118] and CIT-1 [119, 120]. All three molecular sieves belong to the same family of materials which contain different stacking sequences of two related polymorphs (polymorphs A and B) [119, 120]. CIT-1 is the pure polymorph B, while SSZ-33 and SSZ-26 are disordered materials formed mostly from polymorph B, but with fault probabilities of 30 % and 15 %, respectively [116, 117]. The typical SiO_2/Me_2O_3 ratios and templates used during the synthesis of these three materials are summarized in Fig. 20. The two borosilicates are conveniently synthesized from gels with SiO_2/B_2O_3 around 30, while a broader range has been given for the aluminosilicate SSZ-26, viz. SiO_2/Al_2O_3 = 20 to 50. Beside the aluminum content, for zeolite SSZ-26 the following favorable molar ratios of the synthesis gel have been described: template/SiO_2 = 0.15, Na_2O/SiO_2 = 0.15, H_2O/SiO_2 = 27 and OH^-/SiO_2 = 0.30. From such mixtures, zeolite SSZ-26 can be obtained after "several" days at 433 K. It seems that the most critical step in the synthesis of zeolite SSZ-26 is the preparation of the structure directing agent in the desired purity and preventing it from being decomposed ("cracked") by contact with air [116]. Obviously, the critical step during the template preparation is the careful recrystallization in order to separate the diquaternary isomer from the major impurity, viz. the corresponding monoquaternary ammonium compound [116].

Beside the ranges of boron content and the template used, not too many additional details on the synthesis of the borosilicates SSZ-33 and CIT-1 have been published. However, it has been reported by Zones, Davis and coworkers [121–123] that the result of such a synthesis does not only depend on the tem-

SSZ-33
Borosilicate
$n_{SiO_2/B_2O_3} \approx 30$

SSZ-26
Aluminosilicate
$n_{SiO_2/Al_2O_3} = 20-50$

CIT-1
Borosilicate
$n_{SiO_2/B_2O_3} \approx 30$

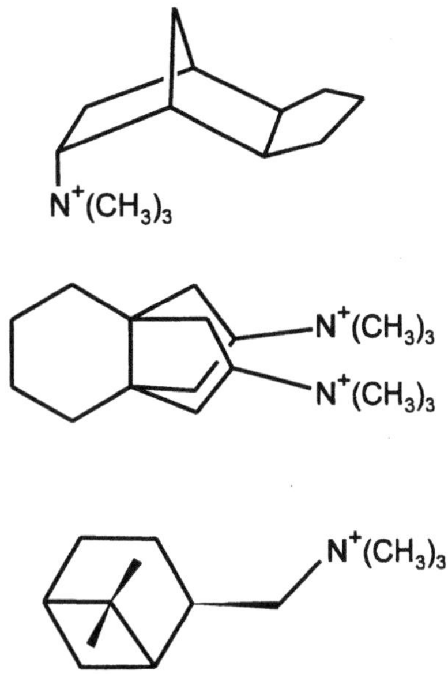

Fig. 20. Templates used for the synthesis of three molecular sieves with intersecting 10- and 12-membered ring channels

plate used and the composition of the synthesis gel, but is also influenced in many cases by the nature of the boron-supplying reagent and the boron concentration in the gel. As a whole, it remains highly desirable to investigate in more detail the influence of the different synthesis parameters on the synthesis of zeolites SSZ-26, SSZ-33 and CIT-1. It seems that, as in the cases of zeolites ZSM-10 and in particular ZSM-18, the laborious synthetic organic work hidden behind the preparation of the structure directing agents has hindered most zeolite scientists from studying the synthesis of these interesting zeolites in more detail.

4.6 Synthesis of Super-Large Pore Aluminosilicates

Probably the first report on the use of a metallocene in zeolite synthesis dates back more than ten years [124]. In the past few years, however, essentially two groups reported on the influence (i.e. structure directing properties) of cobalt metallocene compounds in the synthesis of zeolites and zeolite-like materials [e.g. 125–128]. In none of these studies had a new framework topology been synthesized. However, very recently Balkus and coworkers reported the synthesis of the first fourteen-membered ring zeolite, which they called UTD-1 (University of Texas at Dallas No. 1) [129–132]. UTD-1 was synthesized with a quite unusual template, viz. with bis(pentamethyl-cyclopentadienyl)cobalt(III) hydroxide [129–132]. Both the essentially pure-silica version and high-silica derivatives have been prepared. A typical synthesis of the all-silica version comprises the combination of sodium hydroxide, an aqueous solution of the bis(pentamethylcyclopentadienyl)cobalt(III) hydroxide and fumed silica to form a gel with the molar composition $H_2O/SiO_2=60$; $Na^+/SiO_2=0.1$; $OH^-/SiO_2=0.2$ and $template^+/SiO_2=0.1$. The resulting gel is first aged at room temperature for one hour under stirring and is then crystallized for ca. 2 days at 448 K under static conditions. If an aluminum-containing sample of UTD-1 is to be prepared, an aluminum source, e.g. aluminum isopropoxide dissolved in an aqueous solution of NaOH is added in the last step of the gel preparation. Thereafter, the gel is again aged for one hour at room temperature under stirring and then crystallized as described above [129, 131]. After synthesis, the material obtained has to be calcined in order to remove the organic directing agent. However, since in the case of UTD-1 a transition metal compound is used, cobalt remains in the molecular sieve while the organic component burns off.

UTD-1 is not only a new zeolite with pore openings hitherto unknown for silicate-based materials, but it also demonstrates how the discovery of a new class of templating agents, e.g. the metallocenes, could lead to new framework topologies and perhaps to new applications in catalysis and adsorption. It is expected that the systematic variation of the structure of metallocene-derived templates and their exploration as templates in zeolite synthesis will bring about new insights into the templating ability of these interesting metalloorganic compounds and probably result in the synthesis of further new framework topologies.

5 Conclusion

The development of new or improved processes in catalysis and adsorption were in many cases induced by the development of new catalytic materials and adsorbents. In this context, the synthesis of new aluminosilicates is a continuing challenge in zeolite science. The present review, discussing the synthesis principles of selected more recent zeolites, has shown that there is still much room for innovation in this field. It can be expected that by the use of new classes of templates (one recent example is that of the metallocenes) new structures will be synthesized in the future. Moreover, with the availability of more and more sophisticated tools for modelling zeolite and template structures and their interactions, it will probably be possible to tailor templates for a given (or a theoretical) zeolite structure. Finally, beside the exploration of new templates and new reaction compositions, the influence of the synthesis conditions on the products should not be overlooked, e.g. changing the reaction parameters from subcritical to supercritical conditions could well have an influence on the materials which are formed.

Acknowledgements. Financial support by Deutsche Forschungsgemeinschaft, Fonds der Chemischen Industrie and Max-Buchner-Forschungsstiftung is gratefully acknowledged. Moreover, the author wants to thank Prof. J. Weitkamp for his valuable comments during the preparation of the manuscript of this contribution.

6 References

1. Plank CJ (1984) Chemtech 14:243
2. Dwyer FG (1981) In: Moser WR (ed) Catalysis of organic reactions. Marcel Dekker, New York, p 39
3. Chen NY, Garwood WE, Dwyer FG (1996) Shape selective catalysis in industrial applications, 2nd edn. Marcel Dekker, New York, p 282
4. Chang CD (1983) Catal Rev Sci Eng 25:1
5. Jacobs PA, Martens JA (1987) Synthesis of high-silica aluminosilicate zeolites. Elsevier, Amsterdam. Stud Surf Sci Catal 37
6. Ernst S, Weitkamp J (1994) Catal Today 19:27
7. Chen NY, Maziuk J, Schwartz AB, Weisz PB (1968) Oil Gas J 66(47):154
8. Chen NY, Garwood WE, Heck RH (1987) Ind Eng Chem Res 26:706
9. Ruthven DM (1988) Chem Eng Progr 84(2):42
10. Reiss G (1989) In: Weitkamp J, Karge HG (eds) Zeolites as catalysts, sorbents and detergent builders – applications and innovations. Elsevier, Amsterdam, p 607. Stud Surf Sci Catal 46
11. Jasra RV, Bhat SGT (1988) Separation Science & Technology 26:885
12. Keane Jr M, Sonnichsen GC, Abrams L, Corbin DR, Gier TE, Shannon RD (1987) Appl Catal 32:361
13. Abrams L, Keane Jr M, Sonnichsen GC (1989) J Catal 115:410
14. Nastro A, Giordano G, Trifirò F (1989) Catal Lett 2:369
15. Nawaz S, Kolboe S, Kvisle S, Lillerud K-P, Stöcker M, Ören HM (1991) In: Holmen A, Jens K-J, Kolboe S (eds) Natural gas conversion. Elsevier, Amsterdam, p 421. Stud Surf Sci Catal 61
16. Valyocsik EW, Olson DH, Rodewald PG (1986) Eur. Patent Appl. 193 282 assigned to Mobil Oil Corp
17. Rodewald PG, Valyocsik EW (1987) US Patent 4 665 264 assigned to Mobil Oil Corp
18. Stewart A, Johnson DW, Shannon MD (1988) In: Grobet PJ, Mortier WJ, Vansant EF, Schulz-Ekloff G (eds) Innovation in zeolite materials science. Elsevier, Amsterdam, p 57. Stud Surf Sci Catal 37
19. Ernst S, Weitkamp J (1991) Chem-Ing-Tech 63:748
20. Exter MJ den, Jansen JC, Bekkum H van (1994) In: Weitkamp J, Karge HG, Pfeifer H, Hölderich W (eds) Zeolites and related microporous materials: state of the art 1994. Elsevier, Amsterdam, p 1159. Stud Surf Sci Catal 84B
21. Weitkamp J, Ernst S, Bock T, Kiss A, Kleinschmit P (1995) In: Beyer HK, Karge HG, Kiricsi I, Nagy J (eds) Catalysis by microporous materials. Elsevier, Amsterdam, p 278. Stud Surf Sci Catal 94
22. Gies H (1986) Z Krist 175:93
23. Meier WM, Olson DH, Baerlocher C (1996) Atlas of zeolite structure types 4th edn. Elsevier, London, p 86
24. Gies H (1984) J Incl Phenomena 2:275
25. Meier WM, Olson DH, Baerlocher C (1996) Atlas of zeolite structure types 4th edn. Elsevier, London, p 186
26. Meier WM, Olson DH, Baerlocher C (1996) Atlas of zeolite structure types 4th edn. Elsevier, London, p 122
27. Chu CTW, Chang CD (1985) J Phys Chem 89:1569
28. Sigl M, Ernst S, Weitkamp J, Knözinger H (1997) Catal Letters 45:27
29. Ernst S, Chen NY, Lindner D, Weitkamp J (1989) In: Jansen JC, Moscou L, Post MFM (eds) Zeolites for the nineties, 8th International zeolite conference, July 10–14, 1989, Recent Research Reports, Amsterdam, p 55
30. Argauer RJ, Landolt GR (1972) US Patent 3 702 886 assigned to Mobil Oil Corp
31. Miller SJ, Microporous Materials 2:439
32. Moiweer HH, Jong KP de, Kraushaar-Czarnetzki B, Stork WHJ, Krutzen BCH (1994) In: Weitkamp J, Karge HG, Pfeifer H, Hölderich W (eds) Zeolites and related microporous materials: state of the art 1994. Elsevier, Amsterdam, p 2327. Stud Surf Sci Catal 84C

33. Maerz B, Chen SS, Venkat CR, Mazzone D (1996) Hydrocarbon Technology International (autumn):21
34. Leonowicz ME, Lawton JA, Lawton SL, Rubin MK (1994) Science 264:1910
35. Shannon MD, Casci JL, Cox PA, Andrews SJ (1991) Nature 353:417
36. Unverricht S, Hunger M, Ernst S, Karge HG, Weitkamp J (1994) In: Weitkamp J, Karge HG, Pfeifer H, Hölderich W (eds) Zeolites and related microporous materials: state of the art 1994. Elsevier, Amsterdam, p 37. Stud Surf Sci Catal 84 A
37. Puppe L, Weisser J (1982) DE Patent 3 117 135 assigned to Bayer AG
38. Zones SI (1989) US Patent 4 826 667 assigned to Chevron Corp
39. Millini R, Perego G, Parker Jr WO, Bellussi G, Carluccio L (1995) Microporous Materials 4:221
40. Huss Jr A, Kirker GW, Keville KM, Thomson RT (1991) US Patent 4 992 615 assigned to Mobil Oil Corp
41. Unverricht S, Ernst S, Weitkamp J (1994) In: Weitkamp J, Karge HG, Pfeifer H, Hölderich W (eds) Zeolites and related microporous materials: state of the art 1994. Elsevier, Amsterdam, p 1693. Stud Surf Sci Catal 84C
42. Del Rossi KJ, Huss Jr A (1992) US Patent 5 107 047 assigned to Mobil Oil Corp
43. Chu P, Landis ME, Le QN (1994) US Patent 5 334 795 assigned to Mobil Oil Corp
44. Robert R, Ratnasamy P (1995) J Molec Catal A 100:93
45. Ernst S, Jean B (1997) In: Chon H, Ihm S-K, Uh YS (eds) Progress in zeolite and microporous materials. Elsevier Amsterdam, p 747. Stud Surf Sci Catal 105A
46. Casci JL, Shannon MD, Cox PA, Andrews SJ (1992) In: Occelli ML, Robson H (eds) Synthesis of microporous materials, vol 1. Van Nostrand Reinhold, New York, p 359
47. Nakagawa Y (1993) US Patent 5 254 514 assigned to Chevron Corp
48. Nakagawa Y (1994) In: Weitkamp J, Karge HG, Pfeifer H, Hölderich W (eds) Zeolites and related microporous materials: state of the art 1994. Elsevier, Amsterdam, p 323. Stud Surf Sci Catal 84A
49. Alberti A, Vezzalini G, Galli E, Quartieri S (1996) Eur J Mineral 8:69
50. Galli E, Quartieri S, Vezzalini G, Alberti A (1996) Eur J Mineral 8:687
51. Casci JL, Lake IJS, Maberly TR (1991) US Patent 5 041 402 assigned to ICI
52. Adair B, Chen C-Y, Wan K-T, Davis ME (1996) Microporous Materials 7:261
53. Gläser R, Li R, Ernst S, Weitkamp J, to be published
54. Dessau RM (1993) US Patent 5 254 787 assigned to Mobil Oil Corp
55. Rubin MK, Chu P (1990) US Patent 4 954 325 assigned to Mobil Oil Corp
56. Ravishankar R, Sen T, Ramaswamy V, Sony HS, Ganapathy S, Sivasanker S (1994) In: Weitkamp J, Karge HG, Pfeifer H, Hölderich W (eds) Zeolites and related microporous materials: state of the art 1994. Elsevier, Amsterdam, p 331. Stud Surf Sci Catal 84 A
57. Corma A, Corell C, Pérez-Pariente J (1995) Zeolites 15:2
58. Ravishankar R, Bhattacharya D, Jacob NE, Sivasanker S (1995) Microporous Materials 4:83
59. Mochida I, Eguchi S, Hironaka M, Nagao S, Sakanishi K,and Whitehurst D (1997) Zeolites 18:142
60. Kumar N, Lindfors L-E (1996) Appl Catal A: General 147:175–187
61. Wu P, Lin H, Komatsu T, Yashima T, Chem Commun (1997):663
62. Bennet JM, Chang CD, Lawton SL, Leonowicz ME, Lissy DN, Rubin MK (1993) US Patent 5236 575 assigned to Mobil Oil Corp
63. Lawton SL, Fung AS, Kennedy GJ, Alemany LB, Chang CD, Hatzikos GH, Lissy DN, Rubin MK, Timken H-KC, Steuernagel S, Woessner DE (1996) J Phys Chem 100:3788
64. Kresge CT, Le QN, Roth WJ, Thompson RT (1993) US Patent 5 258 566 assigned to Mobil Oil Corp
65. Huss Jr A, Le QN, Thompson RT (1994) US Patent 5 326 922 assigned to Mobil Oil Corp
66. Roth WJ, Kresge CT, Vartuli JC, Leonowicz ME, Fung AS, McCullen SB (1995) In: Beyer HK, Karge HG, Kiricsi I, Nagy J (eds) Catalysis by microporous materials. Elsevier, Amsterdam, p 301. Stud Surf Sci Catal 94
67. Casci JL, Stewart A (1990) Europ Patent Appl. 377 291 assigned to ICI

68. Moini A, Schmitt KD, Valyocsik EW, Polomski RF (1994) In: Weitkamp J, Karge HG, Pfeifer H, Hölderich W (eds) Zeolites and related microporous materials: state of the art 1994. Elsevier Amsterdam, p 23. Stud Surf Sci Catal 84A
69. Moini A, Schmitt KD, Valyocsik EW, Polomski RF (1994) Zeolites 14:504
70. Zones SI, Harris TV, Rainis A, Santilli D (1992) US Patent 5 106 801 assigned to Chevron Corp
71. Lobo RF, Tsapatsis M, Freyhardt CH, Chan I, Chen C-Y, Zones SI, Davis ME (1997) J Am Chem Soc 119:3732
72. Kumar R, Ramesh Reddy K, Raj A, Ratnasamy P (1993) In: Ballmoos R von, Higgins JB, Treacy MMJ (eds) Proceedings from the ninth international zeolite conference, vol. 1. Butterworth-Heinemann, Stoneham, MA, USA, p 189
73. Ramesh Reddy K, Ramaswamy V, Kumar R, Ramaswamy AV (1994) Zeolites 14:326
74. Weitkamp J (1991) In: Öhlmann G, Pfeifer H, Fricke R (eds) Catalysis and adsorption by zeolites. Elsevier, Amsterdam, p 21. Stud Surf Sci Catal 65
75. Csicsery SM (1995) In: Beyer HK, Karge HG, Kiricsi I, Nagy J (eds) Catalysis by microporous materials. Elsevier, Amsterdam, p 1. Stud Surf Sci Catal 94
76. Weitkamp J, Weiss U, Ernst S (1995) In: Beyer HK, Karge HG, Kiricsi I, Nagy J (eds) Catalysis by microporous materials. Elsevier, Amsterdam, p 363. Stud Surf Sci Catal 94
77. Ciric J (1972) US Patent 3 692 470 assigned to Mobil Oil Corp
78. Hironaka T, Miyake T (1990) JP 2 258 617 assigned to Tosoh Corp
79. Higgins JB, Schmitt KD (1996) Zeolites 16:236
80. Ciric J (1976) US Patent 3 950 496 assigned to Mobil Oil Corp
81. Ciric J, Lawton SL, Kokotailo GT, Griffin GW (1978) J Am Chem Soc 100:2173
82. Lawton SL, Rohrbaugh WJ (1990) Science 247:1319
83. Gale JD, Cheetham AK (1992) Zeolites 12:674
84. Hong, SB, Cho HM, Davis ME (1993) J Phys Chem 97:1622
85. Koelmel CM, Li YS, Freeman CM, Levine SM, Hwang M-J, Maple JR, Newsam JM (1994) J Phys Chem 98:12911
86. Stevens AP, Gorman AM, Freeman CM, Cox PA (1996) J Chem Soc Faraday Trans 92:2065
87. Cox PA, Stevens AP, Banting L, Gorman AM (1994) In: Weitkamp J, Karge HG, Pfeifer H, Hölderich W (eds) Zeolites and related microporous materials: state of the art 1994. Elsevier, Amsterdam, p 2115. Stud Surf Sci Catal 84C
88. Ryan FX (1989) US Patent 4 851 200 assigned to Mobil Oil Corp
89. Tsao Y-YP (1994) US Patent 5 290 534 assigned to Mobil Oil Corp
90. Schmitt KD (1994) US Patent 5 350 570 assigned to Mobil Oil Corp
91. Schmitt KD, Kennedy GJ (1994) Zeolites 14:635
92. Valyocsik EW (1994) WO Patent 95/11196 assigned to Mobil Oil Corp
93. Valyocsik EW (1994) US Patent 5 441 721 assigned to Mobil Oil Corp
94. Zones SI, Rainis A (1995) WO Patent 95/22507 assigned to Chevron Corp.
95. Chen CY, Finger LW, Medrud RC, Crozier PA, Chan IY, Harris TV, Zones SI, Chem Commun, in press.
96. Ernst S, Hunger M, Weitkamp J (1997) Chem-Ing-Tech. 69:77
97. Camblor MA, Corma A, Villaescusa LA, Chem Commun (1997):749
98. Zones SI (1987) US Patent 4 665 110 assigned to Chevron Corp
99. Nordstrand RA van, Santilli DS, Zones SI (1988) In: Flank WH, Whyte Jr TE (eds) Perspectives in molecular sieve science. American Chemical Society, Washington DC, p 236. Am Chem Soc Symp Ser 368
100. Bialek R, Meier WM, Davis ME, Annen MJ (1991) Zeolites 11:438
101. Nordstrand RA van, Santilli DS, Zones SI (1992) In: Occelli M, Robson HE (eds) Synthesis of microporous materials, vol 1. Van Nostrand Reinhold, New York, p 373
102. Santilli DS, Harris TV, Zones SI (1993) Microporous Materials 1:329
103. Ward J (1987) US Patent 4 695 368 assigned to Union Oil Corp
104. Zones SI, Nordstrand RA van, Santilli DS, Wilson DM, Yuen L, Scampavia LD (1989) In: Jacobs PA, van Santen RA (eds) Zeolites: facts, figures, future. Elsevier, Amsterdam, p 299. Stud Surf Sci Catal 49A

105. Zones SI, Yuen LT, Nakagawa Y, Nordstrand RA van, Toto SD (1993) In: Ballmoos R von, Higgins JB, Treacy MMJ (eds) Proceedings from the ninth international zeolite conference, Butterworth-Heinemann, Stoneham, MA, USA, p 163
106. Zones SI, Nakagawa Y, Microporous Materials (1994) 2:543
107. Zones SI, Nakagawa Y (1995) In: Bonneviot L, Kaliaguine S (eds) Zeolites: a refined tool for designing catalytic sites. Elsevier, Amsterdam, p 45. Stud Surf Sci Catal 97
108. Gittleman CS, Watanabe K, Bell AT, Radke CJ (1996) Microporous Materials 6:131
109. Lobo RF, Davis ME (1994) Microporous Materials 3:61
110. Ruiter R de, Jansen JC, van Bekkum H (1992) Zeolites 12:56
111. Pappal DA, Hilbert TL (1996) Petroleum Technology Quaterly (summer):35
112. Meima GR, Aalst MJM van der, Samson MSU, Garces JM, Lee JG (1996) In: Catalysis on Solid Acids and Bases, Proceedings of the DGMK Conference on Catalysis by Solid Acids and Bases, March 14–15, 1996, Berlin, Germany, DGMK Hamburg, p 89
113. Pluth JJ, Smith JV, Howard DG, Tschernich RW (1989) In: Jansen JC, Moscou L, Post MFM (eds) Zeolites for the nineties, 8th International zeolite conference, July 10–14, 1989, Recent Research Reports, Amsterdam, p 55
114. Pluth JJ, Smith JV (1990) American Mineralogist 75:501
115. Smith JV, Pluth JJ, Boggs RC, Howard DG (1991) Chem Commun:363
116. Zones SI, Olmstead MM, Santilli DS (1992) J Am Chem Soc 114:4195
117. Lobo RF, Pan M, Chan I, Li H-X, Medrud RC, Zones SI, Crozier PA, Davis ME (1993) Science 262:1543
118. Lobo RF, Pan M, Chan I, Medrud RC, Zones SI, Crozier PA, Davis ME (1994) J Phys Chem 98:12040
119. Lobo RF, Davis ME (1995) J Am Chem Soc 117:3766
120. Lobo RF, Zones SI, Davis ME (1994) In: Weitkamp J, Karge HG, Pfeifer H, Hölderich W (eds) Zeolites and related microporous materials: state of the art 1994. Elsevier, Amsterdam, p 461. Stud Surf Sci Catal 84 A
121. Zones SI (1994) Microporous Materials 2:281
122. Davis ME, Zones SI (1997) In: Occelli ML, Kessler H (eds) Synthesis of Porous Materials, Marcel Dekker, New York, p 1
123. Lobo RF, Zones SI, Davis ME (1995) J Incl Phenomen Molec Recogn Chem 21:47
124. Valyocsik EW (1986) US Patent 4 568 654 assigned to Mobil Oil Corp
125. Balkus Jr KJ, Shepelev S (1993) Microporous Materials 1:383
126. Balkus Jr KJ, Gabrielov AG, Shepelev S (1995) Microporous Materials 3:489
127. Goor G van de, Freyhardt CC, Behrens P (1995) Z Anorg Allg Chem 621:311
128. Goor G van de, Lindlar B, Felsche J, Behrens P (1995) Chem Commun:2559
129. Balkus Jr KJ, Gabrielov AG (1996) US Patent 5 489 424 assigned to the Board of Regents, The University of Texas System
130. Freyhardt CC, Tsapatsis M, Lobo RF, Balkus Jr KJ, Davis ME (1996) Nature 381:295
131. Balkus Jr KJ, Gabrielov AG, Zones SI, Chan IY (1997) In: Occelli ML, Kessler H (eds) Synthesis of porous materials, Marcel Dekker, New York, p 77
132. Balkus Jr KJ, Biscotto M, Gabrielov AG (1997) In: Chon H, Ihm S-K, Uh YS (eds) Progress in zeolite and microporous materials. Elsevier, Amsterdam, p 747. Stud Surf Sci Catal 105 A

The Synthesis and Properties of M41S and Related Mesoporous Materials

J. C. Vartuli · W. J. Roth · J. S. Beck · S. B. McCullen · C. T. Kresge
Mobil Technology Company, Paulsboro Technical Center, Paulsboro, New Jersey 08066, USA

1 Introduction

Microporous and mesoporous inorganic materials form the backbone of many heterogeneous catalysts and separations media. Because of the extensive commercial applications of these classes of materials, substantial efforts on the part of both academic and industrial researchers have been made to unlock the hidden secrets of the mechanisms of their formation and, through the exploitation of this understanding, to synthesize novel materials with important properties. Much of the mechanistic work has focused on understanding the role of organic directing or templating agents which play a complex, cooperative role of spatial ordering through the filling of void space, balancing charge, and stabilizing structural units.

Templating has been defined, in a general sense, as a process in which an organic species functions as a central structure about which oxide moieties organize into a crystalline lattice [1, 2, 3]. In the simplest case of void filling, the organic merely serves to occupy a space about which the oxide crystallizes or polymerizes. Thus the same organic can be used to synthesize a variety of structures, or vice versa. Structural direction requires that a specific framework is formed from an unique organic compound. However, this does not imply that the resulting oxide structure mimics identically the form of the organic molecule. True "templating" requires, in addition to the structural directing component, that there is an intimate relationship between the oxide lattice and organic form such that the synthesized lattice contains the organic "locked" into position. Thus the lattice reflects the identical geometry of the organic molecule [1].

The use of organic directing agents has a long history in the synthesis of porous inorganic materials. In the early 1960s, Kerr and colleagues revolutionized the field of synthetic zeolites with the introduction of quaternary alkylammonium ions into the reaction mixture [4]. The use of these cations as a replacement for the alkali and alkaline earth cation-based syntheses, developed by Barrer in the 1930s, led to the discovery of some sixty new porous frameworks – the most important being the pentasil structure, ZSM-5 [5]. In these examples, the organic directing agent is functioning as an individual molecule. This concept is illustrated in Fig. 1. Although small organic directing agents have dominated inorganic framework syntheses for nearly fifty years, a fundamental understanding of the ordering mechanism associated with them remains elusive.

The next milestone in the use of organic directing agents, and the subject of this review, occurred some three decades after the work of Kerr. Extending an alkyl chain length of a quaternary ammonium compound to 10–20 carbons changes the solution energetics of these ionic organics, turning them into the well known cationic surfactants that dominate much of the micelle and liquid crystal literature. Now, rather than individual molecular directing agents participating in the spatial ordering of the inorganic reactants, strong organic interactions become a primary driving force in determining the resulting inorganic structures – the M41S family of materials [6, 7]. This supramolecular directing concept is also illustrated in Fig. 1. The implications of these materials and the approach to their synthesis are enormous giving the materials science community the ability, in principle, to tailor the physico-chemical properties of a porous material for a particular application.

Recent work, notably by the groups of Ozin and Mann, has extended these ideas of cooperative spatial ordering of inorganic-organic assemblies into the

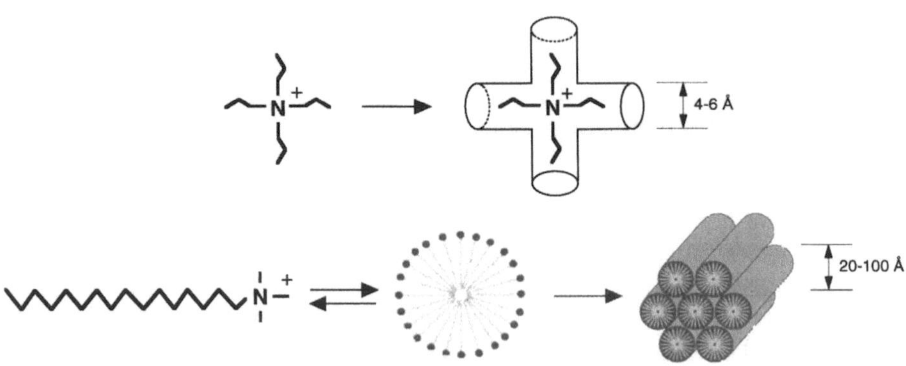

Fig. 1. The formation of microporous molecular sieves using individual small alkyl chain length quaternary directing agents (*top*). The formation of mesoporous molecular sieves using long alkyl chain length quaternary directing agents (*bottom*)

realm of macroscopic, hierarchical materials [8, 9]. Not only do these materials have potential in the traditional fields of application like heterogeneous catalysis, but the growing mechanistic understanding is leading to the generation of materials with striking resemblance to the highly complex forms characteristic of biological structures. It is the biomimetic implications of these materials that may, ultimately, prove to be their ultimate value.

In this review we will cover the synthesis and sorptive properties of these new mesoporous molecular sieves.

2 Structures and Synthesis

The M41S family of mesoporous materials contains several unique structures: those that can be indexed to an hexagonal unit cell, MCM-41 (two dimensional, p6m), SBA-2 (three dimensional, $p6_3/mmc$), and SBA-3 (two dimensional p6m structure similar to MCM-41 but synthesized in an acid media); cubic structures including MCM-48 (Ia3d) and SBA-1 (Pm3n); and lamellar structures including MCM-50 which is a post-stabilized lamellar material. Other materials have been synthesized that are not as readily classified. These materials generally exhibit limited X-ray diffraction information (one peak) and may contain a random array of pores as shown in transmission electron micrographs. However, a common characteristic of all of these materials is that they contain a narrow pore size distribution reminiscent of classical crystalline microporous materials, but within the mesopore dimensions of ~15–200 Å. This characteristic property will be covered later in this review.

MCM-41, the initial member of this family of materials and most extensively investigated, exhibits an X-ray diffraction pattern containing three or more low angle peaks (below 10° 2θ) that can be indexed to an hexagonal hk0 lattice [6, 7]. An example of the characteristic X-ray diffraction pattern is shown in Fig. 2. This particular sample has a unit cell of approximately 46 Å. The transmission electron micrograph also shown in Fig. 2 illustrates the uniform honey-comb like structure with a repeat of approximately 50 Å consistent with the hexagonal indexing of the X-ray diffraction pattern. These materials were first synthesized as silicates and alumino-silicates in basic media using a cationic alkyltrimethylammonium surfactant system where the alkyl chain length varied from ~8 to 16 carbon atoms. Subsequent efforts have shown that this structure can also be prepared in acid media [10, 11], using neutral normal amines [12] (this material has been denoted as HMS by the researchers), nonionic surfactants [13] and dialkyldimethylcationic surfactants [14], with substitution of various heteroatoms in a silicate structure and as well as nonsilicious forms. The heteroatom structures contained Ti, B, Ni, Cr, Fe, Co, Mn, and V [15–24]. The initial nonsilicious materials included oxides of W, Fe, Pb, Mo and Sb [21, 23, 25]. However, many of these materials exhibited very poor thermal stability; upon the removal of the template, the structures collapsed. Stein et al. showed that some of these materials were actually incomplete frameworks more closely resembling those of salts containing Keggin ions of the transition metal and the cationic surfactant molecule [21]. Recently, non-siliceous materials comprising Zr and Ti

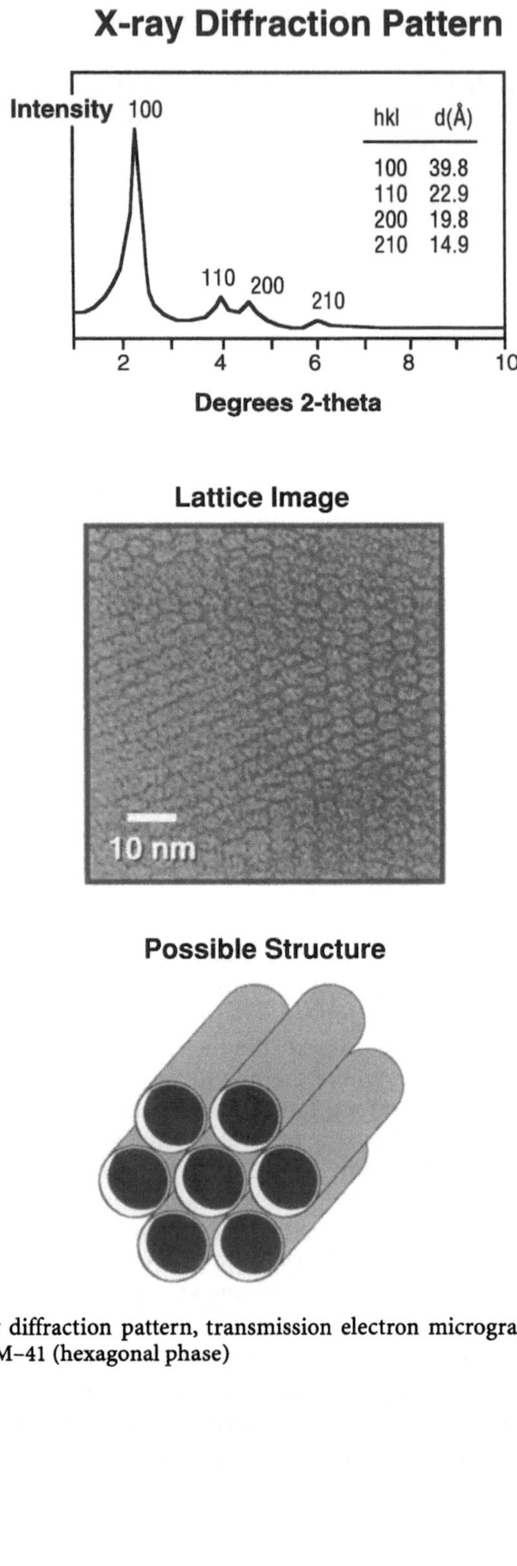

Fig. 2. The X-ray diffraction pattern, transmission electron micrograph, and the proposed structure of MCM-41 (hexagonal phase)

have been prepared that exhibit satisfactory thermal stability and thus are believed to be comprised of extended and complete oxide nets of the elements [26–29].

All of these hexagonal (p6m) materials contain a regular two dimensional array of mesopores which result in a structure having an hexagonal stacking of uniform diameter porous tubes as suggested in transmission electron micrographs. This is illustrated in Fig. 2. The pore size can be varied from about 15 Å to more than 100 Å by the variation of the length of the alkyl chain of the template molecule or by the use of auxiliary solubilized molecules [7]. These techniques will be described in Section 3.

Upon thermal treatment to remove the surfactant template, the X-ray diffraction pattern can exhibit a measurable unit cell contraction (at times, exceeding 5–10 Å) unlike that generally observed for the classical crystalline zeolites. The magnitude of this contraction appears to correlate with both the uniformity of the structure and the ultimate thermal stability. Generally those samples that exhibit little or no unit cell contraction (± 2 Å) upon the removal of the template by thermal treatment also exhibit the highest stability. The contraction of the unit cell has been contributed to the condensation of the silanol groups located within the wall.

Silicon NMR spectra of MCM-41 closely resemble those of amorphous silica suggesting a broad range of Si–O–Si (T–O–T) bond angles, unlike that of classical crystalline zeolites [7, 30] and contain a relatively high concentration of silanols. This silanol concentration appears to be highest for those structures that demonstrate the largest unit cell contraction upon removal of the template by calcination.

Other studies have shown that a related mesoporous silicate can be formed via hydrothermal treatment of a layered silicate such as kanemite and the same alkyltrimethylammonium cationic surfactants [31, 32]. Variation of the synthesis conditions results in a fairly ordered porous material denoted as FSM-16 that has physical properties and diffraction data similar to that of MCM-41 [32].

The hexagonal configuration, $P6_3/mmc$ denoted as SBA-2, has been proposed as a three dimensional hexagonal structure. This silicate was synthesized in both acidic and basic media using Gemini cationic surfactants, which contain two quaternary ammonium head groups separated by a methylene chain of variable length [33]. This framework has been proposed to contain regular supercages resulting from the hexagonal close packing of globular surfactant/silicate moieties. Physisorption of up to ~40 wt.-% nitrogen has been realized for this structure having a pore diameter of approximately 40 Å.

The cubic structures include both the Ia3d member, MCM-48, and the Pm3m material, SBA-1. MCM-48 was also first synthesized in the siliceous form using an alkyltrimethylammonium cationic surfactant system in basic media. Whereas the hexagonal MCM-41 appears to form under a wide range of synthesis conditions and reagents, MCM-48 appears to have a relatively narrow synthesis regime. MCM-48 has been prepared using a silicon alkoxide at slightly higher surfactant concentrations than those that can produce MCM-41. Vartuli et al. have shown that under one set of synthesis conditions using tetraethylorthosilicate and a cetyltrimethylammonium hydroxide solution, MCM-41 was synthesized at surfactant/silica molecular ratios of less than 1, whereas at ratios from 1 to 1.1, MCM-48 could be synthesized [34].

The X-ray diffraction pattern, shown in Fig. 3, consists of several peaks that can be assigned to the Ia3d space group. As in the case of MCM-41, upon thermal removal of the template, a unit cell contraction is noted. This contraction can be as large as 10–20 Å (from ~95 Å for the as-synthesized form to ~80 Å for the calcined version), substantially larger than that observed for well defined MCM-41 [34]. The structure of MCM-48 has been proposed to be bicontinuous with a simplified representation of two infinite three-dimensional, mutually intertwined, unconnected networks of rods as initially proposed by Luzzatti [35]. A more sophisticated and perhaps more realistic model would be based on the concept of an infinite periodic minimal surface of the gyroid form (Q^{230}) [10]. A representative transmission electron micrograph and an X-ray diffraction pattern of MCM-48 and a proposed structure are shown in Fig. 3. Generally both the sorptive properties and the Si-NMR characteristics of MCM-48 are consistent with those of the hexagonal phase.

Less is known about the other cubic structure, SBA-1 (Pm3m) [25]. This material was first synthesized in the siliceous form in acidic media using larger head group cationic surfactants such as alkyltriethylammonium salts. SBA-1 has been proposed to have a large cage structure with a surface area > 1000 m^2g^{-1} for a sample having a pore diameter of 20 Å [33].

Lamellar structures can be prepared using these same surfactant systems. The X-ray diffraction pattern of the as-synthesized materials exhibit several low angle peaks that can be indexed to h00 reflections. These structures are formed under a wide range of synthesis conditions. Initially these lamellar materials were also prepared in the siliceous form but many other compositions under a wide range of synthesis conditions have been reported [6, 7, 25, 33]. Stucky et al. have prepared various compositions under both basic and acidic conditions using cationic as well as anionic surfactants [33]. The compositions include oxides of Pb, Mg, Al, Ga, Mn, Fe, Co, Ni, and Zn as well as ZnP. In another system, using dialkyldimethylammonium bromide, Sayari et al. obtained siliceous lamellar structures when the carbon number of the alkyl chain lengths exceeded 6 [14]. In one system, the siliceous lamellar structures were formed at higher surfactant/silica molecular ratios than those used for the cubic and hexagonal materials [34].

The lamellar structures exhibit poor thermal stability. Upon removal of the template by thermal treatment, the structure collapses resulting in a dense phase with little structural order or porosity. The lamellar phase can be stabilized by subsequent treatment using an alkoxide [36]. Removal of the template after this stabilization treatment resulted in a structure having an X-ray diffraction pattern of only h00 peaks consistent with the lamellar configuration. The X-ray diffraction pattern is illustrated in Fig. 4. The transmission electron micrograph also shown in Fig. 4 is of uniform layers having an interlayer separation (~40 Å) consistent with the X-ray diffraction data.

These lamellar materials have been proposed to form from a bilayer of surfactant molecules with the hydrophilic or charged ends of the surfactant molecules pointed toward the water-inorganic oxide interface while the hydro-

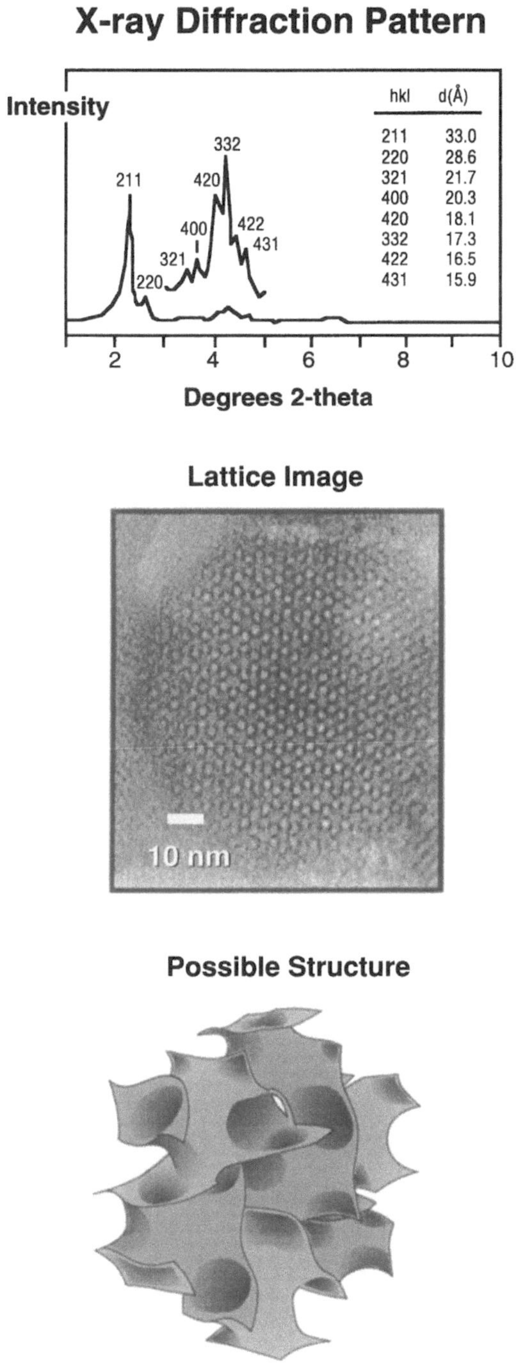

Fig. 3. The X-ray diffraction pattern, transmission electron micrograph, and the proposed structure of MCM-48 (Ia3d cubic phase)

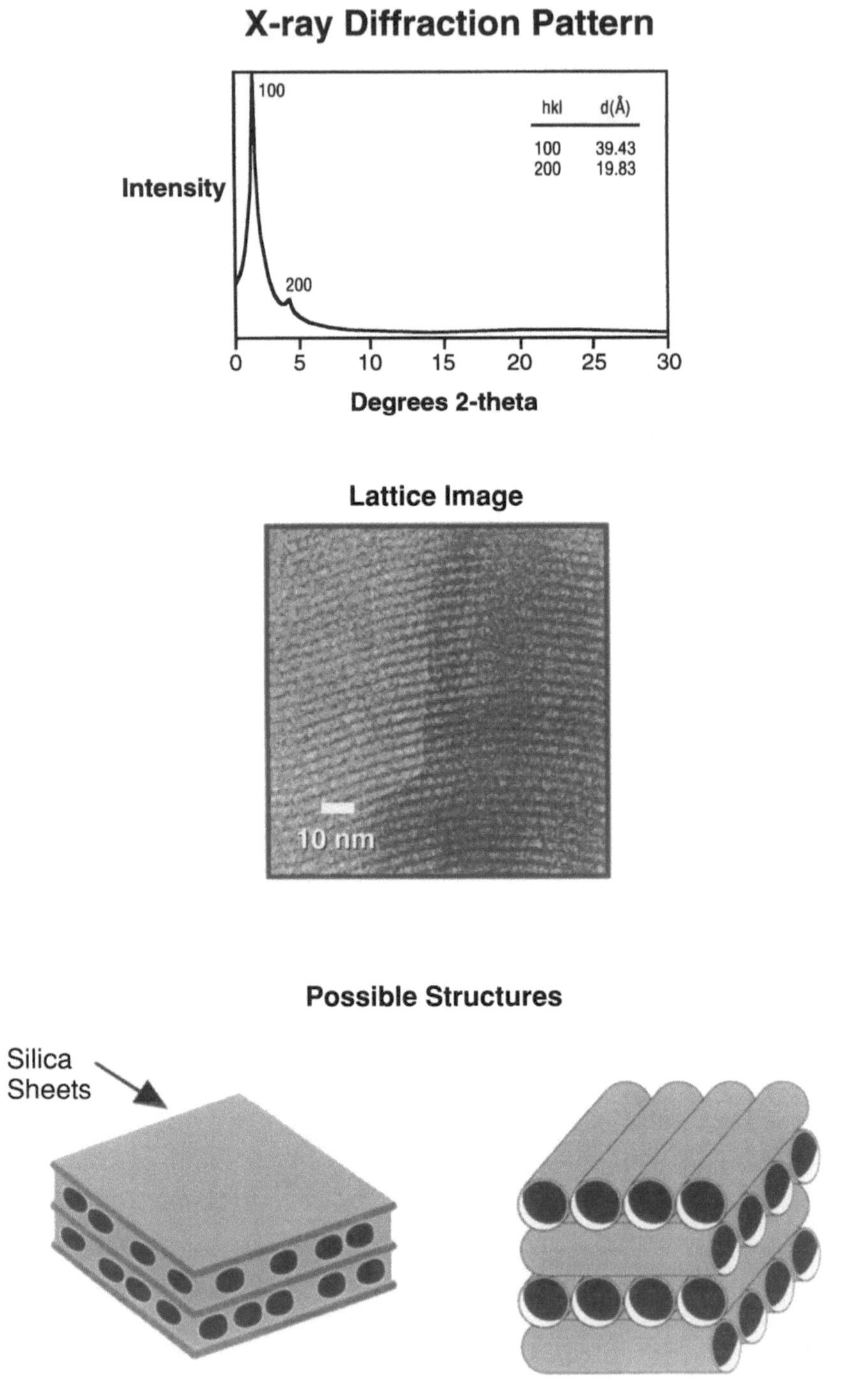

Fig. 4. The X-ray diffraction pattern, transmission electron micrograph, and the proposed structure of MCM-50 (stabilized lamellar phase)

phobic ends of the hydrocarbon molecule face one another. This surfactant structure is then separated by a two-dimensional inorganic oxide sheet similar to that of layered silicates such as magadiite or kenyaite. For the layered configuration, the post-synthesis stabilization would provide stable oxide pillars to separate the oxide sheets and provide porosity. Alternatively, the lamellar phase could be represented by a variation in the stacking of surfactant rods such that the pores of the inorganic oxide product would be arranged in a layered form. For the structure described by the different stacking of rods, the stabilization treatment may provide a method of reinforcing the pore walls. These two proposed structures are shown in Fig. 4.

3
Pore Variation by Direct Synthesis or by Post Functionalization

Pore size tailoring in traditional zeolitic molecular sieve materials has mainly focused on post-synthesis modification. Various organic and inorganic species have been reacted with zeolite surfaces or deposited into the internal pore system with the intent of increasing shape-selectivity and/or altering catalytic activity. Most of this work has focused on reducing the pore size of the molecular sieve. This area has been reviewed by Szostak [37].

In very few cases have methods been devised to increase porosity in zeolites, with a notable exception being the introduction of mesopores via treatment with reagents such as ammonium hexafluorosilicate (AFS) or via steaming of the zeolite. Both types of treatment result in framework dealumination. The removal of framework aluminum results in the formation of defects in the crystal which ultimately lead to a contiguous nest of defects, e.g. the formation of mesopores. The created mesopores are advantageous in improving diffusivity of larger molecules to the active sites of the zeolite. There is much ongoing work in this area and recent results have been reviewed [38]. A recent report by Le Van Mao et al. has shown that mesopores of 120 Å can be generated by treating zeolite A with AFS [39].

Despite the ability to introduce mesopores via the above techniques, there is little control in these processes and the mesopores are disordered in arrangement, randomly scattered and often of wide pore size distribution.

One of the most unique and useful features of the M41S family of materials is the ability to "tailor" both pore size and structure. Through the methods that will be described below, one may, in a systematic fashion, design a mesoporous material with regular, well-defined pore dimensions ranging from ca. 15 Å to >100 Å. The well-documented approaches to accomplishing this pore size variation stand in stark contrast to the ability to accomplish such goals for traditional zeolitic materials. However, it should be pointed out that pore size control in amorphous materials has been successfully achieved in a number of examples [40].

As reported by Mobil, a systematic method to vary the pore size in MCM-41 involves the use of surfactants of varying chain length [6, 7]. When quaternary ammonium surfactants ($C_nH_{2n+1}(CH_3)_3N^+$) with different alkyl chain lengths ($n=8$ to 16) are used, MCM-41 materials exhibiting progressively increasing d_{100}

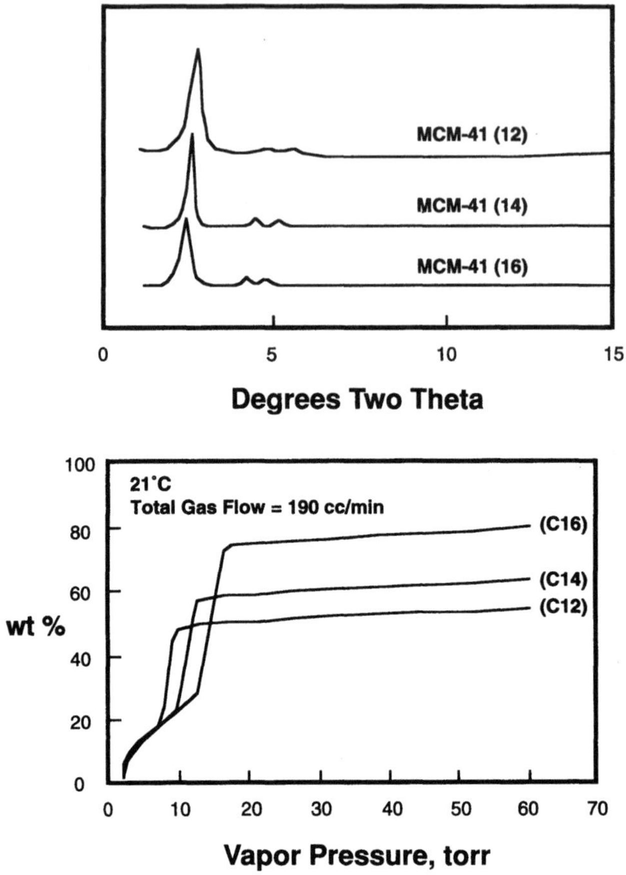

Fig. 5. The X-ray diffraction patterns and corresponding benzene isotherms for MCM-41 materials formed from alkyltrimethylammonium salts having alkyl chain lengths of n = 12, 14, and 16

X-ray diffraction spacings are obtained. Figure 5 illustrates the variation of the d_{100} spacings and the benzene isotherms as a function of alkyl chain length. The changing pore size of these materials was further confirmed by examining the benzene and argon sorption capacities which generally show a shift in the capillary condensation point to higher benzene or argon pressure, with increasing surfactant chain length. This shift is indicative of the filling of larger pores. Furthermore, the total benzene or argon uptakes are generally found to increase with surfactant chain length.

Although surfactant chain length may be used to tailor pore size within a defined range, e.g. 15–50 Å, much more dramatic variation in pore size can be obtained by exploiting the concepts of surfactant/liquid-crystal chemistry. It is well established that a surfactant micellar array is influenced by a number of factors including temperature, pressure, pH and the presence of solvents/solutes in the reaction media.

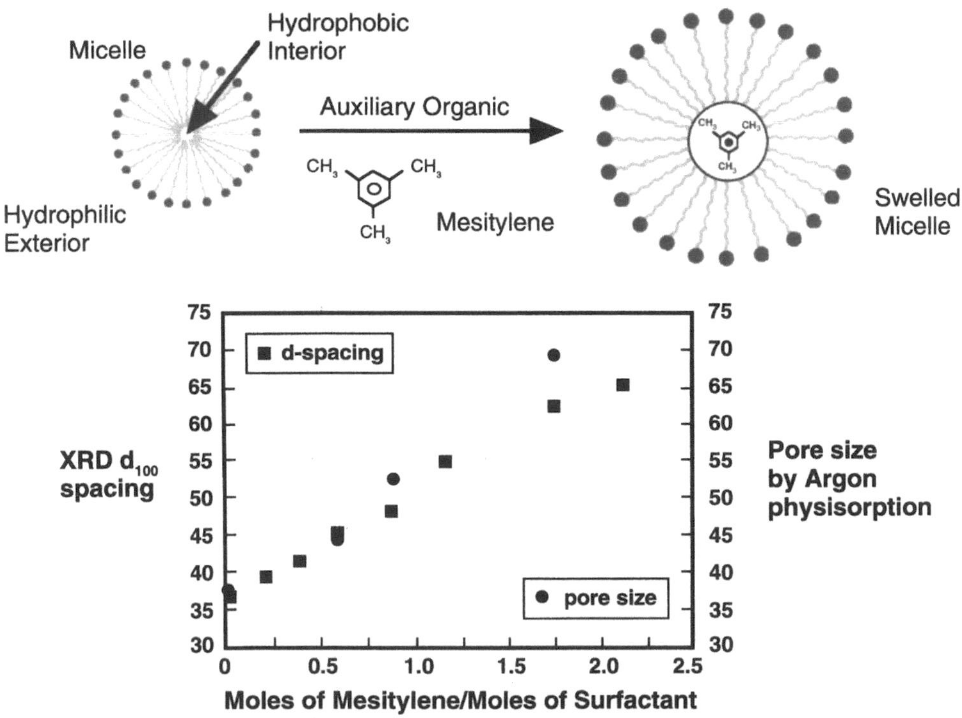

Fig. 6. Powder X-ray diffraction d_{100} spacing and Horvath-Kawazoe pore size as a function of mesitylene/surfactant molar ratio (from [7])

A micellar array in which the core is comprised of hydrophobic hydrocarbon tails is expected to participate in the solubilization of hydrophobic molecules. This process results in the swelling of the micelle with that species. This process is illustrated in Fig. 6. In corroborating the hypothesis that a micellar type arrangement is implicated in the synthesis of MCM-41 type materials, micellar solubilization experiments were performed to try to capture the "swollen" micelles in a silicate cast.

Thus, the typical MCM-41 reaction mixture was modified by the addition of an auxiliary organic, mesitylene (MES) [7]. The molar ratio of MES to $C_{16}H_{33}(CH_3)_3N^+$ in the study ranged from 0 to 2.0. Incremental addition of MES resulted in a proportional increase in d-spacing of the X-ray d_{100} peak. This increase was accompanied by retention of the hexagonal relationship with the X-ray diffraction pattern of each material exhibiting 3–4 peaks related by hexagonal symmetry. A plot of MES/$C_{16}H_{33}(CH_3)_3N^+$ mole ratio versus both d_{100} and pore size, as determined by argon physisorption, results in a linear correlation within this range of solute/surfactant ratios (see Fig. 6).

The benzene sorption isotherms for the several materials prepared with the auxiliary organic show the large sorption capacity characteristic of MCM-41. The isotherms show a gradual shift in capillary condensation to higher benzene pressure with increasing mesitylene content. The material formed with the highest MES/$C_{16}H_{33}(CH_3)_3N^+$ ratio, 2.0, shows only the very beginnings of capillary condensation at the highest benzene pressure examined (60 Torr, $p/p_0=0.6$). Argon isotherms clearly show capillary condensation for these larger pore size materials. Both pore size and pore volume increase with increasing d_{100}.

Materials with even larger pores were prepared from modified reaction mixtures having MES/$C_{16}H_{33}(CH_3)_3N^+$ ratios greater than 3. The products display X-ray powder diffraction patterns exhibiting a single broad line in the extreme low angle region (°2θ); the materials are MCM-41 with a range of d_{100} between 85 and 120 Å. These materials display somewhat irregular, yet essentially hexagonal, pore arrangements. In agreement with these results, argon physisorption derivative uptake curves are significantly broader for these larger pore materials indicating a wider pore size distribution. It is, thus, apparent that MCM-41 materials comprise an extensive family having uniform mesopores of controllable sizes. Figure 7 shows the transmission electron micrographs of MCM-41 materials having pore diameters from 20 to 100 Å.

Functionalization within the mesopores can also affect the pore size. MCM-41 samples contain a large concentration of silanols which can be functionalized via simple elimination reactions. For example, reaction of silanol groups with tetraethylorthosilicate results in the liberation of ethanol and the

Fig. 7. Transmission electron micrographs of several MCM-41 materials having pore sizes of 20, 40, 65, and 100 Å (from [7])

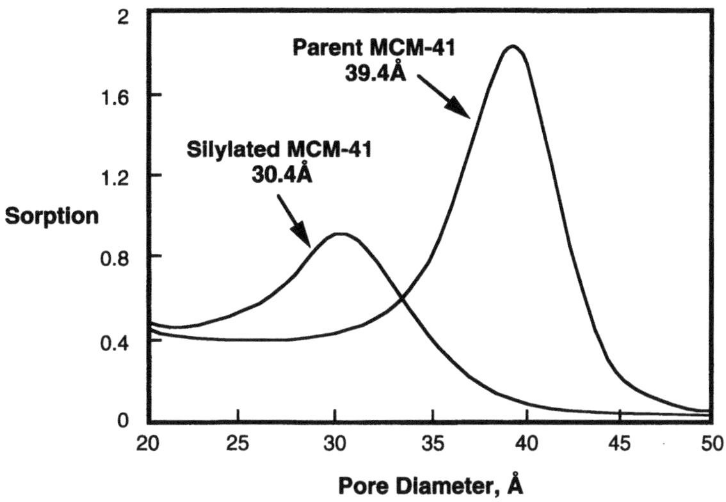

Fig. 8. Horvath-Kawazoe pore size distribution for MCM-41 having a pore diameter of 40 Å and the silylated version (from [7])

attachment of the tetraethoxysilyl group to the interior pore structure. As expected, the resulting internally silylated species has a resultant pore size smaller than the untreated material [7]. This procedure is illustrated in Fig. 8. The material can then be calcined to convert the silylated species to inorganic silica which resides in the pores. The concepts of silylation have been extended to a variety of other species. Some of these materials, such as the silylating agents, have only been used to alter the pore size or affect the hydrophobicity of the pore wall. Other species have been used to anchor new moieties with special catalytic or adsorptive properties. For example, trimethyltin molybdenum carbonyl complexes have been anchored to the walls of MCM-41 via ligand loss (methyl groups + silanol protons) [41]. Upon thermal treatment the complex decomposes to Sn–Mo clusters which reside in the pores. The formation of such species not only alters the pore size, but more importantly, the new species may serve as confined catalysts for catalytic reactions.

4 Mechanism of Formation

This review of the proposed M41S formation mechanism will address the issues of how M41S materials formed, factors controlling this formation and principles of phase selection/transformation. The concepts concerning M41S formation have been formulated based on the studies involving silica (or aluminosilicate)-cationic surfactant systems. Their extension to other compositions will be discussed at the end of this section. As M41S materials are always prepared in the presence of amphiphiles (surfactants), it is safe to conclude that surfactant molecules play a vital role in their formation. Furthermore, the M41S materials appear to be inorganic oxide congeners of lyotropic liquid crystals (LLC) formed by the surfactants in aqueous systems. The salient expressions of this resemblance are the matching structural diversity (hexagonal, cubic and lamellar structures) and size variation, arising from swelling with auxiliary organics or changing surfactant chain length. These similarities, combined with the concept of structure-directing from zeolite science, led to the original proposal of M41S formation due to liquid crystal templating (LCT) [6, 7]. The original proposed mechanistic pathways are illustrated in Fig. 9. Under this general term, two possibilities were envisioned. One reflected the literal meaning of LCT. It invoked pre-existence of surfactant aggregates (e.g. rod-like micelles) or liquid crystal phases in the surfactant precursor solution. Subsequent formation of M41S silica framework occurred due to migration and polymerization of silicate ions into the aqueous region. The second alternative pathway postulated self-assembly of the liquid-crystal like structures due to mutual interaction between dissolved silica and surfactant species. Unlike the first one it was general and did not propose explicitly how M41S formed, and required further study before more details could be provided.

The first pathway became the focus of scrutiny initially as it was relatively easy to verify. It led to formulation of the concept of templating by molecular arrays or aggregates to produce mesopores [42–44]. It was in contrast to templating with single molecules, which could generate only micropores (supramolecular vs monomolecular templating). However, this attractive idea appears to have only a conceptual meaning. It is now quite evident that this simplified LCT model can not be operative in the majority of cases. The appropriate surfactant phase diagrams indicate that most M41S syntheses have been accomplished in the surfactant concentration regime below that required for LLC presence, e.g. down to 1 wt.-% surfactant [10]. It could be argued that the high pH and multi-ion nature of M41S synthesis systems may induce LLC formation. Indeed, increasing ionic strength is known to cause phase changes in amphiphilic systems but it is unlikely to occur to the extent that would be required for M 41S synthesis. This intuitive hypothesis has been substantiated by NMR and low angle neutron scattering [23, 45, 46]. The surfactant precursor solutions used for the synthesis of MCM-41 were found to contain only micellar species without LC phases. Results consistent with the presence of an ordered phase were observed only after the inclusion of silica.

The second mechanism was very general – it only stated what caused M41S formation and required further studies to elaborate mechanistic details. This has been addressed in a series of papers by Stucky et al. [10, 11, 23, 43]. First, however, we would like to analyze two specific proposals concerning attempted elucidation of the self-assembly mechanism. Both pertained to the hexagonal (MCM-41) phase and relied on the identification of putative precursors to MCM-41. Stucky et al. [10] observed the presence of a layered silica/surfactant phase upon mixing of the reactants. With time the hexagonal MCM-41 was produced, which led to the proposal of solid-state transformation, as depicted in Fig. 10.

The second variation of the MCM-41 formation mechanism stemmed from the identification of silica-clad rod-like aggregates in the synthesis mixture [45,

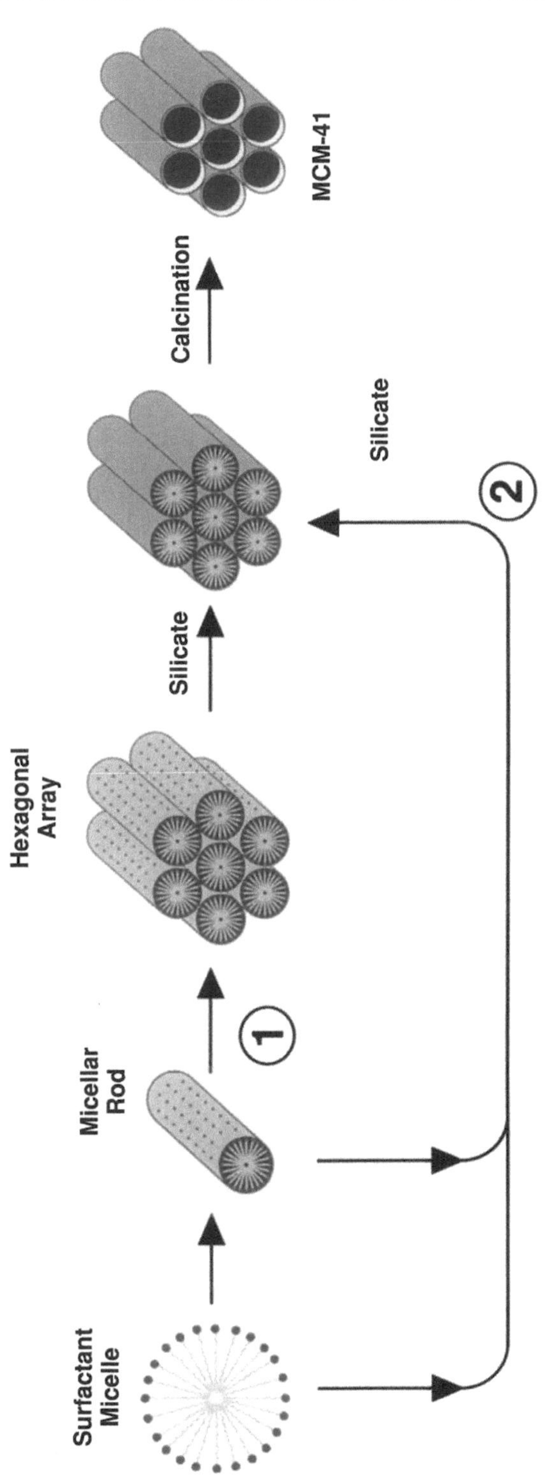

Fig. 9. Possible mechanistic pathways for the formation of MCM-41: (*1*) liquid crystal phase initiated and (*2*) silicate anion initiated (from [6])

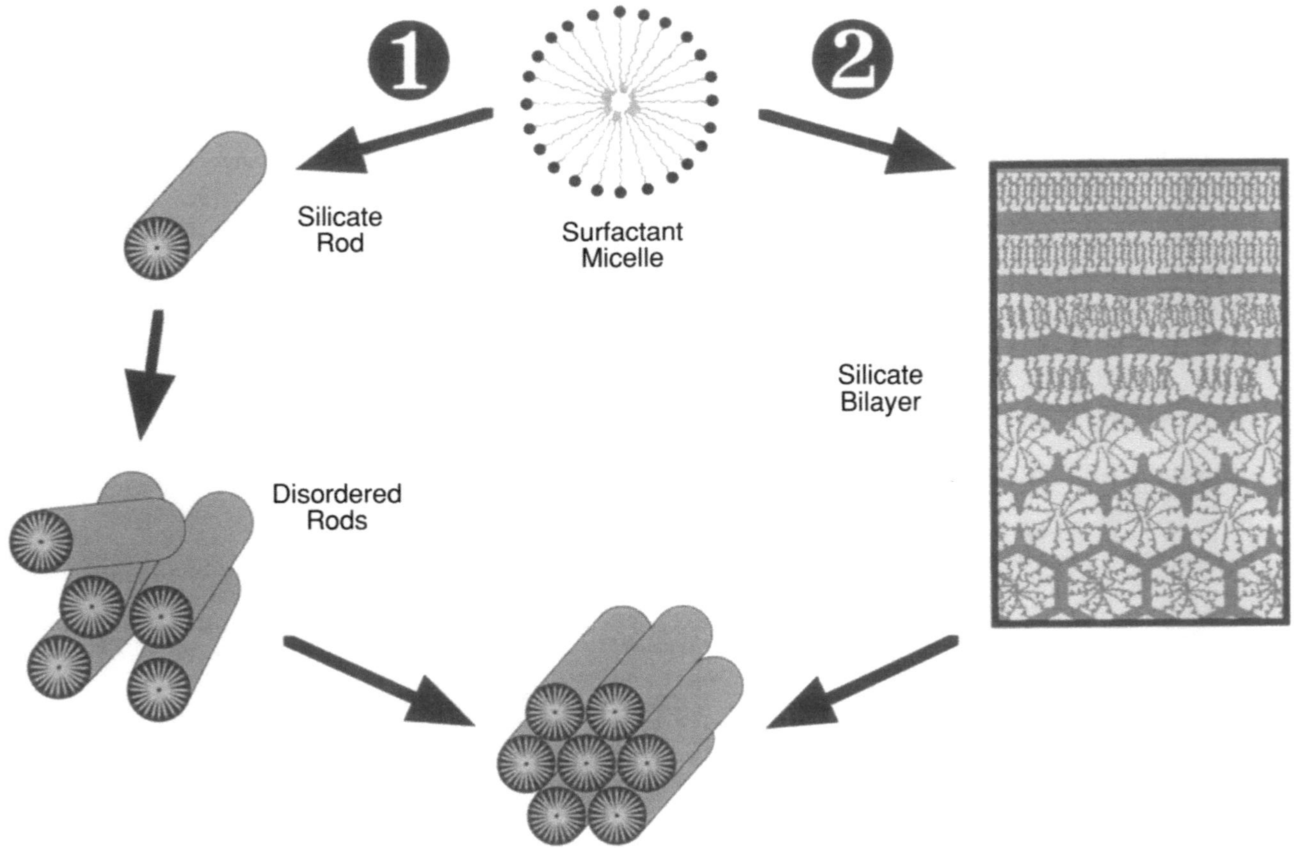

Fig. 10. Proposed formation mechanistic pathways by (*1*) stacking of silicated surfactant rods (from [45]) and (*2*) via the formation of an initial lamellar intermediate (from [10])

46]. The rods were identified by ^{14}N NMR based on the characteristic single isotropic line. Rod-like micelles have been known to form in the presence of electrolytes. The hexagonal MCM-41 could thus arise by close-packing of the silicate-surfactant rods. This mechanism pathway is also shown in Fig. 10. The concept was extended to the cubic (MCM-48) phase. As cubic phase-bearing mixtures were also found to contain these rod micelles [46], it was postulated that "any mechanistic proposal must have the CTMA species assembled into rod-like micelles."

Both of the above hexagonal phase formation mechanisms must be treated as speculative. They are not placed in the broader context and try to account for particular observations without analysis of alternative explanations. In particular, it is quite likely that both precursors (lamellar solid and silica-clad rods) are metastable intermediates that dissolve and serve only as the source of nutrients for the growing MCM-41 framework. The arguments in favor of both pathways are valid essentially only with the hexagonal phase. For example, one can readily justify close-packing of silica-clad rods into hexagonal MCM-41. However, it is not as self-evident to rationalize the formation of cubic and lamellar phases on the same premise. Lastly, neither mechanism addresses the issue of phase transformation.

It is now universally accepted that in the majority of cases the self assembly (second LCT alternative) is the operative mechanism for M41S. As is generally the case with self-assembly, the process is very difficult to describe in detail. The original proposal [6, 7] implied interactions between negative silicate and positive surfactant ions leading to precipitation of the M41S phases. Stucky carried out more detailed analysis of this phenomenon [10, 23, 43]. In a succinct summary [11] this was described as a cooperative mode for the synthesis of mesostructures that involved the following three processes: multidentate binding of silicate oligomers to the cationic surfactant, preferred silicate polymerization in the interfacial region, and charge density matching between the surfactant and the silicate. Some of the rationalization and details are given in the following paragraphs.

While considering the interactions of silicate and surfactant molecules in solution, the monovalent silica monomers were deemed less efficient in competition with the other counterions [10]. However, "oligomeric silica polyanions (existing at high pH) can easily act as multidentate ligands for the cationic head groups of the surfactant, leading to strongly interacting surfactant-silicate interface" [10]. This "preferential multidentate binding of the silicate polyanions causes the interface to quickly become populated by tightly held silicate oligomers, which can subsequently polymerize further" [10]. The polymerization is said to be favored because of the high concentration of the silicate species and partial screening of negative charges due to the presence of surfactant heads. This reasoning led to a two-step pathway to M41S: precipitation of silica-surfactant salts and silica polymerization [10].

The concepts outlined in the preceding paragraph, especially the interaction between silica and surfactant, apparently decoupled from silica polymerization, were explored by Stucky using a discrete silica oligomer, D4R [23]. "To clarify the role of large multiply charged inorganic species in the phase behavior of composite inorganic-organic assemblies a precursor silicate solution (D3R, D4R) can be used to initiate mesophase assembly" [23]. The reaction at room temperature precipitated a (hexagonal) mesophase as a salt with negligible silica polymerization. The latter occurred upon annealing at higher temperature. Varying of the relative surfactant/silica concentration allowed controlling "morphologies of unpolymerized silica-surfactant mesophases within the constraints of a nominal inorganic-organic-water phase space." Thus, at half the original surfactant concentration, a lamellar phase was produced. It was classified as a genuine equilibrium liquid crystal phase as it underwent a reversible first order transformation from a lamellar to an hexagonal phase at 45 °C. NMR showed only intact, unpolymerized D4R silicate species. At higher temperature (70 °C) the reversible lamellar phase becomes quickly and irreversibly polymerized into hexagonal MCM-41. This lamellar-to-hexagonal transformation is reminiscent of the proposed MCM-41 formation pathway described above. However it can not be used to substantiate the latter as there is no direct connection between the two situations. The authors [23] do not observe these equilibrium liquid crystal phases "under the higher temperature conditions used with many mesophase syntheses (for example, 100 °C)" or "for more reactive silicate species (for example, TEOS)." They attribute this to the fact that "polymerization of the interfacial silicate species proceeds at a rate comparable to the silicate-surfactant self-assembly process." Furthermore the D4R structure was not retained during the formation of the higher temperature MCM-41 (as detected by in situ NMR) suggesting that the cubic structure was dissolving during the process and merely providing a source of silicate ions for the mesophase formation.

We now turn to the issue associated with the existence of different structural types of M41S, referred to as the phase selection issue [10, 43]. As mentioned above this structural diversity was considered one of the key elements of the LCT mechanism [6, 7]. The concept of different M41S materials was tied to the phase behavior of lyotropic liquid crystals, in particular the phase transformation with increasing surfactant concentration. M41S systems seemingly showed parallel phase preference with increasing surfactant/silica ratio [34]. However upon close scrutiny no direct physical connection could be demonstrated. LCT must be interpreted as simply indicating that both LC and M41S are governed by the same principles ingrained in the fundamental properties of surfactant systems. This problem has been addressed above in connection with the literal meaning of LCT. The absence of direct physical (as opposed to literal) correlation between the LC phase diagram and M41S phase selection has been noted [23]. This may not be surprising given the higher complexity of the surfactant, silica, and other cation/anion species present. In addition, we now believe that with regard to the phase diagram, the region relevant to M41S phase selection is located in the relatively unexplored region below the Krafft boundary, i. e., where surfactant containing solutions form solid phases.

Stucky et al., stating that long-range order organic arrays (liquid crystals) are not required for M41S nucleation, comment on the phase selection as follows [23]. "The ultimate structure and symmetry are determined by dynamic and often subtle interactions among the organic and inorganic species according to equilibrium thermodynamics. Micelles do rearrange and the product nature depends on inorganic species and its electrostatic and steric interactions with the organic. The conditions such as temperature, ionic strength, pH effect the direction and growth." In general, by analogy to LLC the favored mesophase is that which permits A, the head-group area, to be closest to the optimal value A_0. Quantitatively its value is obtained by minimizing the Gibbs free energy as a function of A. The Gibbs free energy is now described in terms of contributions from different interactions in the system. Although these general considerations suggest the variables affecting the formation of these mesoporous materials, they do not provide a specific recipe to explain the factors required for the phase selection.

A unique mechanism of M41S formation, proposing heterogeneous nucleation has been suggested [47]. It was observed that formation of M41S from a mixture containing TMA–silicate (solution), surfactant and sodium aluminate was contingent upon the presence of a colloidal silica (HiSil) or titania. The paper does not state this but it appears that HiSil contributed additional silica to the system, which therefore possessed different (silica enriched) stoichiometry. The described observation can thus be interpreted as stemming from unfavorable composition regime when HiSil was excluded (not enough silica). In the case of titania no such simple explanation is evident. In view of the scarcity of experimental details and perfunctory presentation of results, no thorough analysis can be attempted. One can envision that titania's adsorptive nature may alter the effective composition of the solution making it more favorable for M41S formation. M41S materials were shown to form from clear solution [48]. This result appears to undermine the validity of the concept of heterogeneous nucleation, assuming that if nuclei formed they were not sufficiently small to avoid detection. Further work is needed for the concept of hetereogeneous nucleation to become acceptable.

The concept of synthesizing mesoporous materials by exploiting surfactant properties has been expanded. Various charge combinations have been used [11] in addition to the original positive surfactant and anionic silicate. These include positive inorganic oxide and negative surfactant, ionic inorganic species with similarly charged surfactant, which is mediated by counterions of opposite charge, as well as neutral surfactant [12].

5
Sorption Properties

In our laboratory, the porosity of M41S materials has been determined by sorption of benzene and argon. The sorption isotherms of M41S materials exhibit four unique features; the material has exceptionally high hydrocarbon sorption capacity, ≥55 wt.% benzene at 50 torr (1 torr = 1.316 mbar) and 25 °C and >50 wt.% hexane at 40 torr and 25 °C; a sharp step in the isotherm for pore filling suggesting a uniform pore size; the p/p_0 position for pore filling occurs at relatively high partial pressure, a result of the mesopore size of MCM-41; pore filling is observed at higher p/p_0 as the pore size increases [6, 7]. These phenomena are illustrated in Fig. 11. Although M41S materials have significant hydrocarbon capacity, the water uptake is lower, 0.05 g/g at $p/p_0 = 0.4$, demonstrating the hydrophobic nature of M41S [45].

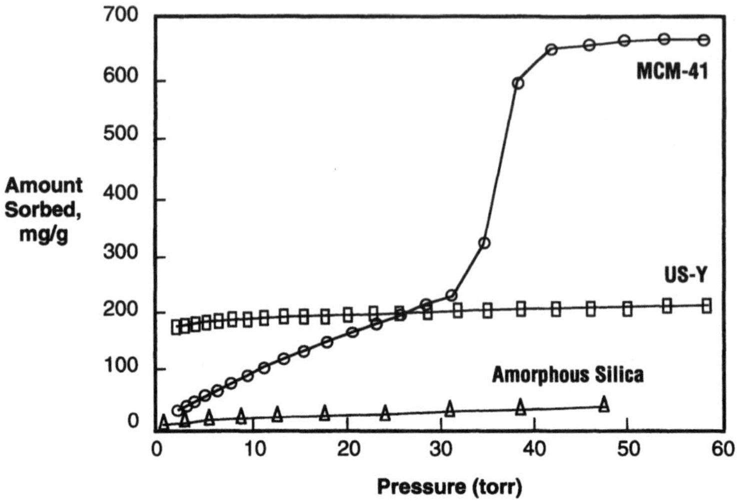

Fig. 11. Benzene adsorption isotherm for MCM-41 having a pore size of 40 Å (from [7])

To probe the unique adsorption characteristics of M41S materials, adsorbates such as N_2, O_2 and argon were examined. These results are illustrated in Fig. 12. The nitrogen sorption properties of the hexagonal structure MCM-41 have shown particularly interesting properties with respect to hysteresis behavior. The nitrogen isotherm displays a sharp uptake at p/p_0 and the isotherm was reversible with no hysteresis. Comparative data on non-porous hydroxylated silica discounted the presence of micropores in MCM-41; the sorption process was described as monolayer-multilayer adsorption followed by capillary condensation [49-51]. Later studies showed reversible Type IV isotherms for materials for pore sizes less than ~40 Å with larger pore materials exhibiting irreversible Type IV isotherms [52, 53]. The physical reasons for this unexpected behavior are still unclear. For the smaller pore materials, a filling process like secondary micropore filling has been proposed. However, the absence of hysteresis in the smaller pore materials conflict with the Laplace and Kelvin theories. Further experimental and theoretical work is necessary for a complete understanding of this unusual behavior. Other adsorbates such as oxygen and argon do exhibit irreversible Type IV isotherms with significant hysteresis loops observed for 40 Å and smaller pore size materials [49, 51, 52]. This is more than likely a result of significant differences in the physical properties of the three adsorbates.

6 Conclusion

Mesoporous molecular sieves with well-defined pore systems are a recent advance in molecular sieve science. However, in the short time span since their discovery, there has been explosive growth in both development of synthetic

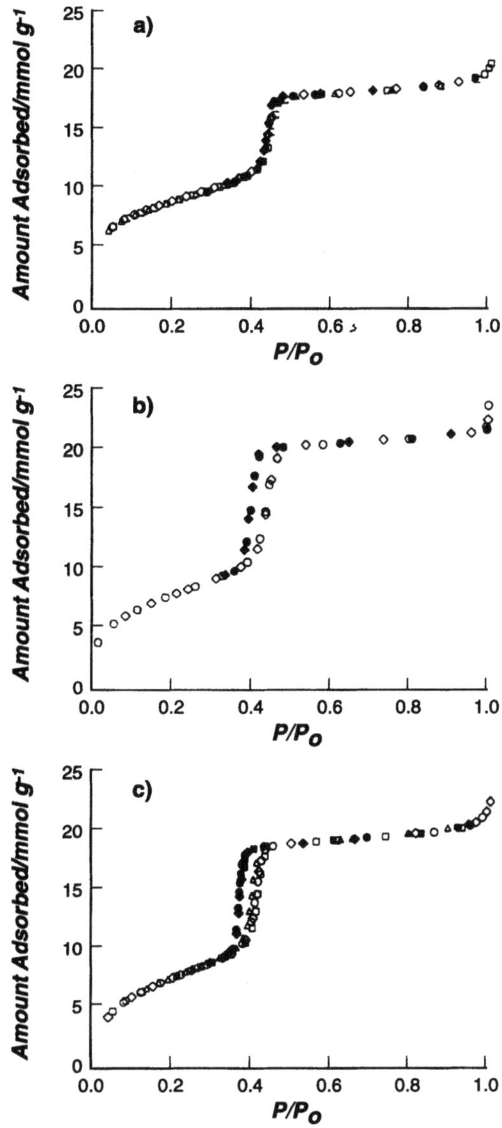

Fig. 12. Sorption isotherms of **a** N_2, **b** Ar, and **c** O_2, on MCM-41 (from [52])

methods to produce these materials and the ability to characterize them. Above we have detailed some of these inroads, ranging from the tunable nature of pore size, structure and sorptive characteristics and texture to the producing of myriad compositional variations, including the production of novel transition metal oxide structures.

Never before in the molecular sieve area, has one been able to control so many of the characteristics of inorganic porous solids. All this appears possible due to an understanding of the most likely mechanisms of formation of these materials. Indeed, one needs only to cleverly combine concepts of inorganic sol gel chemistry with well known surfactant/micelle chemistry to find a niche to explore in this new area.

The mechanistic interplay between micelle/liquid crystal chemistry and the polymerization/crystallization of the inorganic framework provide a rare opportunity to construct an array of new structures based on well established principles of charge density matching at the organic/inorganic interface. With these new structures, pore sizes and compositions at hand, researchers in many fields are armed with a new arsenal of materials with which to attack the nanoscopic problems.

7 References

1. Davis ME, Lobo RF (1992) Chem Mater 4:756
2. Flanigen EM, Patton RL, Wilson ST (1988) Stud Surf Sci Catal 37:13
3. Lok BM, Cannon TR, Messina CA (1983) Zeolites 3:282
4. Kerr GT (1989) Scientific American 261:100
5. Argauer RJ, Landolt GR (1972) US Patent 3702886
6. Kresge CT, Leonowicz ME, Roth WJ, Vartuli JC, Beck JS (1992) Nature 359:710
7. Beck JS, Vartuli JC, Roth WJ, Leonowicz ME, Kresge CT, Schmitt KD, Chu CT-W, Olson DH, Sheppard EW, McCullen SB, Higgins JB, Schlenker JL (1992) J Am Chem Soc 114:10835
8. Oliver S, Combs N, Kuperman A, Lough A, Ozin GA, Nature (1995) 378:47; Oliver S, Combs N, Ozin GA (1995) Adv Mater 7:931; Ozin GA, Oliver S (1995) Adv Mater 7:943
9. Walsh D, Mann S (1995) Nature 377:320
10. Monnier A, Schüth F, Huo Q, Kumar D, Margolese D, Maxwell RS, Stucky GD, Krishnamurty M, Petroff P, Firouzi A, Janicke M, Chmelka BF (1993) Science 261:1299
11. Huo Q, Margolese DI, Ciesla U, Feng P, Gier TE, Sieger P, Leon R, Petroff P, Schüth F, Stucky GD (1994) Nature 368:317
12. Tanev PT, Pinnavaia TJ (1995) Science 267:865
13. Bagshaw S, Prouzet E, Pinnavaia TJ (1995) Science 269:1242
14. Karra VR, Moudrakovski IL, Sayari A (1996) J of Porous Mat 3:77
15. Tanev PT, Chibwe M, Pinnavaia TJ (1994) Nature 368:321
16. Corma A, Navarro MT, Pérez-Pariente J, Sánchez F (1994) In: Weitkamp J, Karge HG, Pfeifer H, Hölderich W (eds) Zeolites and related microporous materials: state of the art. Elsevier Science, p 69
17. Franke O, Rathousky J, Schulz-Ekloff G, Stárek J, Zukal A (1994) In: Weitkamp J, Karge HG, Pfeifer H, Hölderich W (eds) Zeolites and related microporous materials: state of the art. Elsevier Science, p 77
18. Llewellyn PL, Ciesla U, Decher H, Stadler R, Schüth F (1994) In: Weitkamp J, Karge HG, Pfeifer H, Hölderich W (eds) Zeolites and related microporous materials: state of the art. Elsevier Science, p 2013
19. Corma A, Navsarro MT, Pariente JP (1994) J Chem Soc Chem Comm 147
20. Corma A, Fornés V, Navarro MT, Pérez-Pariente J (1994) J Catal 148:569
21. Stein A, Fendorf M, Jarvie TP, Mueller KT, Benesi AJ, Mallouk TE (1995) Chem Mater 7:304
22. Sayari A, Danumah C, Moudrakovski IL (1995) Chem Mater 7:813
23. Firouzi A, Kumar D, Bull LM, Besier T, Sieger P, Huo Q, Walker SA, Zasadzinski JA, Glinka C, Nicol J, Margolese D, Stucky GD, Chmelka BF (1995) Science 267:1138
24. Zhao D, Goldfarb D (1995) J Chem Soc Chem Comm 875

25. Huo Q, Margolese DL, Ciesla U, Demuth DG, Feng P, Gier TE, Sieger P, Firouzi A, Chmelka BF, Schüth F, Stucky GD (1994) Chem Mater 6:1176
26. Luca V, MacLauchlan DJ, Hook JM, Withers R (1995) Chem Mater 7:2220
27. Abe T, Taguchi A, Iwamoto M (1995) Chem Mater 7:1429
28. Antonelli DM, Ying JY (1995) Angew Chem Int Ed Engl 34:204
29. Schüth F, Ber. (1995) Bunsenges Phys Chem 99:1306
30. Feuston BP, Higgins JB (1994) J Phys Chem 98:4459
31. Inagaki S, Fukushima Y, Kuroda K (1993) J Chem Soc Chem Comm 680
32. Inagaki S, Fukushima Y, Kuroda K (1994) In: Weitkamp J, Karge HG, Pfeifer H, Hölderich W (eds) Zeolites and related microporous materials: state of the art. Elsevier Science, p 125
33. Huo Q, Leon R, Petroff P, Stucky GD (1995) Science 268:1324
34. Vartuli JC, Schmitt KD, Kresge CT, Roth WJ, Leonowicz ME, McCullen SB, Hellring SD, Beck JS, Schlenker JL, Olson DH, Sheppard EW (1994) Chem Mater 6:2317
35. Husson F, Mustacchi H, Luzzatti V (1960) Acta Cryst 13:668
36. Vartuli JC, Kresge CT, Roth WJ, McCullen SB, Beck JS, Schmitt KD, Leonowicz ME, Lutner JD, Sheppard EW (1995) 209th National ACS meeting, Anaheim, CA, to be published
37. Szostak R (1991) Stud Surf Sci Catal 58:153
38. Chen NY, Degnan Jr. TM, Smith CM (1994) Molecular transport and reaction in zeolites. VCH Publishers, New York, p 32
39. Le Van Mao G, Denes NTC (1995) Mat Res Soc Symp Proc 371:123
40. Komarov VS, Kuznetsova TF (1978) Vestsi Akad Navuk BSSR 2:22
41. Huber C, Moller K, Bein T (1994) J Chem Soc Chem Comm 2619
42. Beck JS, Vartuli JS, Kennedy GJ, Kresge CT, Roth WJ, Schramm SE (1994) Chem Mater 6:1816
43. Stucky GD, Monnier A, Schüth F, Huo Q, Kumar D, Margolese D, Krishnamurthy M, Petroff P, Firouzi A, Janicke M, Chmelka BF (1994) Mol Cryst Liq Cryst 240:187
44. Chenite A, Lepage Y, Sayari A (1995) Chem Mater 7:1015
45. Chen CY, Burkett SL, Li HX, Davis ME (1993) Microporous Materials 2:27
46. Davis ME, Chen CY, Burkett SL, Lobo RF (1994) Mat Res Soc Symp Proc 346:831
47. Liu A, Kim AY, Virden JW, Bunker BB (1995) Langmuir 11:689
48. Elder KJ, White JW (1995) J Chem Soc Chem Comm 155
49. Franke O, Schulz-Ekloff G, Rathousky J, Starek J, Zukal A (1993) J Chem Soc Chem Comm 724
50. Branton PJ, Hall PG, Sing KSW (1993) J Chem Soc Chem Comm 1257
51. Branton PJ, Hall PG, Sing KSW, Reichert H, Schüth F, Unger KK (1994) J Chem Soc Faraday Trans 90:2965
52. Llewellyn PL, Grillet Y, Schüth F, Reichert H, Unger KK (1994) Microporous Materials 3:345
53. Schmidt R, Stöcker M, Hansen E, Akporiaye D, Ellestad OH (1994) Microporous Materials 3:443

Approaches for the Synthesis of Ultra-Large and Ultra-Small Zeolite Crystals

Eric N. Coker* · Jacobus C. Jansen

Department of Chemical Technology, Delft University of Technology, Julianalaan 136, NL-2628 BL Delft, The Netherlands. *E-mail: J. C. Jansen@stm.tudelft.nl*

* Present address: BP Chemicals Limited, Chertsey Road, Sunbury-on-Thames, Middlesex TW16 7LL, UK. *E-mail: CokerEN@bp.com*

1 General Introduction

Ultra-large crystals: crystals of which the unit cell of framework atoms can be regularly repeated in three directions are the object of this section. The perfectness of the crystal and the framework are only subject to the ability to elucidate the fractional co-ordinates of the atoms comprising the unit cell. The size of

Molecular Sieves, Vol. 1

those crystals clearly depends on the type of zeolite. Of zeolite beta, even the smallest single crystal is not known, while analcime forms single crystals of a few millimeters. However, crystals may also be considered ultra-large in cases where one or two dimensions are well developed, such as elongated prismatic MFI of 60 μm × 60 μm × 350 μm [1] and ferrierite tablets of 1000 μm × 1000 μm × 100 μm [2]. The availability of ultra-large crystals of zeolites and other microporous materials enables fundamental studies to be carried out, providing insights and achieving results which would hardly be possible using small crystals. In particular, measurement of the diffusional properties of species which diffuse quickly through the porous network and accurate determination of zeolite structures, down to the atomic positions, with single-crystal X-ray techniques are possible. Measurement of micropore diffusivities is essential for the optimization and design of catalytic reactors and adsorptive separation processes. The construction of ordered composites of zeolite crystals containing guest species allows non-linear optical devices, quantum-confined semiconductors, and other such materials to be realized. Continuous, ordered membranes of zeolites grown onto porous supports may be prepared more easily and reproducibly once ways are found to grow large enough single crystals. In case the zeolite is to be applied as a membrane, the crystal form might be important. For example, ultra-large ferrierite crystals were grown in tablet form, [2] as depicted in Fig. 1. Access to the relatively small number of long eight- and ten-ring channels parallel to the large crystal face would be poor in a membrane set up where the crystals lie flat on a support surface.

Ultra-small crystals do not have to be, and often are not single crystals. The ratio of external to internal surface areas is of particular relevance; it may be an order of magnitude higher for ultra-small crystals compared to ultra-large ones. The ultra-small crystals are desirable for many industrial processes, particularly catalytic ones. The larger the number of pore openings per unit length of pore, the greater is the accessibility of the catalytic sites and the faster is the attainment of equilibrium in adsorptive and catalytic applications.

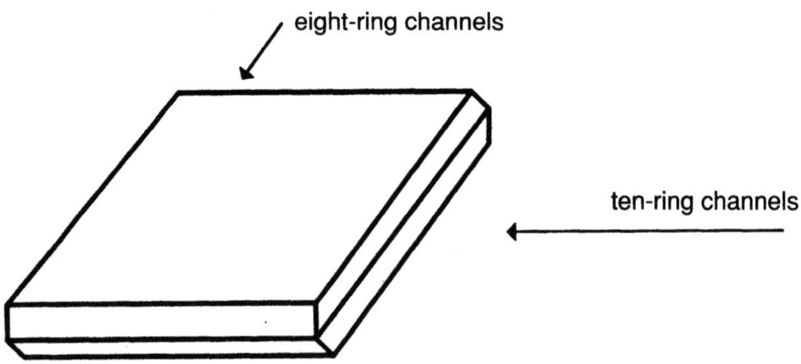

Fig. 1. Typical tablet form of ferrierite. The tablet-shaped crystal face, with a surface area of 1 to 2 mm^2, comprises no channel entrances. The eight- and ten-ring channels run parallel to the large crystal plane

Naturally-produced zeolites often far surpass those made in the laboratory in terms of their large size and sheer beauty. However, the time-frame of nature's synthesis procedure is measured in millennia – clearly not an option for even the most patient scientist. Hence it is up to the scientist to use his ingenuity in creating methods to prepare ultra-large or ultra-small crystals in the laboratory.

1.1 Aspects of a Zeolite Synthesis Solution

Classical zeolite syntheses are carried out in aqueous media under hydrothermal conditions and autogenous pressure. The sequence of events thought to occur in a typical zeolite synthesis experiment is given below. Once a synthesis solution has been mixed, a number of processes occur, some at ambient temperature and some preferentially at elevated temperature. An incubation period may exist, during which no apparent nucleation or growth occurs; it is assumed that reorganization of the nutrients into some "pre-nucleation" state occurs (Fig. 2). During the early stages of synthesis, the nutrients organize themselves around inorganic or organic cations through van der Waals or electrostatic interactions, forming small units possessing some of the characteristics of a zeolitic structure. Such units may be considered as nuclei; they remain unstable with respect to dissolution until they have reached a critical size [3]. Once the nucleus exceeds that size, the probability that it will grow into a crystal is relatively high. In many cases, the rate of crystal growth has been shown to be linear over most of the growth period, but tailing-off as the nutrients eventually become depleted and growth stops.

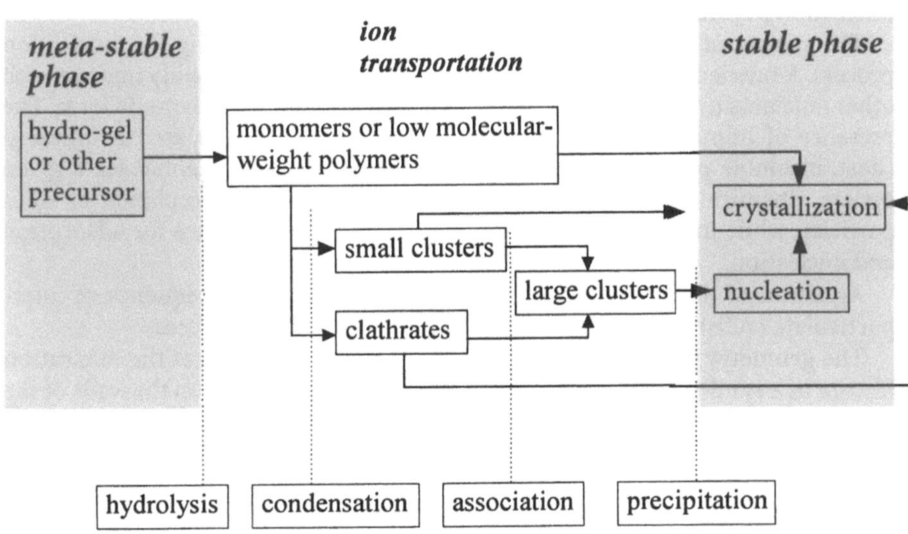

Fig. 2. A simplified view of some of the major processes occurring in a zeolite synthesis reaction

While some compositions produce gel-like amorphous aluminosilicate precursor phases early in the synthesis, others remain essentially clear. The amorphous gel phase may be regarded as a reservoir of nutrients, continually dissolving to replace nutrients removed from solution by condensation onto growing crystals. Nucleation occurs in the gel phase or near its surface, where nutrient concentrations are high, and the growing crystals may be partially suspended by the gel network. In clear-solution syntheses, nucleation may be homogeneous in solution, or heterogeneous at the vessel walls or dust particles. Once the crystals reach a certain size, they begin to settle out of solution under the influence of gravity.

1.2 Parameters Influencing Nucleation

The degree of supersaturation of the solution is the principal driving force for nucleation; the higher the supersaturation, the greater the tendency for nucleus formation. The degree of supersaturation is determined by two main factors: nutrient concentration and temperature. As the nutrient concentration increases, so does the degree of supersaturation, until the formation of an amorphous gel phase occurs. As the temperature of the system increases, so does the solubility of the nutrients; by definition, the degree of supersaturation decreases. However, increasing the temperature also increases reaction kinetics, thus accelerating nucleation and crystal growth. The specific effect of a rise in temperature with respect to crystal size will depend upon the particular chemistry and conditions of the synthesis. Theoretical modeling of zeolite syntheses have shown that the rate of nucleation can be very sensitive to changes in the temperature of the system [4]. The rate of heat-up of the system can also be critical; a slow heating rate (i.e. a reactor of high heat capacity) will result in a spreading-out of the nucleation period.

The nature of the reagents may be critical in determining the nucleation process. A nutrient present in a monomeric form may more readily interact with other nutrients to form nuclei than one which is present in a polymeric form. The presence of impurities, either chemical (Fe^{3+}, alkaline earths, etc.) or physical (dust, insoluble particulates) can increase the nucleation potential for a given system. Chemical impurities may, for instance, encourage flocculation of silica particles, while insoluble material presents a convenient surface for adsorption and nucleation.

Agitation of the solution during synthesis increases the frequency of interparticulate collisions, which causes secondary nucleation to occur.

The geometry and size of the synthesis vessel also influences the nucleation history of a synthesis batch. Heterogeneous nucleation occurs on the walls of the reactor, thus vessels with a high ratio of solution volume to wall area may produce fewer nuclei per unit volume of solution.

1.3
Parameters Influencing Crystal Growth

The availability of nutrients, or, more specifically, the amount of nutrient available to each crystal nucleus, is a major limiting factor in the growth of large crystals. The presence of excess nutrients in a given solution volume, or their addition during the synthesis, while avoiding excessive nucleation is a challenge to the researcher.

As one raises the temperature at which a particular synthesis is carried out, the rate of crystal growth increases. However, the solubility of the reagents and also the crystalline zeolite is also increased, so a lower equilibrium conversion of nutrients into zeolite crystals can be expected.

Increasing the concentration of nutrients brings about a similar increase in crystal growth rate, while also increasing the nucleation potential of the mixture.

As the density and viscosity of a solution is increased, a reduction in crystal growth rate occurs, while the concomitant reduction of sedimentation allows the crystals to reach a larger size before falling to the bottom of the reactor.

For the growth of large crystals, a long synthesis time is needed since one is dealing with a system containing relatively few nuclei; the cumulative surface area of crystalline material available for condensation of nutrients (i.e. crystal growth) is low, thus requiring a long time to deplete the nutrient pool. Similarly, a short time is required for the synthesis of small crystals as the crystal surface area is much higher. One problem with the long synthesis times for large crystals is the metastability of zeolites under such conditions. Examples of this will be seen later in the chapter.

1.4
Comparison of Large and Small Crystal Syntheses

The following paragraph lists the parameters which may be used to promote the growth of larger or smaller crystals. Because zeolite synthesis solutions are so complex and vary in their composition tremendously depending upon the type of zeolite, an absolute method for the preparation of ultra-large or ultra-small crystals cannot be given. Rather, a suitable combination of these parameters needs to be chosen. Table 1 summarizes the most important parameters which may influence the preparation of ultra-large, or ultra-small crystals.

2
Synthesis of Ultra-Large Crystals

2.1
Case Study: Zeolites LTA and FAU

In this case study, a few examples of systematic studies on the factors affecting crystal size will be given, followed by a survey of attempts to produce large crystals of zeolites LTA (Linde Type A) and FAU (faujasite).

Table 1. Parameters which influence the growth of ultra-large and ultra-small zeolite crystals

Parameter	Effect	Result
Towards Ultra-Large Crystals		
removal of contaminants	cleaner mixtures	fewer nucleation centers
gel phase or microgravity	static mixture, reduced particle settling	nucleation and rate of mass transfer reduced
larger vessel	larger nutrient pool and volume to surface ratio	extended growth and fewer nuclei per volume
polymeric reagents	time-released nutrients	reduced number of nuclei
organic solvents with stoichiometric water	new mineralizer with specific properties	extended growth period
temperature spike	induce nucleation in short time	reduced number of nuclei
Towards Ultra-Small Crystals		
clear solutions	homogeneous solution, no sol or gel particles which determine the crystallite size and distribution	explosive nucleation
lower temperature	homogeneous solution	explosive nucleation
colloidal sol	homogeneous small precursor	explosive nucleation
short crystallization time	interrupted growth	reduced crystal growth
monomeric reagents	largest interface between silica and template	high nucleation rate

Since the crystallization fields for zeolites A and faujasite (X and Y) lie next to each other on the crystallization phase field diagram (Fig. 3) [5], they often grow in the same, or similar solutions. This is especially true when attempting to increase crystal sizes through the addition of complexing agents to the solutions (see Sect. 2.1.1). Hence, these two types of zeolites will be discussed together.

The effects of alkalinity, silica source, potassium content [6], method of gel formation [7] and solubility of the gel [8] on the size of zeolite A crystals have been investigated. Decreasing the alkalinity, using a silica source of lower specific surface area (measured according to Carman's permeability method [9]), and replacement of some Na^+ by K^+ each independently brought about an increase in the average size and the dimensions of the largest zeolite A crystals. The effect of replacing some Na^+ by K^+ can be understood if one considers that hydrated Na^+ acts as a structure director for zeolite A while hydrated K^+ does not. Compositions with less Na^+ form fewer zeolite A nuclei. The most influential of these parameters was found to be the choice of silica source. The silicas were introduced to the reaction mixture in their solid forms, thus their solubilities played an important role in the course of the syntheses. Similarly, the solubility of the amorphous nutrient gel influences crystal size and number. Crystal growth was

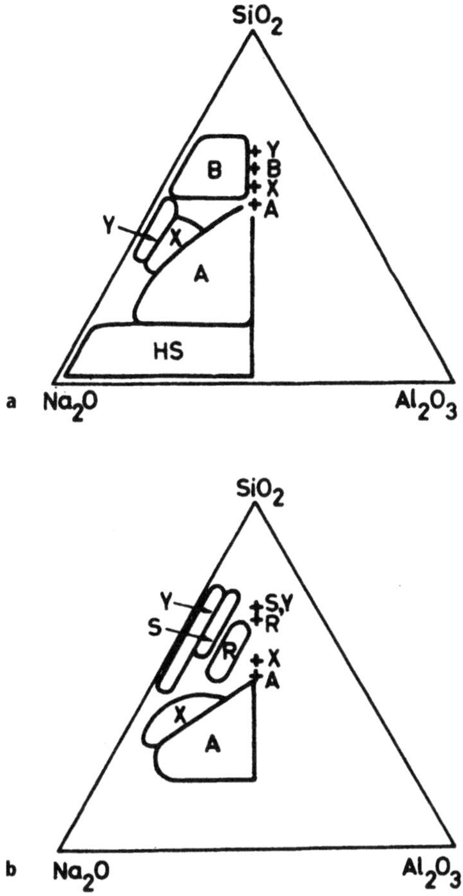

Fig. 3. A projection of the Na_2O, Al_2O_3, SiO_2, H_2O reaction composition system at 100 °C and 90–98 mol-% H_2O; source of SiO_2 is **a** sodium silicate and **b** colloidal silica. The areas enclosing a letter represent the compositions which yield the corresponding phase, while the + marks the typical composition of the product. *A* = zeolite A; *X* = zeolite X; *Y* = zeolite Y; *B* = zeolite P; *R* = chabazite; *S* = gmelinite and *HS* = hydroxysodalite. Reprinted with permission from [5]

reported to commence earlier from nutrient gels which were prepared quickly than from slowly mixed gels, although final crystal sizes were larger in the latter case. These results were ascribed to differences in the solubility of the gels prepared by different methods. In order to influence the gel solubility without influencing the alkalinity of the solutions, variable concentrations of glycerol were introduced to the synthesis mixtures [8]. It was supposed that glycerol alters the gel solubility through changing solution viscosity without changing the alkalinity. Higher concentrations of glycerol produced larger crystals, and the resulting reduction in mass transport lengthened the growth period of the crystals.

In 1967, Ciric published the preparation of large single crystals of zeolites A, P and X [10]. His method was based upon that for growing large crystals of low solubility, low stability crystals at low temperatures (ambient) which was introduced by Henisch et al. a few years earlier [11]. In essence, a 12.5-mm diameter, 300-mm long Tygon tube was filled with a gel made from an acrylic acid polymer dissolved in sodium hydroxide. The tube was bent into a vertical loop, and aqueous slurries of sodium aluminate and sodium di-silicate nonahydrate were introduced from opposite ends. The ends of the tube were then sealed and the nutrients were allowed to diffuse through the gel at 80 to 90 °C for two weeks. The nutrient-containing slurries presumably dissolved due to the presence of sodium hydroxide (ca. 1 mol dm^{-3}) in the supporting gel. At the boundary between the two nutrient solutions, crystals nucleated and grew. Because of the uncontrollable nature of the composition (Si/Al ratio, for instance), nucleation of various phases occurred. The mixtures of products were recovered only in low yield. In one experiment, crystals of zeolite A up to 75 μm and spheres of zeolite P of around 110 μm as well as gibbsite up to 250 μm were observed. A similar experiment, using a Pyrex tube in place of the Tygon tube, produced zeolite A (small) in the Al-rich zone, zeolite P (up to 150 μm) in the Si-rich zone and faujasite (up to 40 μm) in-between. Reaction of the alkaline solutions with the glass vessel could be responsible for he formation of numerous, small zeolite A crystals.

The formation of larger crystals using the supporting gel medium instead of the usual solution phase growth was presumably due to the low degree of nucleation which occurred when the first portions of silicate and aluminate mixed, followed by the prolonged growth period during which the nuclei were fed with continuous streams of the nutrients. The high density and viscosity of the acrylic gel medium allowed the crystals to remain suspended in the stream of nutrients instead of settling to the bottom of the tube.

Table 2 summarizes the solution compositions and resulting crystal sizes from various researchers attempting to grow large zeolite A crystals, while Table 3 presents corresponding data for zeolite X.

2.1.1
Charnell's Method – Synthesis in the Presence of Triethanolamine

Research into the synthesis of large crystals of zeolites A and X has been largely dominated by efforts which use triethanolamine (2,2′,2″-nitrilotriethanol), or other additives, in the synthesis mixture. Triethanolamine has been used extensively as an aluminum corrosion inhibitor [12] and as an aluminum chelating agent [13], and in 1971, Charnell introduced its use as a "stabilizing and buffering or complexing agent" in the synthesis of large zeolite crystals [14]. According to his article, triethanolamine was added to both a sodium meta-silicate solution and a sodium aluminate solution, which were filtered through 0.2 μm membranes and subsequently mixed together and hydrothermally treated at 75 to 85 °C in a covered 1.9 liter polypropylene jar. Compositions producing zeolite A were fully crystalline after 2 to 3 weeks, while zeolite X compositions took 3 to 5 weeks. The products contained crystals "approaching 100 and 140 μm in

Table 2. Compositions of synthesis solutions for the preparation of large zeolite A crystals

Solution composition[a]	Temperature and time	Crystals obtained[b]	Remarks	Ref.
not specified	80–90 °C, 14 d	A (60–75 μm) +P (110 μm) +gibbsite (250 μm)	acrylic gel medium in Tygon tubing. Low yield, many impurities	[10]
$10 SiO_2 : 13.89 Al_2O_3 : 23.89 Na_2O : 42.92 TEA : 2389 H_2O$	75–85 °C, 14–21 d	A (< 100 μm), broad p.s.d	triethanolamine added to synthesis mixture	[14]
$10 SiO_2 : 13.89 Al_2O_3 : 23.89 Na_2O : 42.92 TEA : 2389 H_2O$	not specified	A (< 40 μm), broad p.s.d.	repeat of recipe from [14]	[16]
$10 SiO_2 : 8.33 Al_2O_3 : 20.75 Na_2O : 36.50 TEA : 1967 H_2O$	85 °C, "some weeks"	A (< 50 μm), broad p.s.d.	$[Na_2O]$ and Si/Al ratio optimized	[17]
$10 SiO_2 : 13.89 Al_2O_3 : 23.89 Na_2O : 42.92 TEA : 2389 H_2O$	75 °C, 14–56 d	A (< 40 μm)	various reagents used	[18]
$10 SiO_2 : 11.18 Al_2O_3 : 21.81 Na_2O : 64.31 TEA : 2389 H_2O$	70 °C, 40 d	A (< 75 μm), P and X impurities	systematically changed $[NaAlO_2]$ and [TEA]	[19]
$10 SiO_2 : 11.90 Al_2O_3 : 23.10 Na_2O : 119.0 TEA : 2310 H_2O$	95 °C, 7 d	A (< 50 μm) + impurity	nature of TEA-Al interaction investigated	[24]
$10 SiO_2 : 11.90 Al_2O_3 : 29.88 Na_2O : 119.0 TEA : 2310 H_2O$	95 °C, 3 d	A (< 45 μm) + LOS	Increased $[Na_2O]$ reduced impurity level	[27]

[a] TEA = triethanolamine. The precise composition of the solutions used to synthesize large crystals is often not given in the literature. A certain amount of interpolation was required to derive the above compositions, and the authors apologize for any inaccuracies.

[b] Size of largest observed crystals in parentheses; p.s.d. = particle size distribution. The definition of crystal size varies between authors; some may mean the distance between opposite corners (maximum diameter), while others may mean edge-length or equivalent spherical diameter.

Table 3. Compositions of synthesis solutions for the preparation of large zeolite X crystals

Solution composition[a]	Temperature and time	Crystals obtained[b]	Remarks	Ref.
not specified	80–90 °C, 14 d	X (40 µm) + A (small) + P (150 µm)	acrylic gel medium in Pyrex tubing	[10]
$10SiO_2 : 6.94 Al_2O_3 : 16.94 Na_2O : 42.92 TEA : 2458 H_2O$	75–85 °C, 21–35 d	X (<140 µm)	triethanolamine added	[14]
$10SiO_2 : 4.44 Al_2O_3 : 11.11 Na_2O : 26.67 TEA : 1822 H_2O$	not specified	X (< 35 µm)	more dilute; better reproducibility	[16]
$10SiO_2 : 3.64 Al_2O_3 : 9.09 Na_2O : 21.82 TEA : 1200 H_2O$	not specified	X (< 70 µm)	higher SiO_2 concentration	[16]
$10SiO_2 : 6.94 Al_2O_3 : 16.94 Na_2O : 42.92 TEA : 2458 H_2O$	not specified	X (< 80 µm) + P impurities	repeat of recipe from [14]	[16]
$10SiO_2 : 4.90 Al_2O_3 : 16.86 Na_2O : 42.11 TEA : 2221 H_2O$	85 °C, "some weeks"	X (< 80 µm)	$[Na_2O]$ and Si/Al ratio optimized	[17]
$10SiO_2 : 6.94 Al_2O_3 : 16.94 Na_2O : 42.92 TEA : 2458 H_2O$	75 °C, 21–84 d	X (< 70 µm) + ≥ 15% A	various reagents used	[18]
$10SiO_2 : 2.86 Al_2O_3 : 13.60 Na_2O : 5.71 TEA : 1297 H_2O$	115 °C, 4 d	X (< 90 µm) + chabazite (40%)	size strongly dependent on silica source	[25]
$10SiO_2 : 6.94 Al_2O_3 : 16.94 Na_2O : 42.92 TEA : 2458 H_2O$	65 °C, 200 d	X (< 500 µm) + poly-disperse powder	special autoclave : zone-heating	[38]
$10SiO_2 : 6.94 Al_2O_3 : 16.94 Na_2O : 42.92 TEA : 2458 H_2O$	75 °C	X (< 135 µm) + A	aluminate added dropwise to silicate	[21, 22]

[a] See footnote a to Table 2.

[b] Size of largest observed crystals in parentheses. The definition of crystal size varies between authors; some may mean the distance between opposite apices (tip-to-tip length), while others may mean edge-length or equivalent spherical diameter.

diameter" for zeolites A and X, respectively, although the distribution of crystal sizes from such a preparation is large. Performing the synthesis at higher temperature (85 to 100 °C) yielded a product of narrower size distribution, but of smaller mean size.

Many researchers have attempted to reproduce Charnell's results (see below), but with varying degrees of success. Unfortunately, too little detail is given describing the experimental procedure which was originally used. If one assumes that the "diameter" of zeolite A quoted by Charnell was measured between opposite corners, through the center of the cube, then cubes of edge-lengths of approximately 60 μm should be expected. This is in accordance with the crystals shown in his figure 1 [14]. The role of triethanolamine in zeolite syntheses has been the subject of a number of articles, and it has been shown that triethanolamine forms a complex with Al^{3+} which is stable in zeolite synthesis mixtures (Fig. 4). The complexation of Al^{3+} reduces the effective concentration of "free" aluminum in solution, and thus reduces the degree of supersaturation of the synthesis mixture. It is the supersaturation of the solution phase (with respect to silicate and aluminate) which provides the driving force for nucleation and growth of zeolite crystals. It is to be expected then, that reduction of the level of supersaturation through removal of "free" aluminum species will reduce the tendency for nucleation and crystal growth. Thus fewer crystals nucleate in the presence of triethanolamine, and the growth period is extended. It can be assumed that the interaction between triethanolamine and aluminum is an equilibrium one. Thus, as the concentration of aluminum in solution is depleted due to crystal growth, the triethanolamine-aluminum complexes break up to replenish the aluminum in solution and allow crystal growth to continue. The nuclei which are formed may grow into relatively larger crystals since the available nutrients are shared between fewer crystals than in the case of synthesis without suppression of nucleation.

The ability of triethanolamine to complex to aluminum has been shown to result in an increase in the effective Si/Al ratio in solution, and thus the formation of crystals of higher Si/Al ratio [15]. As the amount of triethanolamine which was added to a zeolite A mixture was gradually increased from 30 up to 70 wt.-%, the product changed from pure zeolite A to a mixture of zeolites A

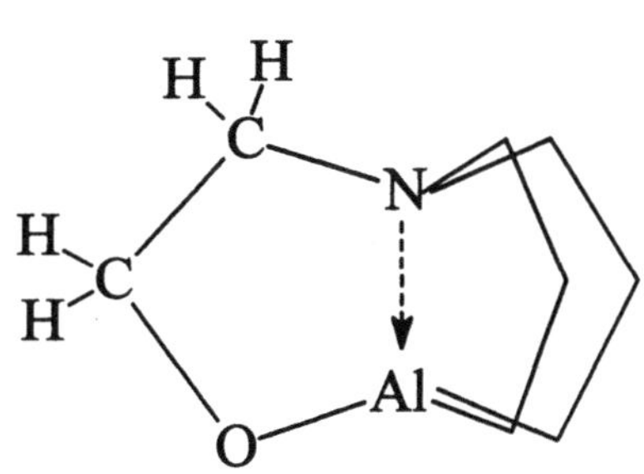

Fig. 4. The proposed interaction between triethanolamine and Al^{3+} which reduces the availability of aluminum in solution. For simplicity, atoms are labeled on only one arm of triethanolamine

and X, and finally to pure zeolite X. Similarly, a zeolite X mixture to which 40 wt.-% triethanolamine was added, yielded zeolite Y.

Attempts to reproduce Charnell's results yielded smaller crystals and significant level of impurities, mainly zeolite P. To achieve a reproducible synthesis and reduction in impurity levels, higher dilution was recommended [16]. The size of zeolite X crystals was limited to approximately 35 µm in a dilute system, and about twice the size in a more concentrated system. A narrower distribution of particle sizes was reported. By following Charnell's procedures, broad size distributions of zeolite A crystals up to 40 µm and zeolite X crystals up to 80 µm were obtained together with impurity phases [16]. An optimum concentration of Na_2O was found for the synthesis of zeolite A (maximum 40 µm) and zeolite X (maximum 60 µm) [17]. At the optimum concentration, "the growth rate is highest and... supersaturation is removed by growth of the crystals." At Na_2O concentrations below the optimum value, the synthesis was growth-determined, and the synthesis times increased, while at concentrations above the optimum value, poly-nuclear crystallization occurred, yielding crystal aggregates. Subsequent adjustment of the Si/Al ratio of the mixtures resulted in a further increase in crystal dimensions up to edge lengths of 50 µm for zeolite A (Table 2) and 80 µm for zeolite X (Table 3).

Zeolite A commonly formed rhombic dodecahedra (truncated cubes with {110} faces evident, Fig. 5), sometimes in mixtures with cubic crystals in syntheses containing triethanolamine [18]. Depending upon the method of gel formation, zeolite X compositions were found to produce varying proportions of zeolite A, with inter-growths or over-growths of the two phases common. The

Fig. 5. A zeolite A crystal which exhibits rhombic dodecahedral morphology. The narrow faces appearing along the edge of each cube face are the {110} crystallographic facets

zeolite A crystals formed in solutions of zeolite X composition were mostly in the form of pure cubes, in contrast to the rhombic dodecahedra formed in solutions of zeolite A composition.

Systematic studies have been carried out in which the silica and alumina sources, the gel preparation procedure and the chemical composition of the gels for zeolite A synthesis were varied [19, 20]. Changing the composition of the gel phase was the only factor which significantly affected the size of crystals produced. Higher dilution and reduction of the ratio of "reactive" Al to Si were two factors recommended in order to grow larger crystals. The "reactive" Al may be regarded as that which is not bound to triethanolamine. Decreasing the concentration of "reactive" aluminum in a zeolite A preparation brought the solution composition closer to that of zeolite X (Fig. 3), and larger and larger A crystals were obtained up until zeolite X began to form. From this point on, no further increase in size of zeolite A crystals was observed; the proportion of zeolite A in the product decreased steadily while that of zeolite X increased. Rhombic dodecahedral zeolite A crystals (with {110} faces) were common, but as the solution composition was altered and zeolite X began to form, the {110} faces disappeared, and pure cubes of zeolite A grew. In an attempt to reproduce Charnell's original preparation of zeolite A, pure zeolite A was obtained, but only up to a maximum size of 40 μm [19]. By adjusting the solution composition and using a reactor of larger volume, an optimum zeolite A size of up to 75 μm was achieved, however, the bulk of the product consisted of 15 to 60 μm zeolite A while zeolites X and P were also present in small quantities (Table 2). All compositions which were not close to that given by Charnell yielded significant quantities of zeolite P as a by-product. It seems generally true that attempting to increase the size of zeolite A crystals leads to a compromise in product purity.

The nature and quality of the starting materials and the method of mixing the nutrient solutions were two parameters which were found to be dominant in determining the size of zeolite X crystals obtained in the presence of triethanolamine [21, 22]. This was ascribed to the fact that these two factors have a strong influence on the nucleation stage of the synthesis. Only through control of nucleation could large crystals be reproducibly produced. Other factors, such as dilution, triethanolamine content and aluminum content, which, it was claimed, effect mostly the growth process, did not have so much influence on the crystal size. Zeolite X crystals of up to 135 μm were obtained using silica sol and $Al(OH)_3$ as reagents, and adding the aluminate solution to the silicate solution drop-wise with stirring. The importance of performing ultra-filtration of all solutions in the preparation of large crystals was stressed.

The group at Worcester Polytechnic Institute, Massachusetts, USA, has carried out a large volume of work on the synthesis of zeolites in the presence of triethanolamine, and other organic additives [23–28]. Their work is largely concerned with developing experiments for the successful growth of zeolites under conditions of micro-gravity (US space shuttle in low Earth orbit). The subject of growing zeolites in micro-gravity will not be addressed in this chapter, but the reader is referred to a recent review of this subject [29]; only the results of ground-based experiments will be presented here. The effectiveness of triethanolamine as a suspending agent, i.e. a means to increase solution viscosity, thereby hindering the settling of growing crystals has been considered [23]. This allows longer growth periods and reduces the high shear which occurs on particle-settling. The suspension of the growing crystals in the solution also helps to keep the supersaturation at a lower level due to their continued consumption of nutrients, thus reducing further nucleation. Calculations were performed to estimate the velocity of settling from solution under gravity for particles of various sizes, and it was found that a 1 μm particle would settle-out in about 3 h. Considering that the total synthesis time for that composition was of the order of 30 h, a 1 μm particle would spend almost 90% of the synthesis time on the floor of the reactor, where its growth would be hindered. As the particle size increased, the time taken to settle was reduced. The calculations were performed assuming a constant particle size throughout settling. In reality, the particles would continue to grow as they settled, thus accelerating in their rate of sedimentation. The benefits of using triethanolamine in the preparation of larger crystals was demonstrated, but this result was most probably due to a combination of increased viscosity and interaction between triethanolamine and aluminum (hence reduced supersaturation).

In some studies, the triethanolamine content of a zeolite A batch mixture was systematically varied [24, 26, 27]. Without addition of triethanolamine, crystals of 2 μm were obtained, while crystal sizes increased smoothly on increasing the molar ratio of triethanolamine to Al_2O_3 to 10, at which point crystals up to 50 μm were obtained. On increasing the triethanolamine content further, increasing amounts of zeolite X were formed, and at 30 moles per mole of Al_2O_3, the product contained zeolite X (70%) and hydroxysodalite (30%) only. No further increase in size of the zeolite A crystals was observed once zeolite X was present in the product. Figures 6 and 7 illustrate the increase in maximum crystal size and the increasing contamination of the product (zeolite A) with zeolite X and hydroxysodalite as the concentration of triethanolamine in the synthesis mixture is increased. Note how the {110} facets on the zeolite A crystals diminish with increasing concentration of triethanolamine.

A systematic study of the effect of varying the silica source on the particle size of zeolite X growing in the presence of triethanolamine has been reported [25]. A triethanolamine to aluminum ratio of 1 : 1 was used throughout the study, since higher ratios lead to higher levels of contaminating phases. Pronounced differences were observed, with maximum sizes varying between 7.5 and 90 μm, depending upon the silica source employed with all other parameters equal (Fig. 8). Each of the eleven different silica sources were introduced to the mixtures in a dissolved, filtered form, thus differences in surface areas, solubilities, etc. should not have influenced the product. It was concluded that any of a number of impurities present in the silica sources may be responsible for promoting nucleation of zeolite X. The impurities for which analysis was performed were transition metals and alkali- and alkaline-earth metals. Those silica sources which contained few impurities (e.g. Cab-O-Sil and silicic acid) produced large crystals, while those containing relatively greater amounts of impurities produced much smaller crystals. Since the yield of zeolite from each silica source (% silica converted into zeolite) remained effectively constant, the number of crystals which formed in the small-particle system was much higher than that in

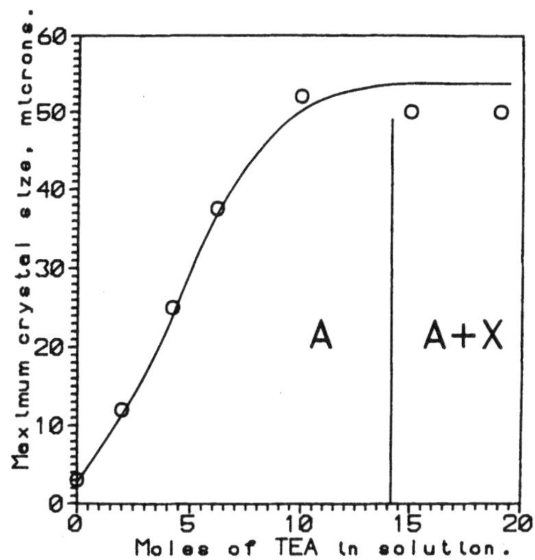

Fig. 6. The effect of varying the concentration of triethanolamine on the size of the largest crystals of zeolite A. Synthesis at 95 °C, composition 0.84 $SiO_2:Al_2O_3:1.94\ Na_2O:194\ H_2O:y$ TEA. The x-axis represents the number of moles of triethanolamine "y" in the composition. Reprinted with permission from Zeolites, vol. 13, Morris M, Dixon AG, Sacco A, Thompson RW, Investigations on the relative effectiveness of some tertiary alkanolamines in the synthesis of large-crystal zeolite NaA, (1993) pp. 113–21, Elsevier Science Inc.

the large-particle system. No one particular impurity was shown to be responsible for the promotion of nucleation of zeolite X; increasing the concentration of most impurities brought about a decrease in the size of the crystals. Additionally, the nature of, and amount of other zeolitic phases which formed were dependent upon the silica source. Generally, as the size of the largest particles increased, so did the degree of contamination from other phases, as has been observed in zeolite A systems. Zeolite R (chabazite) constituted nearly 40% of the product prepared using Cab-O-Sil (zeolite X crystals up to 90 μm), while zeolite P contaminated the product from a metasilicate pentahydrate to a level of 10% (zeolite X crystals up to 50 μm). Deliberate contamination of a relatively pure silica source by addition of various metal salts, however, did not significantly influence the size of the zeolite X crystals obtained. The crystallization phase field diagrams in Fig. 3 also illustrate how changing the source of silica may affect the products which form from a given composition.

2.1.2
Mechanism of Triethanolamine's Action

The presence of a triethanolamine-aluminum complex has been demonstrated by ^{13}C NMR of synthesis solutions which contained the organic [26]. Two sets of peaks were observed, one for free triethanolamine (which was present in excess with respect to aluminum), and one for the proposed triethanolamine-alumi-

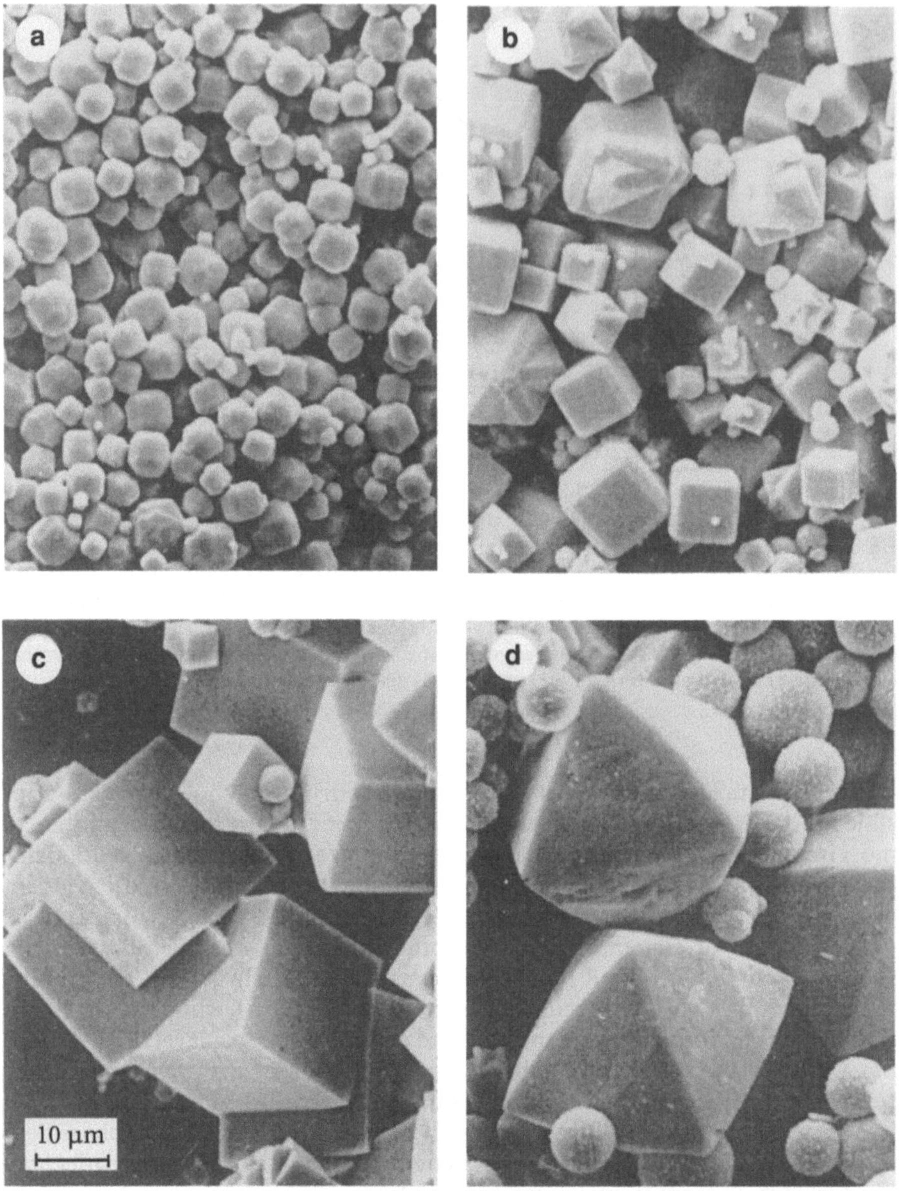

Fig. 7. The effect of varying the concentration of triethanolamine on the size and purity of zeolite A. Synthesis at 95 °C, composition 0.84 SiO_2 : Al_2O_3 : 1.94 Na_2O : 194 H_2O : y TEA, **a** y = 2.12; **b** y = 4.23; **c** y = 6.34; **d** y = 30.0. At high triethanolamine concentrations, the products are predominantly zeolite X and hydroxysodalite. All four electron micrographs are shown at the same magnification. Reprinted with permission from Stud Surf Sci Catal, vol. 49A, Scott G, Dixon AG, Sacco A, Thompson RW, Synthesis of zeolite NaA in the presence of triethanolamine, (1989), pp. 363–72, Elsevier Science Inc.

Fig. 8. The effect of varying the silica source, with all other parameters remaining unchanged, on the synthesis of zeolite X. Three sodium metasilicates with varying degrees of hydration and fumed silica are represented. All four electron micrographs are shown at the same magnification. Reprinted with permission from Zeolites, vol. 13, Hamilton KE, Coker EN, Sacco A, Dixon AG, Thompson RW, The effects of the silica source on the crystallization of zeolite NaX, (1993), pp. 645–53, Elsevier Science Inc.

num complex. A plot of relative ^{13}C-NMR peak intensity for the complex versus the content of triethanolamine in solution showed that the complex is formed from an equal number (assumed 1 each) of triethanolamine and aluminum moieties (Fig. 9). A study was made to determine which components of the triethanolamine molecule were responsible for its strong interaction with aluminum. Triethanolamine possesses a trivalent nitrogen atom (with a lone pair of electrons) to which three 2-hydroxyethyl groups are attached. Triethyl-substituted amines possessing zero, one, or two 2-hydroxyethyl groups had no influence on the nucleation or growth of zeolite A. Thus, the presence of three hydroxyl groups appears to be necessary for formation of the complex (see Fig. 4). Whether the hydroxyl groups are directly involved in binding to aluminum, or simply act to solubilize the complex could not be unequivocally determined from these experiments. Comparison of the relative effectiveness in the synthesis of zeolite A of tris(hydroxymethyl)ethane and tris(hydroxymethyl)*amino*ethane proved that the amino group was necessary for effective

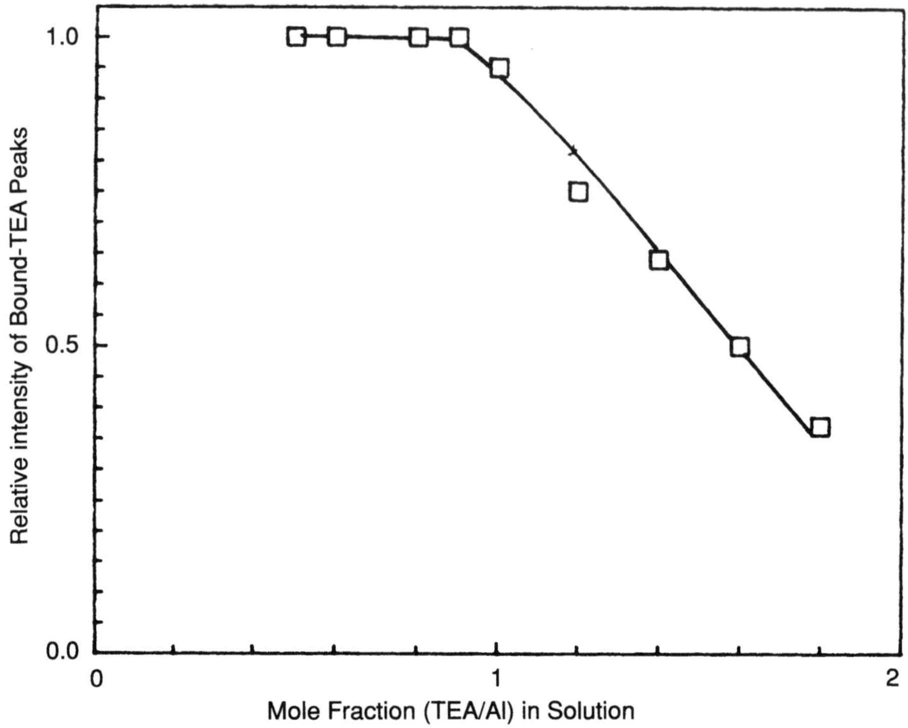

Fig. 9. Results from ^{13}C NMR which indicate that, in an excess of aluminum, all triethanolamine is complexed to Al^{3+}, while above a triethanolamine to Al^{3+} ratio of 1, uncomplexed triethanolamine exists in solution. The stoichiometry of the complex is thus 1:1. Reprinted with permission from Zeolites, vol. 10, Scott G, Thompson RW, Dixon AG, Sacco A, The role of triethanolamine in zeolite crystallization, (1990), pp. 44–50, Elsevier Science Inc.

aluminum complexation. It seems clear that the presence of three hydroxyl groups and an amino nitrogen (i.e. a lone pair of electrons) is necessary for triethanolamine to bind effectively to aluminum. Triethanolamine's ability to complex aluminum is evident from the above work, however little work had been performed investigating the effect of triethanolamine in all-silica systems. Triethanolamine often has little or no effect on the synthesis of high-silica ZSM-5 (silicalite-1) [30], however, when the sodium content of the system is very low, triethanolamine may assist in solubilizing the silica, possibly through a pH effect, thus shortening reaction times and increasing the crystal size [26]. No complex between triethanolamine and silicon moieties in zeolite synthesis mixtures has been detected.

Aluminum speciation in zeolite synthesis mixtures containing triethanolamine and several other organic compounds has been investigated using ^{27}Al NMR [28]. It was demonstrated that all aluminum in sodium aluminate solutions containing an excess of triethanolamine was present in complexed form, probably with tetrahedral symmetry. Virtually no tetrahydroxy aluminate ions were observed by NMR, even when the solution was heated to a typical reaction temperature of 85 °C, showing that the triethanolamine-aluminum complex is relatively stable under these conditions.

One interesting side-effect resulting from the use of triethanolamine (or other chelating agents) in zeolite synthesis solutions is the potential to prepare zeolites which contain unusually small numbers of framework impurity atoms. Triethanolamine may form complexes with ions of metals such as iron [31] and copper [32] in solution, and thus prevent or reduce their incorporation into zeolite frameworks. This effect has been demonstrated for the synthesis of virtually iron-free zeolite X [33]. After intentional introduction of solubilized iron salts to the synthesis solution, triethanolamine reduced the concentration of iron in the final products significantly when compared to syntheses without the organic additive. When not intentionally added to solution, iron could not be detected by electron spin resonance spectroscopy in samples synthesized in the presence of triethanolamine, while samples prepared without the organic additive showed clear signals for iron in both framework and extra-framework positions.

As a final comment on the use of triethanolamine in zeolite syntheses, its effect on the viscosity of the reacting solution should be considered. Too high a concentration of triethanolamine will increase the viscosity to a point where diffusion of the nutrients to the growing crystals will be severely hindered. At this point, crystal sizes should begin to drop.

2.1.3 *Synthesis Using Organic Additives Other than Triethanolamine*

Some of the first reports on the synthesis of zeolites in the presence of complexing agents were aimed at influencing not the crystal size, but the silicon-to-aluminum ratio of FAU zeolites [34, 35]. A range of inorganic and organic complexing agents were investigated, principally phosphates. No mention is made of the size of crystals which were obtained from such reactions, but it is interesting to see the change in Si/Al ratio of the products, which presumably reflects to some degree the complexing ability of the various additives. Increasing the concentration of any one complexing agent generally resulted in faujasite crystals of increasing Si/Al ratio. If the concentration were increased beyond a certain point, the composition did not crystallize within the time frame of the experiment. Faujasites with Si/Al ratios up to 2.4 were obtained from compositions containing phosphate, compared to 1.4 without phosphate. Phosphate and phytate gave the greatest increase in Si/Al ratio, implying that their complexing ability towards aluminum under the prevailing conditions was better than that of the other additives investigated. Part of the success of these two complexing agents is attributed to their good buffering capabilities.

Charnell [14], as well as introducing the use of triethanolamine to the zeolite synthesis field, investigated the influence of other water-miscible organic bases, including 2-dimethylaminoethanol, ethanolamine, diethylenetriamine, 2-amino-2-methyl-1-propanol, morpholine and ethylenediamine. None of these were found to be as effective as triethanolamine.

A number of organic compounds were tested for their influence on the synthesis of zeolites A, X and Y [15]. Between 0.5 and 200% by weight of each com-

Table 4. Relative effectiveness of various complexing agents in the synthesis of zeolite A. Gel composition: $10\,SiO_2 : 11.90\,Al_2O_3 : 23.10\,Na_2O : 2310\,H_2O$: x (complexing agent), from [27]

Acronym	Chemical formula	Product formed (max./average size)	
		x=59.5	x=119
TEA	$N(CH_2CH_2OH)_3$	NaA, 30/– µm	NaA, 50/– µm
TPA	$N(CH_2CHOHCH_3)_3$	NaA, 24.4/16.8 µm	NaA, 25.4/15.8 µm
BEP	$N(CH_2CH_2OH)_2(CH_2CHOHCH_3)$	NaA, 33.7/18.4 µm	NaA, 36.3/22.0 µm
DIPSO	$N(CH_2CH_2OH)_2(CH_2CHOHCH_2SO_3Na)$	NaA, 37.8/23.4 µm	NaA + NaX, 30.3/23.4 µm
BIS	$N(CH_2CH_2OH)_2(C(CH_2OH)_3)$	NaA, 33.2/21.2 µm	NaA + NaX, 32.2/20.2 µm

pound were added to synthesis mixtures. Besides triethanolamine, the additives investigated were: 1,4-dioxane, tetrahydrofuran, acetone, ethanol, *n*-propanol, *n*-butanol, *n*-nonanol, ethyleneglycol, glycerol, 1-arabitol, *d*-sorbitol, *d*-glucose, cane sugar, pyridine, α-methylpyridine, piperidine, phenol, dimethylformamide, cyclohexanone and anisole. Of all compounds added, only triethanolamine was found to have a positive influence on crystal size.

An investigation into the effectiveness of ethylenediamine, ethyleneglycol, glycerol and glycine in the preparation of large zeolite A crystals from solutions of various Si/Al ratios revealed that none were as effective as triethanolamine [17].

Tertiary alkanolamines which contain less than three hydroxyalkyl groups have been shown to be ineffective in the production of large zeolite A crystals [27,28]. Investigations of sodium aluminate solutions containing the various organics by ^{27}Al NMR confirmed the ineffectiveness of certain additives to bind to Al^{3+}. Five additives which each contained three hydroxyalkyl groups were chosen for further study (Table 4) [27]. The additives all possessed hydroxyl groups on the second carbon atom, thus preserving somewhat the geometry of the triethanolamine molecule, while altering the length of the alkyl chains. While bis-(2-hydroxyethyl)-amino-2-propanol (BEP) was found to act similarly to triethanolamine, 1,1′,1″-nitrilotri-2-propanol (TPA) was found to be slightly less effective in chelating aluminum. Crystallization of a mixture containing 10-parts of sodium-3-[*N,N*-bis-(2-hydroxyethyl)-amino]-2-hydroxypropane sulfonate (DIPSO), produced a product containing zeolite X in addition to zeolite A, indicating its greater aluminum-chelating ability compared to triethanolamine. The zeolite A crystals reached a size of 30 µm between opposite faces, and interestingly, the corners of all of the cubes were rounded; none of the other chelating agents used in this study showed the same effect under the prevailing conditions. One of the chelating agents, viz. 2,2-bis(hydroxymethyl)-2,2′,2″-nitrilotriethanol (BIS), contained five hydroxyl groups and was concluded to be the most efficient of all complexing agents studied. Again, a product containing both zeolites A and X was achieved using 10 parts of the additive. From Fig. 11 of reference [27], it appears that the zeolite A crystals prepared with 10 parts BIS only reach about 10 µm, while zeolite X exceeds 30 µm (tip-to-tip). Although the maximum size of the observed zeolite A crystals was smaller than those produced in the presence of triethanolamine, the co-formation of zeolite X at lower concentrations of additive indicates that complexing of aluminum is more efficient and that, under refined conditions, larger crystals than those obtained in the presence of triethanolamine may be achievable.

2.1.4
Addition of Nutrients During Synthesis

Recently, a method of adding nutrients to a zeolite A batch during the synthesis has been described [36, 37]. The composition studied was one which produced an amorphous gel phase upon mixing the reagents. The time at which the gel suddenly shrinks in volume is taken as the point when nucleation is complete and crystal growth dominates. Addition of extra nutrients (doubling the volume of the mixture while maintaining the same concentration) at this point was reported to increase the size of the crystals by 25 to 30% without producing a second population of crystals. This method shows promise, but may be restricted to systems with rapid crystal growth rates which crystallize relatively quickly.

2.1.5
Synthesis of Zeolite X Using a Zone-Heating Technique

Zeolite X crystals with tip-to-tip lengths reaching 500 μm (0.5 mm) have been prepared in relatively low yield using an autoclave of unusual design and carefully aged solutions [38]. The authors proposed three factors to be of key importance in the preparation of large zeolite crystals. These were 1) optimization of solution primary treatment (i.e. composition) and gel formation, 2) reduction of the number of heterogeneous nuclei through filtering of solutions prior to synthesis, and 3) reduction of settling of the growing particles. Charnell's recipe was used throughout their study, and the filtered solutions were allowed to age for up to 22 d prior to hydrothermal treatment. Both the size of the largest crystals and the time required for complete conversion passed through a maximum at approximately 15 d of aging. Crystals of up to 250 μm were obtained in 200 d after this optimum aging time. On increasing the aging time to 22 d, the maximum crystal size decreased to about 50 μm, while with no aging, crystals of 90 μm were recovered. In order to reduce the settling of particles under gravity, the researchers turned to a method similar to zone melting. A 100 mm tall, 80 mm diameter polyethylene flask was filled with aged nutrient solution and was embraced by a ring heater, which would locally heat the gel to 65 °C. The heater was slowly moved from the top of the flask to the bottom (0.5 mm d^{-1}). Nucleation and growth of crystals occurred in the heated portion of the gel, and because of the higher density and viscosity of the cooler gel below, the crystals did not settle out of solution so easily. They sat on the surface of the cooler, denser gel, where they were surrounded by nutrients and could grow to larger sizes than if they had fallen to the bottom of the reactor, and had been covered by other crystals. Besides forming large zeolite X crystals, agglomerates of crystals and poly-disperse powder resulted from such an experiment. The authors make no comment about the presence of impurity phases in their products.

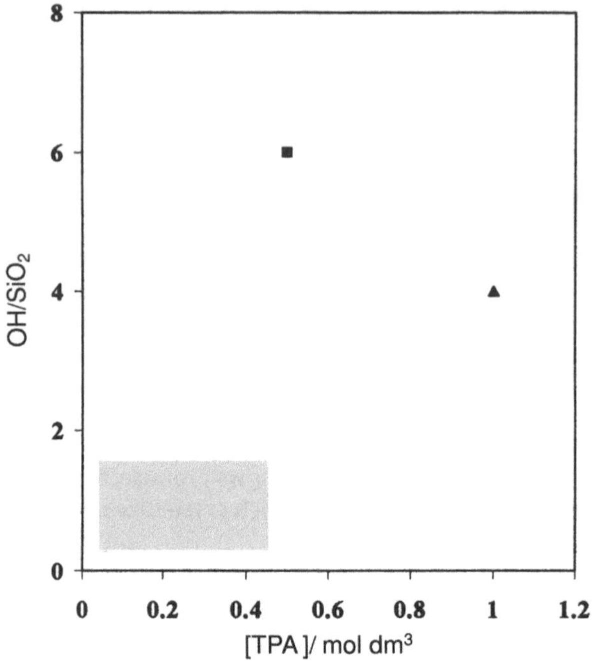

Fig. 10. Compositional diagram of regular and extreme recipes for MFI-type zeolites. *Shaded area* represents "normal" recipes for intermediate sized MFI crystals, while compositions ■ [1] and ▲ [41] yield ultra-large crystals

2.2 Case Study: Zeolites MFI and Analcime

Single crystals of 500 µm × 350 µm × 350 µm and incomplete single crystals of up to 750 µm in the c-direction of MFI with a Si/Al ratio of 300 have been made [39, 40]. The synthesis formulation and procedure developed [41] deviate significantly from the regular synthesis recipes used for the preparation of MFI-type materials as illustrated in the compositional diagram of reactants in Fig. 10.

After preparation of the mixture, a "clear" solution is observed which does not change after treatment at 180 °C for 60 h. The next event is the formation of gel-spheres that reach diameters up to 2 mm. During in situ observation with a look-through autoclave at 180 °C, the first few crystallites were seen only on the external surface of the gel-sphere. The single crystallites proceeded to grow *into* the gel-sphere. Ultimately, relatively few large single crystals are formed from one gel-sphere [40]. A number of aspects mentioned in Table 1 are relevant here. The formation of a large gel-sphere provides a large homogeneous nutrient pool for silica and alumina. Simultaneously the environment of the growing crystal is rather static as it grows into the gel-sphere. Another aspect is the relatively low nucleation rate at the interface of the liquid and the gel-sphere, as shown in Fig. 11. Furthermore, it has been observed that small gel-spheres produce small

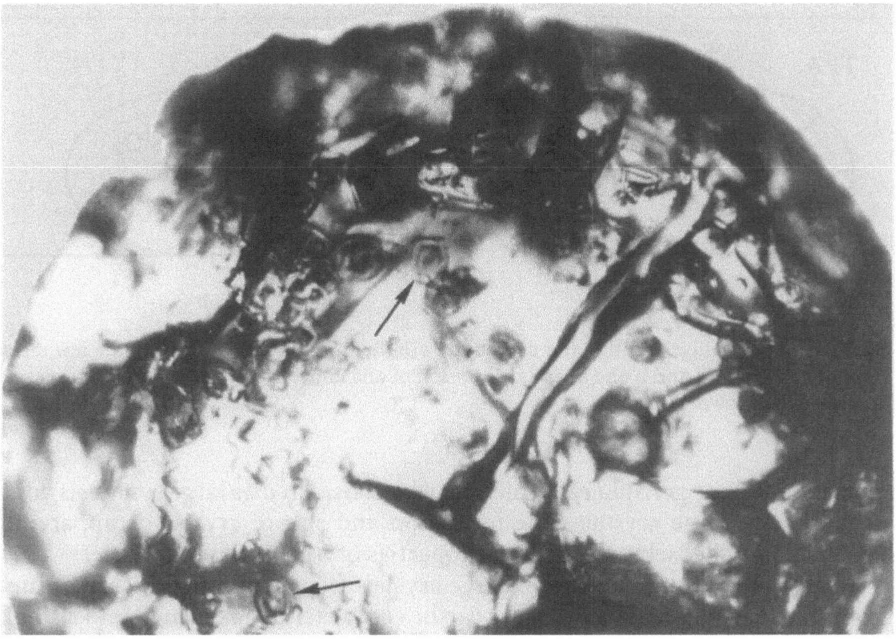

Fig. 11. In-situ observation of MFI-type single crystallites on the external surface of a "self-assembled" gel-sphere. The gel-sphere is comprised of silica sol particles, sodium and water. As no TPA is present in the gel-sphere, interaction of TPA with silica and nucleation have to begin on the external surface of the sphere. Because the external surface is normally covered with Na^+, competition for silica results in only a few crystallites forming; some are indicated by arrows

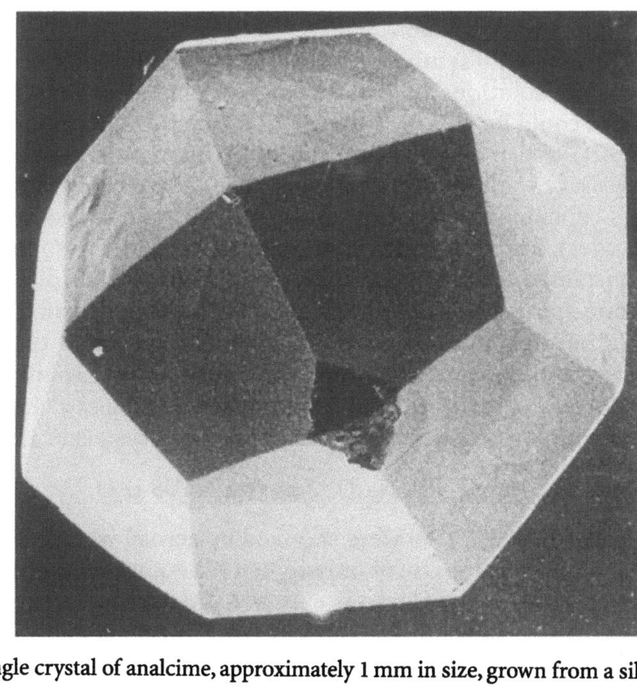

Fig. 12. A single crystal of analcime, approximately 1 mm in size, grown from a silica gel-sphere

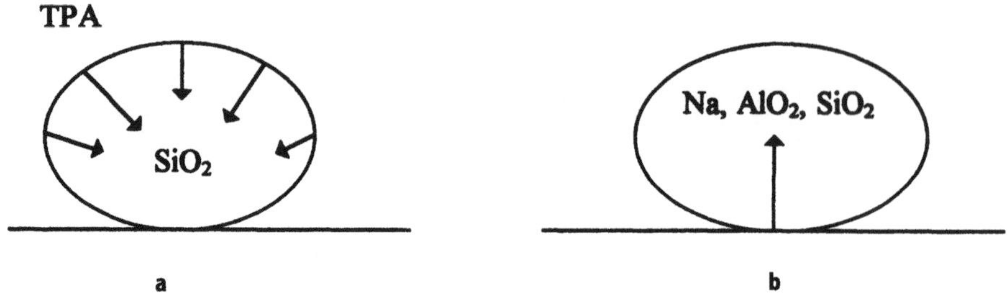

Fig. 13. The nucleation point(s) and crystal growth direction in a gel-sphere of **a** a few ultra-large crystals of MFI and **b** one ultra-large crystal of analcime. Larger gel-spheres allow larger crystals to be grown

crystals and large gel-spheres, which may be considered as large synthesis volumes, produce large crystals. The formation and size of crystals is apparently dependent on the "self-assembling" properties of a gel precursor phase.

Analcime was studied as well. Large crystals, depicted in Fig. 12, are formed from large gel-spheres. A gel-sphere can be completely transformed into a single crystal of analcime. The nucleation and the growth history are certainly different from MFI; in the case of analcime, only one crystal is formed from a precursor phase (gel-sphere) comprising all of the necessary ingredients. Figure 13 illustrates the differences in nucleation and growth of MFI and analcime in gel-spheres.

2.3
Mordenite

Sun et al. [42] reported the synthesis of large single crystals of mordenite from clear homogeneous solutions in the absence of templates. Crystals of up to 185 μm × 125 μm could be obtained under the correct conditions. A mixture of two silica sources, aerosil and sodium silicate (in an approximately 4:1 molar ratio), was necessary to obtain large crystals. The authors suggest that the sodium silicate is consumed first, followed more slowly by the less active aerosil (fine silica powder). Thus, crystallization was initiated by the sodium silicate, and was then sustained over a long period of time by the gradual dissolution and release of silicate species from the aerosil. For these experiments, aerosil was added to a solution of sodium silicate and NaOH in water and the whole was then homogenized under ambient conditions until a clear solution was obtained. The syntheses were carried out in Teflon-lined reactors at 150 °C for 7 to 30 d. The molar composition which yielded the largest crystals in 15 d was:

$$100\ SiO_2 : 1.33\ Al_2O_3 : 20\ Na_2O : 5.33\ NaCl : 733\ H_2O$$

Of the 75 equivalents of SiO_2, 60 were supplied by aerosil while the remaining 15 came from sodium silicate. Use of only aerosil (75 equivalents) in the composition produced crystals only as large as 8 μm × 3 μm after 5 d, while use of only sodium silicate (75 equivalents) resulted in 2 μm × 1 μm crystals after 5 d. The

Fig. 14. Mordenite crystals prepared from mixtures of composition 100 SiO_2 : 5.26 Al_2O_3 : 22.74 Na_2O : 1545 H_2O at 175 °C. Pre-treatment of the silicic acid (silica source) in air for 20 h at: **a** un-treated; **b** 300 °C; **c** 550 °C; **d** 850 °C. Reprinted with permission from Zeolites, vol. 16, Warzywoda J, Dixon AG, Thompson RW, Sacco A, Suib L, The role of the dissolution of silicic acid powders in aluminosilicate synthesis mixtures in the crystallization of large mordenite crystals, (1996), pp. 125–37, Elsevier Science Inc.

nature of the salt added had a marked effect upon the resulting crystals, for example, substitution of Na-acetate for NaCl decreased the crystal size to 27 μm × 30 μm, changing significantly the aspect ratio of the crystals.

Warzywoda et al. [43, 44] recently introduced a method of controlling mordenite crystal size by heat-treating porous silica reagents in air at different temperatures prior to synthesis. Syntheses were carried out at 175 °C in Teflon-lined reactors for 2 to 6 d using the following composition:

$$100\ SiO_2 : 5.26\ Al_2O_3 : 22.74\ Na_2O : 1545\ H_2O$$

The silicas were, in all cases, introduced to the synthesis mixture in their solid form, thus their solubilities played a key part in the synthesis itself. Silicas which possessed small, medium or large pores were investigated. Crystals up to 175 μm in length were obtained after optimal treatment of a large-pore silica gel, while crystals up to 250 μm were obtained from a heat-treated silicic acid [44]. Thermal treatments were most effective in the range of 700 to 900 °C (Fig. 14), varying slightly between silica sources. The effect of thermal treatment was shown to be a decrease in the surface area and pore volume of the materials, while the diameter of the remaining pores was assumed to be unaffected. Thus, heat treatment would reduce the contact of the internal volume of the silica particles with the caustic synthesis medium. A slow release of silicate species into the reaction medium was thus realized. Silicas which possessed small (2.2 to 2.5 nm) or medium pores (6 nm) were not suitable for preparing large crystals of mordenite, even after heat treatment, while heat-treated large pore (14 to 15 nm) silicas produced the best results. With small pore silicas, heat treatment was postulated to reduce the surface areas too much, i. e. the solubility of the silica became so low that solutions of significantly lower Si/Al were formed, which resulted in the more ready formation of the more aluminous zeolite P. Generally, on increasing the temperature of heat treatment of a large pore silica, an increase in size of the mordenite product was observed until a certain threshold temperature was reached after which impurity phases (zeolite P and quartz) co-crystallized, reducing the size of the mordenite crystals. Medium pore silicas did not show much improvement in crystal size upon heat treatment. Without heat treatment, all of the silica sources investigated produced pure mordenite (30 to 60 μm) after 2 d of hydrothermal treatment.

2.4 Non-Aluminosilicate Materials

The aluminophosphate and related analogues of zeolites can often be grown to sizes exceeding those attainable with the aluminosilicate zeolites. This reflects the fundamentally different nature of the solutions used for the syntheses. Non-aluminosilicate molecular sieves often crystallize from solutions which are less alkaline than zeolite synthesis mixtures. The problems associated with instability of the microporous phases under hydrothermal conditions may be smaller with non-aluminosilicate materials.

Of the non-aluminosilicate microporous materials, the AFI ($AlPO_4$-5) structure-type is the one most frequently grown in large crystal form. These materials are of interest for the preparation of non-linear optical devices, where "molecular wires" of ordered guest molecules reside within the one-dimensional pore system [45, 46]. Large crystals are required for their easy alignment, thus producing a coherent non-linear optical effect. Another important consideration for such applications is the structural purity of the crystals, so that losses of

Table 5. Synthesis mixture compositions for large AFI crystals

Material	Composition	Template	Conditions	Maximum crystal size (μm)	Ref.
$AlPO_4$-5	$Al_2O_3 : P_2O_5 : 0.5\ Pr_3N : 300\ H_2O$	tripropylamine	150 °C, 3 d	500	[1]
$AlPO_4$-5	$Al_2O_3 : 1.03\ P_2O_5 : 1.55$ TEA $: 1000\ H_2O$	triethylamine	192 °C, 6 d	920	[47]
$AlPO_4$-5	$Al_2O_3 : P_2O_5 : 1.55$ TEA $: 1000\ H_2O$	triethylamine	190 °C, 6 d	750	[48]
$AlPO_4$-5	$Al_2O_3 : P_2O_5 : 2.2\ Pr_4NOH : 1.7\ NH_4F : 318\ H_2O$	tetrapropyl ammonium-fluoride	170 °C, 10 d	500	[49]
SAPO-5	$Al_2O_3 : P_2O_5 : \leq 0.6\ SiO_2 : 1.5$ TEA $: \geq 600\ H_2O$	triethylamine	180–200 °C, 10 d	580	[51, 52]
CoAPO-5	$Al_2O_3 : P_2O_5 : 0.2–0.25$ DAE $: 0.045\ CoCl_2 : 18\ H_2O$	2-diethylamino-ethanol	50 °C, 1 d, then 200 °C, 3 d	400	[54]

signal due to light scattering may be minimized. A number of publications describe syntheses of large crystals of $AlPO_4$–5 (aluminophosphate) [1, 47, 48, 49, 50], SAPO–5 (silicoaluminophosphate) [51, 52, 53] and also CoAPO-5 (cobalt aluminophosphate) [54]. Some representative compositions and conditions are given in Table 5.

Müller et al. prepared $AlPO_4$-5 crystals 500 μm in length in relatively low yield following a series of factorial experiments in which certain synthesis parameters were systematically varied [1]. Under the conditions studied, decreasing the dilution lead to a decrease in crystal size, while drastically increasing the yield. Temperature and concentration of template (tripropylamine) had little effect on the crystal size. Crystals of $AlPO_4$-5 up to 920 μm have been prepared in higher yield, using triethylamine as the template, although the distribution of sizes was broad and the proportion of intergrowths large [47]. Better crystal morphology and higher yields were achieved from less dilute solutions, but the maximum size was 650 μm. Increasing dilution has been found to lead to an increasing aspect ratio (crystal length to width ratio) as well as an increase in the contamination of $AlPO_4$-5 by other products [48]. Synthesis in a fluoride medium may yield 500 μm crystals after 10 d at 170 °C [49], or crystals up to 130 μm in length using microwave heating to 175 °C for about 60 s [50].

Crystals of SAPO-5 up to 580 μm in length may be synthesized (see Table 5), although phase purity deteriorates with increasing size [51, 52]. The yield of SAPO-5 was 60 to 80% and separation of the unidentified impurity phase was possible by decantation due to the difference in size. Increased solution dilution brought about an increased particle size (Fig. 15), while increasing the template concentration increased the width of the crystals without changing their length. Large optically clear SAPO-5 crystals have been prepared and studied using FTIR microscopy [53, 55].

The cobalt-containing CoAPO-5 has been prepared as single crystals up to 400 μm in length after carefully controlled aging of the synthesis mixture prior to hydrothermal treatment [54]. The temperature of the mixture during aging was critical and yielded the largest crystals if aging was performed for 1 d at 50 °C. When the ratio of template (2-diethylaminoethanol) to alumina was approximately 0.25, the crystal size reached a maximum.

2.5 Synthesis in Non-Aqueous Media

Non-aqueous solvent systems may, under appropriate conditions, yield large crystals of microporous materials [2, 56]. The structure types grown in large-crystal form include 1) all silica: ferrierite, silicalite-1 and dodecasil-3C; 2) aluminosilicate: ferrierite; and 3) aluminophosphate: $3(C_2H_5)_3NH(Al_3P_4O_{16})$ (novel structure). The concentration of silica in organic solvents is generally much lower than that in aqueous systems under similar conditions. In these organothermal systems, water is only present in reagent quantities and solvents such as pyridine, triethylamine, tetra- or di-ethylene glycol, tetrahydrofuran, dimethylformamide, dimethylsulfoxide, sulfolane, lutidine and mineral oil may

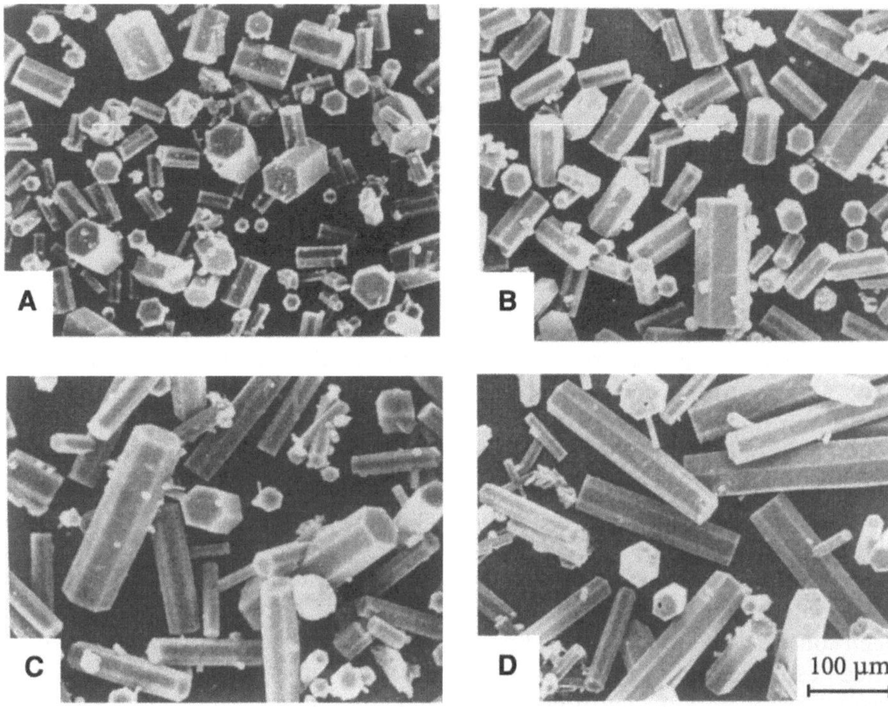

Fig. 15. Variation of the size and aspect ratio of SAPO-5 crystals with the water content of the synthesis mixture. Synthesis at 190 °C from composition Al_2O_3:P_2O_5:0.2 SiO_2:1.55 TEA: x H_2O; **A** x = 300; **B** x = 450; **C** x = 600; **D** x = 750. Reprinted with permission from Stud Surf Sci Catal, vol. 65, Finger G, Kornatowski J, Richter-Mendau J, Jancke K, Bülow M, Rozwadowski M, On controlled growth of SAPO-5 molecular sieve crystals of different sizes and shapes, (1991), pp. 501 – 509, Elsevier Science Inc.

be employed to varying degrees of success. Solvents with a strong affinity for hydrogen bonding interact strongly with the surface of the silica aggregates (from fumed silica reagent) forming a hydrophobic shell around the silica which prevents or reduces the interaction between the mineralizer (OH^- or F^-) and silica; in extreme cases, crystallization does not occur. Solvents with an intermediate hydrogen-bonding ability allowed restricted access of the mineralizer to the silica, thus controlling the dissolution of silica and the degree of supersaturation of the solution, leading to large crystals. The use of gel-forming solutions allowed crystal growth to occur without immediate sedimentation of the crystals, and with greatly reduced mass transfer rates due to reduced convection. Thus, crystals remained suspended in the nutrient pool for longer periods of time and the number of collision-bred nuclei formed was reduced, compared to similar non-gel-forming systems. Hydrogen fluoride acts as the mineralizer, and, when combined with alkylamines and particularly pyridine as the solvent, forms stable poly-hydrogen fluorides through ion pairing. The ion-pairing provides a reservoir of HF, which is released into solution gradually and controls the dissolution and mass transfer of silica. The lower levels of supersaturation and reduced rates of nucleation in these organothermal syntheses lead to the growth of larger crystals. Temperatures of 180–200 °C and reaction periods of between one and several weeks are required for the growth of large crystals. Dodecasil-3C crystals up to 2.1 mm, silica ferrierite up to 1.3 mm, aluminosilicate ferrierite up to ca. 1 mm and $3(C_2H_5)_3NH(Al_3P_4O_{16})$ of a few mm were reported.

2.6 Growth onto Seed Crystals

Klemperer and Marquart reported the synthesis of crystals of ZSM-39 (dodecasil-3C) up to 10 mm in size by seeded growth [57]. Seeds (sections cut from 2 to 3 mm single crystals of ZSM-39 [58]) were suspended in reaction mixtures containing silica, hydrofluoric acid, pyridine and water and treated at 190 °C under autogenous pressure for up to 14 d. Rods of fused silica as well as fumed silica acted as nutrient sources. Fumed silica, possessing a high surface area, dissolved quickly and provided rapid growth in the early stages (0.8 mm d^{-1}), while the fused silica rods provided a slow but steady supply of silica, with a correspondingly slower growth rate (0.15 mm d^{-1}) once the fumed silica had been consumed. Fumed silica was found necessary to bring the concentration of silica in solution up to a high level before dissolution of the seed crystals could occur. Cristobalite and an unidentified silica polymorph co-crystallized with ZSM-39, and were found attached to the reactor walls. The ZSM-39 crystals grown from the seeds were essentially transparent, with highly reflective surfaces, while some twinning could be seen under polarized light. Powder XRD and ^{29}Si and ^{13}C CP-MAS NMR showed them to be free from impurities. The use of multiple silica sources has also been used in the un-seeded synthesis of large-crystal mordenite, as described in Sect. 2.3 of this chapter [42].

3 Synthesis of Ultra-Small Crystals

Based upon experimental work, various parameters can be optimized in order to achieve small crystals, as described in Table 1. Examples of the successful preparation of ultra-small crystals are listed in Table 6.

Table 6. The synthesis of ultra-small crystals

Zeolite type	Synthesis Variable	Ref.
A	aging, sol colloid formation	[61]
Y	addition of TMA	[62]
A and Y	clear solution	[61]
Silicalite-1	clear solution	[59]
ZSM-2	colloid of lithium aluminosilicate glass	[63]
HS	clear solution	[64]

3.1 Case Study: Zeolite MFI

Excellent studies have been performed on the synthesis of ultra-small crystals of MFI-type zeolite [59,60]. In particular the preparation of suspensions of colloidal silicalite-1 crystals (less than 100 nm) with a narrow particle-size distribution have been extensively studied.

The following recipe was used to prepare a colloidal suspension of TPA-silicalite-1:
The main chemicals used were:

- Either tetraethoxysilane (TEOS, Aldrich, < 3 ppm Al) or silica sol, Ludox TM (DuPont, 50 wt.-% SiO_2, 0.39 wt.-% Na_2O, 160 ppm Al, particle size 22–24 nm) or Ludox SM (DuPont, 30 wt.-% SiO_2, 0.66 wt.-% Na_2O, particle size 7–8 nm)
- Tetrapropylammoniumhydroxide (TPAOH, Sigma, 1 M aqueous solution, 195 ppm Na, 27 ppm K, < 10 ppm Al)
- NaOH pellets (Eka Nobel AB, p. a.)
- Additional chemicals and water used were as pure as possible.

In a typical experiment, the alkali sources were added to the TEOS under conditions of vigorous stirring, and the solution was shaken for 12h to promote the hydrolysis of TEOS. The synthesis mixture which was a clear solution, was then heat-treated at 98 °C in polypropene bottles with a reflux set up and under static conditions.

The molar composition of the synthesis mixture is given below:

$$9\ \text{TPAOH} : 0.05\ \text{NaOH} : 25\ SiO_2 : 480\ H_2O : 100\ \text{EtOH}$$

After 24h the colloidal suspension was separated from the liquid phase by repeating the following procedure 4 times:

- centrifuging the sample with a relative centrifugal force of 34,000g for 4h
- decanting the liquid carefully from the zeolite phase
- re-dispersing the solid in distilled water by shaking for at least 2h

The size of the crystals is about 100nm while the form is sometimes too rough to recognize the well known cubic shape. The elongated prismatic shape is not observed in the studies referred to above. In contrast to elongated prismatic crystals, cubic shaped crystals grow from gel or sol particles as has been frequently established by in situ observations [40]. The suggestion is therefore made that even in the clear solution systems discussed, the crystals form in a precursor phase of a colloidal sol consisting of particles not larger than 150 nm. Table 7 summarizes the experimental parameters and results from two separate groups which produced ultra-small silicalite-1 crystals.

The crystal growth rate of MFI in (I) is 0.6 nm min^{-1} and in (II) is 6.6 nm min^{-1}. Although there is a distinct difference in the Na concentration in the two studies the main effect is most probably the substantial amount of ethanol present in (I) retarding the crystal growth rate. The energy of activation for crystallization of silicalite-1 is about 95 kJ mol^{-1} as reported in several papers. This clearly indicates that the growing process of the crystals is not controlled

Table 7. Comparable studies from the literature, regarding the crystallization of ultra-small silicalite-1 crystals from clear solutions [59, 60]

	I, [59]	II, [60]
SiO_2 (moles)	25	25
SiO_2 source	TEOS	silicic acid
Na_2O (moles)	1	0.1
TPAOH (moles)	9	9
H_2O (moles)	480	450
EtOH (moles)	100	-
Temperature (°C)	96	98
Time to achieve 100 nm particles of zeolite (h)	25	5

by diffusion of building units to the crystal, but to chemical interaction of the silicate species involved. In cases where ethanol is present, the energy of activation is approximately 45 kJ mol^{-1}. Apparently the ethanol inhibits the chemical interaction on the surface of the growing crystal.

3.2 Case Study: Zeolite Y

The following procedure was used to prepare a batch of ultra-small crystals of zeolite Y free of other solid phases [61].

The ingredients used were:

- Silica sol Ludox SM (DuPont, 30 wt.-% SiO_2, 0.66 wt.-% Na_2O, particle size 7–8 nm)
- Aluminum sulfate ($Al_2(SO_4)_3 \cdot 18H_2O$, Kebo Lab, Sweden, puriss.)
- Tetramethylammoniumhydroxide ($TMAOH \cdot 5H_2O$, Sigma)
- NaOH pellets (Eka Nobel AB, Sweden, p.a.)

The mixtures were prepared as follows:

- Ammonia was added to an aqueous solution of $Al_2(SO_4)_3$, the precipitate of $Al(OH)_3$ was filtered, re-dispersed in water and filtered again. This procedure was repeated until all ammonia and sulfate were removed
- The wet $Al(OH)_3$ filter cake was added to an 8.4 M solution of TMAOH and stirred until a clear solution of TMA-aluminate was achieved
- To this solution, calculated amounts of water and silica sol were added to obtain the molar composition of the reaction mixture for zeolite Y as given below:

$$2.43\ TMA_2O : 0.04\ Na_2O : 1.0\ Al_2O_3 : 3.4\ SiO_2 : 370\ H_2O$$

The experiment was carried out in a polypropene bottle with a reflux set up at 98 °C. Pure zeolite Y was formed after an induction time of 60 h followed by a crystal growth time of 40 h. The crystals were recovered by centrifugation for 2 h at 49,000 g. After carefully decanting the mother liquid, the crystals were re-dispersed in water and the procedure was repeated until a pure phase of zeolite Y was left. Next, the crystals were dried for 2 h at 105 °C and equilibrated over saturated $CaCl_2$ for at least 16 h. The elemental analysis of the zeolite Y crystals is not given in the paper.

4 Conclusions

Ultra-large and ultra-small crystals of zeolites and zeolite-like materials may be prepared, ranging from all silica to Si/Al = 1, and including $AlPO_4$, SAPO and MeAPO materials. Ultra-large crystals, up to several mm may be prepared in an environment where nucleation is low, mass transfer is high and the rate of chemical reaction at the crystal surface is maximal. Some specific examples of how to achieve such conditions are: (a) the use of ultra-pure and dust-free solutions; (b) control of nucleation; (c) slow release of reagents; (d) extended growth period; and (e) high alkalinity (high nutrient mobility). Synthesis mixtures which form hydro-gels provide a stable growth medium with reduced convective motion and crystal sedimentation, thus reducing collision-bred nucleation and particle settling compared to clear-solution systems. Control of nucleation through chemical additives may be effective in producing large crystals, while increasing the tendency for the formation of competing impurity phases. Synthesis in non-aqueous systems, with carefully chosen solvents and mineralizers may lead to ultra-large crystals due to precise control of the nucleation process, through regulation of the solubility of the nutrients.

In the case of explosive nucleation, the nutrient pool is rapidly depleted and further substantial crystal growth is avoided. Thus ultra-small crystals, down to 30 nm, with a small crystal size distribution can be obtained. Generally, monomeric reagents and low synthesis temperatures are used for preparing ultra-small crystals.

5 References

1. Müller U, Brenner A, Reich A, Unger KK (1989) ACS Symp Ser 398:346–59
2. Nadimi S, Oliver S, Kuperman A, Lough A, Ozin GA, Garces JM, Olken MM, Rudolf P (1994) Stud Surf Sci Catal 84A:93–100
3. Barrer RM (1982) Hydrothermal Chemistry of Zeolites. Academic Press, London
4. den Ouden CJJ, Thompson RW (1992) Ind Eng Chem Res 31:369–73
5. Breck DW, Flanigen EM (1968) In: Molecular Sieves. Society of Chemical Industry, London, p. 47
6. Meise W, Schwochow FE (1973) Adv Chem Ser 121:169–78
7. Finger G (1976) Z Chem 16:414–6
8. Finger G (1977) Z Chem 17:197–8
9. a) Carman PC, Malherbe PR (1950) J Soc Chem Ind 69:134–43; b) Carman PC (1954) J Oil Colour Chem Ass 37:165–73
10. Ciric J (1967) Science 155:689
11. Henisch HK, Dennis J, Hanoka JI (1965) J Phys Chem Solids 26:493–500
12. Scott JA, Anthony WH (1967) French Patent 1,475,582
13. Baker AS (1964) British Patent 1,101,427

14. Charnell JF (1971) J Cryst Growth 8:291–4
15. Wolf F, Horn S (1973) Z Chem 13:109–10
16. Finger G (1975) Z Chem 15:117–8
17. Neels H, Schmitz W, Berger EM, Lutz D (1978) Krist Tech 13:1345–50
18. Gutsze A, Kornatowski J, Neels H, Schmitz W, Finger G (1985) Cryst Res Technol 20:151–8
19. Kornatowski J, Finger G, Schmitz W (1987) Polish J Chem 61:155–64
20. Schmitz W, Kornatowski J, Finger G (1987) Cryst Res Technol 22:35–41
21. Schmitz W, Kornatowski J, Finger G (1988) Cryst Res Technol 23:K25–8
22. Kornatowski J, Finger G, Schmitz W (1990) Cryst Res Technol 25:17–23
23. Sand LB, Sacco A, Thompson RW, Dixon AG (1987) Zeolites 7:387–92
24. Scott G, Dixon AG, Sacco A, Thompson RW (1989) Stud Surf Sci Catal 49A:363–72
25. Hamilton KE, Coker EN, Sacco A, Dixon AG, Thompson RW (1993) Zeolites 13:645–53
26. Scott G, Thompson RW, Dixon AG, Sacco A (1990) Zeolites 10:44–50
27. Morris M, Dixon AG, Sacco A, Thompson RW (1993) Zeolites 13:113–21
28. Morris M, Sacco A, Dixon AG, Thompson RW (1991) Zeolites 11:178–83
29. Coker EN, Jansen JC, Martens JA, Jacobs PA, DiRenzo F, Fajula F, Sacco A. Microp Mesop Materials, in press
30. Mostowicz R, Sand LB (1983) Zeolites 3:219–25
31. a) Böhme KG (1964) German Patent 1,181,644
b) Ross TK, Pearson C (1964) Corrosion Sci 4:449–52
32. Korshunov IA, Malyugina NI (1958) Uch Zap Gor'k Gos Univ in NI Lobachevskogo, Ser Khim 32:21–4 [Chemical Abstracts 54:5319e]
33. Coker EN, Thompson RW, Dixon AG, Sacco A, Nam SS, Suib SL (1993) J Phys Chem 97:6465–9
34. Kühl GH (1968) "Molecular Sieves" Soc Chem Ind, London, 85–91
35. Kühl GH (1970) Preprints, 2nd International Conference on Molecular Sieve Zeolites, pp. 59–70
36. Zhang H, Kamotani Y, Ostrach S (1993) J Cryst Growth 128:1288–92
37. Zhang H, Kamotani Y, Ostrach S (1995) J Spacecraft and Rockets 32:110–6
38. Bogomolov VN, Petranovsky VP (1986) Zeolites 6:418–9
39. Jansen JC, unpublished work
40. Jansen JC, Engelen CWR, van Bekkum H (1989) Am Chem Soc Symp Ser 398:257–73
41. Lermer H, Draeger M, Steffen J, Unger KK (1985) Zeolites 5:131–4
42. Sun Y, Qiu S, Song T, Pang W, Yue Y (1993) J Chem Soc Chem Commun 1993:1048–50
43. Warzywoda J, Dixon AG, Thompson RW, Sacco A (1995) J Mater Chem 5:1019–25
44. Warzywoda J, Dixon AG, Thompson RW, Sacco A, Suib SL (1996) Zeolites 16:125–37
45. Cox SD, Gier TE, Stucky GD, Bierlein J (1988) J Am Chem Soc 110:2986–7
46. Caro J, Finger G, Kornatowski J, Richter-Mendau J, Werner J, Zibrowius B (1992) Adv Mater 4:273–6
47. Kornatowski J, Finger G (1990) Bull Soc Chim Belg 99:857–9
48. Finger G, Richter-Mendau J, Bülow M, Kornatowski J (1991) Zeolites 11:443–8
49. Qui S, Pang W, Kessler H, Guth J-L (1989) Zeolites 9:440–4
50. Girnus I, Jancke K, Vetter R, Richter-Mendau J, Caro J (1995) Zeolites 15:33–9
51. Finger G, Kornatowski J (1990) Zeolites 10:615–7
52. Finger G, Kornatowski J, Richter-Mendau J, Jancke K, Bülow M, Rozwadowski M (1991) Stud Surf Sci Catal 65:501–9
53. Demuth D, Stucky GD, Unger KK, Schüth F (1995) Microp Mater 3:473–87
54. Yokomori Y, Kawachi Y (1995) Zeolites 15:637–9
55. Schüth F, Demuth D, Zibrowius B, Kornatowski J, Finger G (1994) J Am Chem Soc 116:1090–5
56. Kuperman A, Nadimi S, Oliver S, Ozin GA, Garces JM, Olken MM (1993) Nature 365:239–42
57. Klemperer WG, Marquart TA (1994) Mat Res Soc Symp 346:819–24
58. Chae HK, Klemperer WG, Payne DA, Suchicital CTA, Wake DR, Wilson SR (1991) Am Chem Soc Symp Series 455:528–40

59. Persson AE, Schoeman BJ, Sterte J, Otterstedt JE (1994) Zeolites 14:557–67
60. Twomey TAM, Mackay M, Kuipers HPCE, Thompson RW (1994) 14:162–8
61. Schoeman BJ, Sterte J, Otterstedt JE (1994) Zeolites 14:110–6
62. Bodart P, Nagy JB, Gabelica Z, Derouane EG (1986) J Chim Phys 83:777
63. Schoeman BJ, Sterte J, Otterstedt JE (1995) J Colloid Interface Sci 170:449–56
64. Schoeman BJ, Sterte J, Otterstedt JE (1994) Zeolites 14:208–16

Synthesis of Molecular Sieve Phosphates

R. Szostak

Department of Chemistry, Clark Atlanta University, 223 James P. Brawley Dr. SW, Atlanta, Georgia 30314, USA. *E-mail: rszostak@cau.edu*

1 Phosphates

The chemistry of phosphate complexation in inorganic systems is important in agriculture and soil science and more recently in remediation technology because of the low solubility product of many of the metal phosphate salts. A large number of inorganic phosphate phases have been identified [1, 2]. In the area of catalysis the aluminophosphates have been investigated both as catalysts and catalyst supports due to their surface acidity and thermal stability. Researchers searching for zeolite-like materials in the late 1970s turned to the phosphates in their quest for new compositions with potentially novel structures. The phosphate-based molecular sieves proved to be an area rich with new compositions having topologies and building units not previously imagined [3, 4]. This chapter will cover the synthetic advances in the area of new microporous molecular sieves with an update of developments in the aluminophosphate systems.

1.1 Phosphate Structures vs Zeolite Structures

Though many zeolite topologies have been formed with aluminophosphate compositions, there are some significant differences between these two systems. At low pH and in the presence of phosphate as a ligand, aluminum more readily expands its coordination to six with four positions occupied by O^{-2} of the framework phosphate and two sites occupied by additional ligands. To date, the commonly found ligands include water, hydroxide and fluoride ions. Thus the aluminophosphate literature abounds with new porous structures which are not classically zeolite analogs but $AlPO_4$-hydrates, $AlPO_4$-hydroxides and $AlPO_4$-fluorides. An additional phosphate may also coordinate to the framework as observed in the structure of $AlPO_4$-22 [5]. A summary of the more common types of coordination found for these ligands in the microporous aluminophosphate systems studied is listed below:

H_2O monodentate
OH^- mono- or bidentate
F^- mono- or bidentate

The ability of each of these species to act as a mono- or bidentate ligand ultimately plays a key role in the nature of the phase produced. Many of these materials, however, do not exhibit thermal stability. Some exhibit molecular sieving properties but these properties are rarely reported. A number of the aluminophosphates have been found to undergo solid state rearrangements to other structures at relatively low temperatures. In the zeolites, only a few topotactic rearrangements have been noted (materials with the EAB, CHA, and CAN topologies) and only at temperatures exceeding 500 °C [6–8].

Exploring other parts of the periodic table for microporous phosphates has led to the discovery of many open structures, most notably in the gallophosphate, beryllophosphate, cobalt phosphate, zinc and iron phosphate systems. The key synthesis feature for the successful formation of phosphate complexes lies in the rich gel chemistry which occurs with these metals at low pH.

1.2 Natural Phosphate Minerals and Molecular Sieves

Aluminophosphates with 1:1 ratio of aluminum to phosphorus exist in nature. These include the minerals: berlinite, which is the quartz analog, variscite, metavariscite, augelite, senegalite, wavellite, trolleite, bolivarite, and evansite. The minerals variscite and metavariscite both have the chemical composition $AlPO_4-2H_2O$. For both phases the framework aluminum exists in octahedral coordination with aluminum connected to four phosphate groups and two coordinated water molecules [9, 10]. This type of framework typifies many of the aluminophosphate materials subsequently discovered. In variscite and metavariscite, the alternating aluminum and phosphorus form six-member rings throughout the structure. These structures are thermally unstable with structural collapse occurring with loss of the water molecules of coordination. It was the early attempts to synthesize these phases that resulted in the discovery of several novel aluminophosphate materials.

Phosphorus is not readily found as a substitutional element in the natural aluminosilicates due in part to the lack of phosphate sources in the vicinity of the aluminosilicate minerals. The feldspars generally only contain less than 0.1% [11, 12]. Griphite, however, is an example of an aluminosilicate garnet containing phosphate [13].

In nature two beryllophosphate minerals are known. Both contain zeolite topologies. Tiptopite is analogous to cancrinite and pahasapaite is analogous to zeolite rho [14, 15]. The negative framework charge is balanced by alkali and alkaline earth ions.

A very unusual natural microporous iron aluminophosphate which exhibits reversible adsorption properties called cacoxenite ($AlFe_{24}O_5(OH)_{12}(PO_4)_{17}$-$(H_2O)_{24}\cdot$ ca. 51 H_2O) contains a unidimensional channel system with 14.2 Å pore opening [16, 17]. In this structure the iron exists in octahedral coordination with aluminum occupying a distorted octahedral position within the framework. This phosphate mineral was the first known example of a microporous material with a pore size larger than 7 Å. It undergoes structural collapse when heated.

1.3
Inorganic Routes to the $AlPO_4$s

In the early 1960s, a surprising wealth of phases were prepared in the pure aluminophosphate system from a batch composition of $Al_2O_3:P_2O_5:xH_2O$ (T=100 °C) by d'Yvoire [18–20]. By simply varying the concentration of water in his synthesis mixture he was able to prepare the synthetic analogs to variscite, metavariscite and new phases which he called H1, H2, H3, H4, H5, and H6 along with their thermal products A, B, C, D and E. The interrelationship between these phases is shown in Table 1. A key feature to note is the relative difference in the ease of dehydration for some of these phases. The dashed arrows indicate that the reverse (rehydration) does occur but with difficulty. All materials prepared from this batch composition contain an Al:P ratio of unity. Because the conditions of synthesis were very similar differing only in the amount of water in the pre-crystallization mixture, it is not surprising that similarities exist in the structures of these phases. A comparison of the framework structures of dehydrated H1(VPI-5), hydrated H2, and dehydrated H3 ($AlPO_4$-C) is shown in Fig. 1. H1 consists of 6-rings alternating with double 4-rings to produce a large 18-member ring channel. H2 also contains double 4-rings alternating with

Table 1. A summary of the hydrates (prepared by d'Yvoire) and their anhydrous and thermal counterparts

Hydrates		Reversible anhydrous counterparts		Irreversible anhydrous phases
MV	20 °C(P_2O_5) or 70–120 °C ⇄	A	450 °C →	Tridymite
V	20 °C(P_2O_5) or 80–100 °C ⇄	B	400 °C →	Tridymite (+ Quartz)
H5	110 °C ⇄			
H3	100 °C–180 °C ⇄	C (↓ 200 °C to D)		
H6	110 °C ⇄	D	600–850 °C →	Tridymite + Cristobalite
H2	20 °C (P_2O_5) ⇄	E	110 °C →	Tridymite
H1	⇄	VPI-5	100 °C →	$AlPO_4$-8
H4	250 °C →			Tridymite

6-rings giving rise to a 10-ring channel. H3 contains single 4-rings and single 6-rings resulting in an 8-ring pore. H2 dehydrates at room temperature using P_2O_5 as a desiccant, H3 dehydrates between 100 and 180 °C. H1 partially dehydrates around 100 °C undergoing a simultaneous topotactic transformation to $AlPO_4$-8.

From this simple batch composition, however, many of these phases could not be isolated appearing only as mixtures. The use of mineral acids such as HCl, other acids such as HNO_3 as well as the addition of organic amines allowed later researchers to isolate several phases in high purity [21–23]. The most notable being H1 which was found to contain an 18-member ring channel [24]. Structural studies confirm that, even though amines were added to the synthesis mixture in order to obtain suitable purity of this phase, none of the organic is incorporated into the very large channel system [25].

In general, it appears that temperature plays a critical role in the synthesis of these hydrates [26–29]. In the di-*n*-propyl amine system the structure of the final product is highly dependent on the temperature to which the gel is heated [28]. The effect of temperature is shown in Fig. 2. At low temperatures the

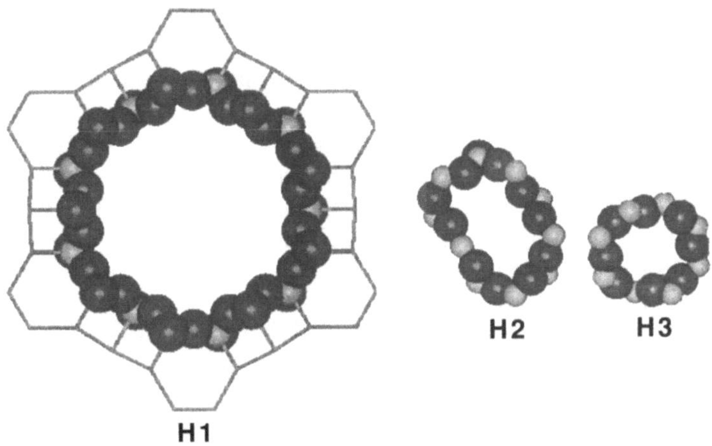

Fig. 1. A pore size comparison of three of the $AlPO_4$ hydrates: H1 (*left*), H2 (*center*), H3 (*right*). All three structures contain four- and six-member rings

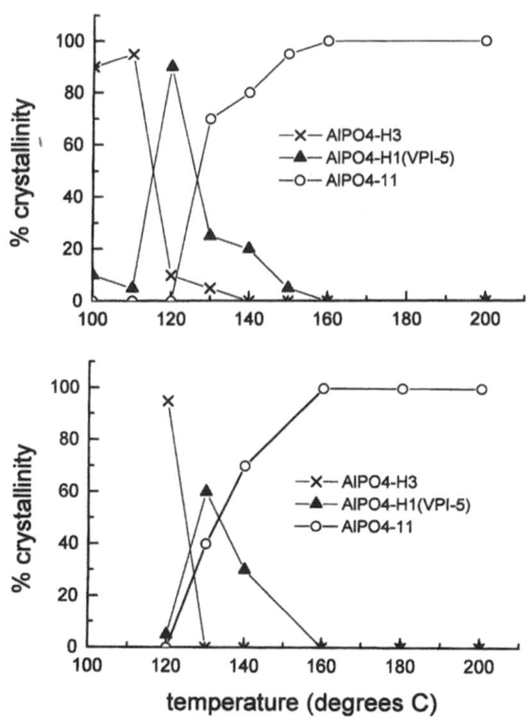

Fig. 2. A comparison of the products obtained from the synthesis mixtures containing di-*n*-propylamine (DPA) as a function of temperature: (*top*) 0.9 DPA : 1 Al_2O_3 : 1 P_2O_5 : 30 H_2O (*bottom*) 1.0 DPA : 1 Al_2O_3 : 1 P_2O_5 : 40 H_2O

preferred product is H3. Intermediate temperatures around 120 °C result in the formation of higher purity H1 (referred to as VPI-5 in the organic system) while temperatures above 160 °C produce the $AlPO_4$-11 phase in high purity. $AlPO_4$-11 has only been prepared from organic amine containing gels.

D'Yvoire used highly acidic conditions in the synthesis of these phases reporting a pH commonly below three. The addition of other acids such as HCl also maintains a very acidic pH through the course of crystallization. In the presence of organic amines the pH can change substantially depending on the basicity of the amine. For the synthesis of H1 (VPI-5) in the presence of organic amines such as di-*n*-propyl amine, the pH of the gel initially begins below 4. As crystallization proceeds, an increase in pH is noted with a final neutral pH attained at the completion of the crystallization [26].

Like the natural aluminophosphates variscite and metavariscite, these as-synthesized aluminophosphates are hydrates and contain octahedral aluminum where two of the coordination sites are occupied by water. Dehydration converts the octahedral aluminum to tetrahedral aluminum as shown in Fig. 3. This process, however, is limited in reversibility. For some structures such as H1, an irreversible topotactic transformation occurs resulting in the generation of a new phase, $AlPO_4$-8 [30–35]. H2 can be reversibly dehydrated at room temperature but with further heat treatment converts to $AlPO_4$-tridymite [27]. H3, too, can be dehydrated to form $AlPO_4$-C. Thermal treatment results in a topotactic transformation to $AlPO_4$-D which can be hydrated to produce a different hydrate, H6. None of these phases have counterparts in the aluminosilicate zeolite system.

D'Yvoire also explored the addition of inorganic cations to his aluminophosphate batch composition. Low levels of potassium added to the mixture did result in the formation of metavariscite. Increasing the ratio of K/Al suppressed the formation of metavariscite. Some variscite and H2 could be formed, but the majority of what precipitated was amorphous. A competing phase in this system

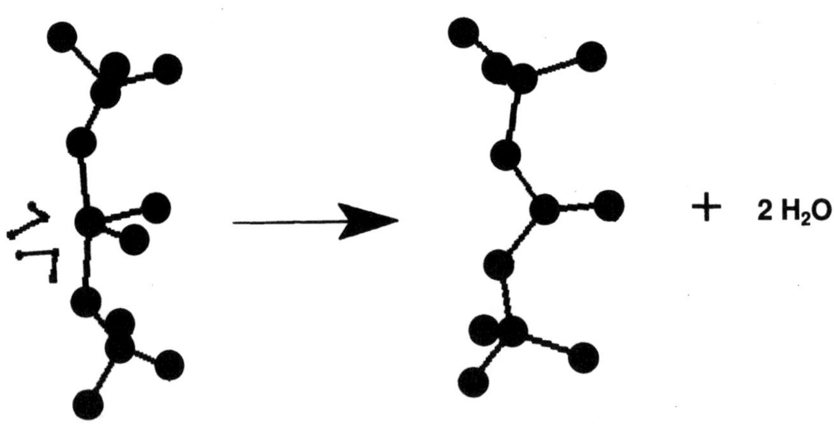

Fig. 3. The transformation of octahedral aluminum to tetrahedral aluminum through loss of two waters of coordination

was taranakite, $KAl_3H_8(PO_4)_6 \cdot 6H_2O$. Only recently has the addition of potassium salts been reinvestigated and novel microporous phases produced [36].

1.4
Phosphorus in the Zeolite Framework

Barrer was one of the first to explore the incorporation of phosphorus into aluminosilicate structures. Substitution would occur through the replacement of two framework Si ions with an Al+P in order that equivalence be maintained. He reported little success [37–39]. Flanigen and Grose also explored the synthesis of aluminosilicates containing substituted phosphate [40]. A high hydroxide content and significant amounts of inorganic cations were used in the synthesis. In order to encourage the uptake of phosphate by the aluminosilicate gel, a very slow copolymerization method was employed. Several zeolite topologies were found to incorporate phosphate, though in widely variable amounts ranging from 5 to 25% P_2O_5. These included LTA, LTL and GIS topologies.

2
Organic Amines and Aluminophosphate Molecular Sieves

The use of organic additives represents a major breakthrough in the generation of new microporous phosphates. Unlike the materials prepared by d'Yvoire, some of the phases obtained were analogous to the known zeolites. Table 2 provides a listing of the aluminophosphate materials that have been crystallized to date and for which structures have been determined. Included in the list are materials analogous to known silicate-based phases as well as phases in which the structure is unique.

These organic containing aluminophosphates are preferentially crystallized at temperatures ranging from 125 to 200 °C [41]. This is distinctly different from the zeolites which can form, albeit over a substantial period of time, at temperatures between room temperature and 100 °C. In the aluminophosphate system, at 100 °C, metavariscite, variscite, and H3 are generally the preferred products formed. Above 200 °C, the aluminophosphate analogs of cristobalite, tridymite, and quartz are formed. This temperature dependence is illustrated in Fig. 2 for the di-*n*-propylamine containing gel system and has been previously discussed. For both batch compositions examined in this study, $AlPO_4$-H3 represented the primary product at temperatures below 100 °C and is relatively insensitive to the presence of amine in the gel. $AlPO_4$-H1 only appears in high purity over a narrow range of temperatures (ca. 120 °C) regardless of other factors. Slightly adjusting the di-*n*-propyl amine concentration and decreasing the water content has a greater effect on increasing the purity of this phase as seen in the difference in % crystallinity of this phase for the trace at the bottom compared to the top trace in this figure. Temperatures above 150 °C result in the crystallization of $AlPO_4$-11 which is also not very sensitive to the water concentration or slight changes in the amine/Al ratio.

The crystallization of aluminophosphate phases is much more rapid than that of the aluminosilicate materials. It is not unusual to complete crystallization in

Table 2. List of known aluminophosphate phases with new microporous or known zeolite-like structures

Common name	Code	Ring size	Common name	Code	Ring size
No silicate analog			*Silicate analog*		
$AlPO_4$-18	AEI	8,8	$AlPO_4$-5	AFI	12
$AlPO_4$-11	AEI	10	$AlPO_4$-16	AST	6
$AlPO_4$-8	AET	14	$AlPO_4$-34	CHA	8
$AlPO_4$-41	AFO	10	SAPO-37	FAU	12
SAPO-40	AFR	12,8	MAPO-43	GIS	8,8
MAPSO-46	AFS	12,8	SAPO-42	LTA	8
$AlPO_4$-52	AFT	8	$AlPO_4$-20	SOD	6
SAPO-56	AFX	8			
MgAPO-50	AFY	12,8			
$AlPO_4$-H2	AHT	10			
$AlPO_4$-C	APC	8,8			
$AlPO_4$-D	APD	8,8			
MAPO-39	ATN	8			
$AlPO_4$-31	ATO	12			
MAPO-36	ATS	12			
$AlPO_4$-33	ATT	8,8			
$AlPO_4$-25	ATV	8			
$AlPO_4$-22	AWW	8			
DAF-1	DFO	12, 12			
VPI-5	VFI	18			
ZAPO-Ml	ZON	8,8			
STA-1-D	SAO	12, 12			
UIO-6	OSI	12			

several hours as opposed to several days or weeks for the zeolites. $AlPO_4$-5 crystals have been observed to form after 30 min when a reactive source of aluminum is used [42].

It is obvious from the plethora of microporous aluminophosphate phases that the organic cation or neutral amine strongly influences the course of crystallization. Without the organic amines, dense phases or the phases reported by d'Yvoire commonly result. Though the amine many times appears to promote a specific topology, it is unclear if parallels can be drawn to the role of organic templates in zeolite synthesis or to specific structure-directing abilities of a particular organic species [43]. For example, $AlPO_4$-5 crystallizes in the presence of a substantial number of different types of amine additives including quaternary ammonium cations, neutral amines, diamines, cyclic amines and alkanolamines [44]. Conversely, a wide variety of different topologies and pore sizes and shapes can be prepared using a single amine additive.

The concentration of the amine is critical in the generation of a particular microporous structure. Cyclohexylamine produces $AlPO_4$-17(ERI) when the organic/(Al-P)O_2 ratio is 0.45, $AlPO_4$-5(AFI) when the ratio is 0.64, and $AlPO_4$-34(CHA) when the ratio is nearer to unity (0.82). The amount of amine in the gel also helps in the incorporation of other framework ions such as silicon increasing its solubility by increasing the pH [45].

Gel aging is also found to influence the outcome of the final phase. VPI-5 is obtained in pure form generally from organic containing $AlPO_4$ gels that have been aged at room temperature before proceeding to the final crystallization temperature [26]. The ATS topology is readily prepared in zinc, iron, magnesium, cobalt, manganese, and titanium aluminophosphate gels through direct synthesis [46,47] but the pure $AlPO_4$ end member, $AlPO_4$-36, requires a two stage heating process where aging the gel at 120 °C is necessary before heating to the final crystallization temperature of 142 °C [48].

2.1 Practical Considerations in Aluminophosphate Synthesis

Unlike the zeolites, the aluminophosphate molecular sieves do not exhibit very high stability in their mother liquor at room temperature. Upon cooling of the autoclave, immediate filtration or centrifugation is prudent, otherwise pitting of the crystal surface is observed [49]. Though many of the $AlPO_4$ phases are thermally stable to temperatures over 500 °C, heating water solutions containing crystals to 200 °C results in collapse of the structure and the formation of dense phases or amorphous materials [50]. The pH is also critical in the stability of the aluminophosphate molecular sieves. Both acidic and basic pH will destroy the structure. Salt solutions will also encourage the dissolution of the crystalline $AlPO_4$s [51]. Leaving the amines in the structure improves the solution stability of the material [52]. It has been suggested that these materials should be stored dry and in the as-synthesized form [51].

2.2 Synthesis Using Microwaves

Microwave technology as a way to induce rapid heating of solutions has been applied to the crystallization of zeolite phases. Zeolites A, Y, and ZSM-5 can all be prepared under microwave radiation [53, 54]. Aluminophosphate gels which, under normal thermal heating procedures, produce $AlPO_4$-5 were also found to produce the same phase under microwave heating [55, 56]. Crystal formation was observed as early as 60 seconds after irradiation. The introduction of an additional framework component such as cobalt to produce CoAPO-5 has also been demonstrated utilizing microwave synthesis methods [57].

2.3 Structural vs Non-Structural Hydroxide Coordination

Synthesis of some aluminophosphate structures under basic conditions can result in the formation of the neutral aluminophosphate structure containing trapped organic hydroxide salts. $AlPO_4$-5 for example contains encapsulated TPAOH where the hydroxide anion is nestled in the base of the tripod formed by three of the propyl groups constrained in the unidimensional 12-ring channels [58]. In structures containing organic cations that do not have the ability to nest the hydroxide counterion, the hydroxide can interact with the surrounding framework aluminum under the proper pH conditions. In $AlPO_4$-17, the hydroxide balances the organic piperidinium cation, but the hydroxide bridges between two adjacent aluminum atoms producing an apparent distorted five-fold coordinated aluminum in the structure [59]. Other structures that have been shown to contain bridging hydroxyl groups with aluminum in five-fold coordination include: $AlPO_4$-12 [60], -14 [61], -14 A [62], -21 [63, 64], -18 [65], -31 [66], and -EN3 [67]. $AlPO_4$-15 contains octahedral aluminum in two different configurations, one containing two hydroxyl groups, the other a water and a hydroxyl ligand [68].

2.4 Hydration, Dehydration, and Topotactic Transformation

The ability of the aluminum to attain six coordination when associated with phosphate ligands results in the aluminophosphate molecular sieves exhibiting some unusual thermal properties. As discussed previously, D'Yvoire's H_n ($n=1-6$) materials were hydrated species with some of the aluminum in octahedral coordination, four points attached to the phosphate in the framework, and the two remaining coordination sites occupied by molecules of water. The lability of these water molecules is dependent on the structure. $AlPO_4$s prepared from organic-containing systems can also exhibit expanded coordination around the aluminum.

$AlPO_4$-11 is a 10-ring structure containing five crystallographically different tetrahedrally coordinated aluminum ions. Apparent structural changes which were observed in the X-ray diffraction pattern concurrent with water adsorption led researchers to examine this structure more closely in order to understand the source of this unusual adsorptive behavior [69]. Initial NMR studies were inconsistent with the structure, and the presence of an octahedral aluminum was indicated for the hydrated material [70]. Further detailed studies confirmed that there is one tetrahedral aluminum site in $AlPO_4$-11 that undergoes preferential and reversible hydration [71–73].

The transformation of the anhydrous $AlPO_4$-C to $AlPO_4$-D involves more than the simple relaxation of the framework observed in the dehydration/rehydration examples described above [74]. The aluminophosphate hydrate, $AlPO_4$-H3, can be dehydrated at 100 °C to produce $AlPO_4$-C. This process can be reversed, however, when the temperature is further raised to 250 °C, $AlPO_4$-C undergoes a topotactic transformation to produce $AlPO_4$-D. This transformation converts the double-crankshaft chain found in $AlPO_4$-C to an UDUD or narsarsukite-type arrangement of $AlPO_4$-D. This type of transformation is illustrated in Fig. 4.

$AlPO_4$-H1(VPI-5) dihydrate can be transformed under careful conditions to the monohydrate [75]. The topotactic transformation of $AlPO_4$-H1(VPI-5) reduces the 18-member ring opening of VPI-5 to a 14-member ring of $AlPO_4$-8 where one third of the aluminum ions in the new structure occupy octahedral sites coordinated to two water molecules [76–78]. This is an unusual example of an incomplete dehydration occurring during topotactic transformation. This anomaly may be a result of the strain within the VPI-5 structure generated upon

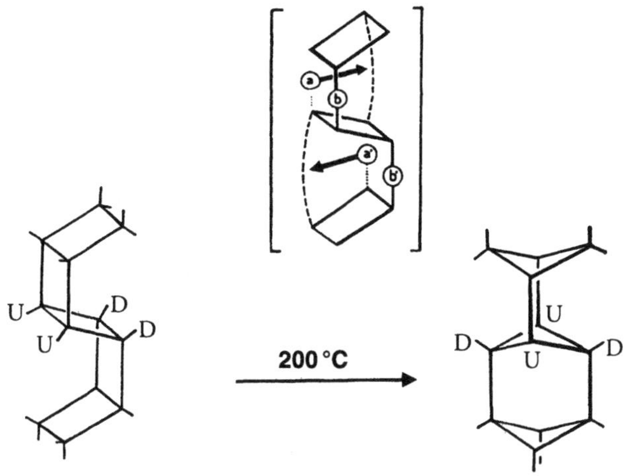

Fig. 4. The topotactic transformation of $AlPO_4$-C to $AlPO_4$-D. The UUDD (double-crankshaft) chain in $AlPO_4$-C and the UDUD (narsarsukite-type) chain in $AlPO_4$-D are shown. (Reprinted with permission of Elsevier Publishers, Amsterdam)

removal of the pore water as the crystals are heated to 100 °C inducing structural rearrangement before complete dehydration can be attained. This transformation, though facile, does result in the generation of numerous stacking faults within the structure which block the pores and give rise to adsorption properties more characteristic of smaller pore material than that expected of the 14-member ring structure [79, 80].

A similar type of transformation is observed with the thermal conversion of $AlPO_4$-21 to $AlPO_4$-25. This transformation is catalyzed by the loss of a bridging OH group. In the structure of $AlPO_4$-21 two different aluminum coordinations are observed: an $Al(OP)_4$ tetrahedron and two five-coordinated distorted $Al(OP)_4(OH)$ moieties. Bridging of the OH between two aluminum sites gives rise to a mix of three- and five- member rings within the structure as shown in Fig. 5 a. Thermal treatment disrupts this arrangement and additionally causes the UD configuration in $AlPO_4$-21 to flip similar to that which has been observed for $AlPO_4$-C and -D (see Fig. 5 b) [63, 64, 81].

2.5
Fluoride in $AlPO_4$ Synthesis

Though crystallization of the aluminophosphate materials generally prefers acidic conditions, the use of fluoride as a mineralizing agent has been found to offer further flexibility in the synthesis of $AlPO_4$ phases. Generally, the introduction of fluoride to an aluminophosphate gel results in a more rapid formation of the crystalline product. Fluoride ion addition may also increases the size of the crystals produced [82]. In some cases centimeter long crystals have been obtained [83].

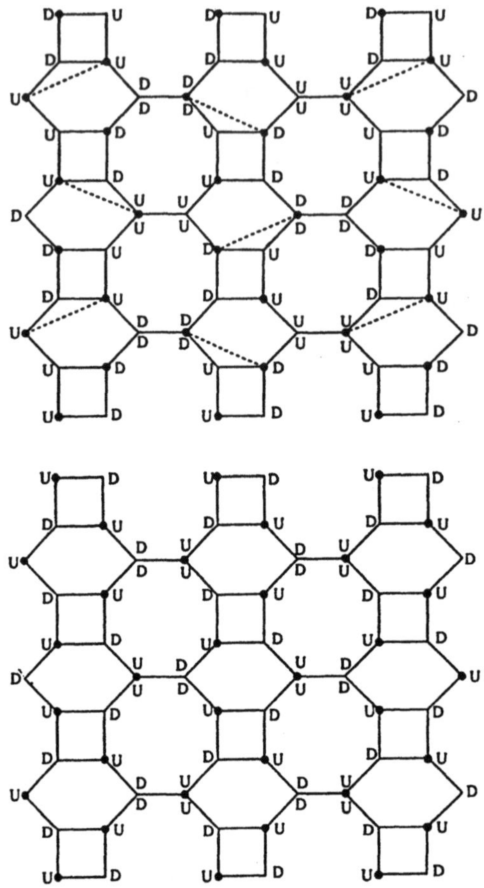

a b

Fig. 5. **a** Idealized diagram of $AlPO_4$-21. The aluminum ions are marked by the solid circles, and the phosphorus are located at the vertices. U and D represent upward and downward pointing connections. There is a bridging oxygen in the midpoint of each solid line and a bridging OH group in the middle of the dashed lines which results in two distinct five-coordinated Al sites. **b** Idealized diagram of the $AlPO_4$-25 framework. (Reprinted with permission of the American Chemical Society, Washington D.C.)

Fluoride coordinates readily to aluminum ions and like water and hydroxide results in aluminum expanding the coordination sphere from four to six. Unlike coordinated water, fluoride ions prefer to behave as bidentate ligands linking two aluminum ions in the structure. The addition of fluoride ions to the synthesis requires a cation to balance the charge. Generally this counter-ion is the organic amine. $AlPO_4$-5 can be prepared in both a fluoride media and in the presence of hydroxide as a counter-ion and therefore provides a good example of the effect the synthesis conditions have on the nature of the final structure even though the basic topology is the same. The location of the fluoride counter-ion to the organic template in $AlPO_4$-5 does differ from that of the corresponding hydroxide – an indication of the strong attraction between fluoride ions and the framework aluminum [84]. In $AlPO_4$-5, the fluoride occupies sites associated with the four rings of the framework and is not intimately associated with the organic cation unlike the hydroxide counter-ion which is nestled in the tripod of the tetraproylammonium cation described previously [58].

When working in gel systems containing variable quantities of fluoride (and additionally alkali metals) changes may be observable in the X-ray diffraction pattern which are not associated with the generation of new topologies. With very high loadings of fluoride ions the distortion of the framework due to the presence of bridging fluoride ions is very symmetric [84]. With lesser amounts of fluoride, the distortion can be observed in the changing of the X-ray powder diffraction pattern. This is illustrated in Fig. 6 for $AlPO_4$-5. The X-ray diffraction pattern at the top of this figure is characteristic of $AlPO_4$-5 prepared under normal synthesis conditions. The bottom trace was taken from a sample recovered from a synthesis containing potassium fluoride [85].

The ability of the fluoride ion to occupy framework associated sites can lead to further difficulties in differentiating new phases from known phases based on the presence of a different X-ray diffraction pattern. A triclinic form of $AlPO_4$-chabazite is produced in the presence of 1-methylimidazole and HF [82]. It contains a F/T atom ratio of 1/6. Other ratios of F/T can also be synthesized for this

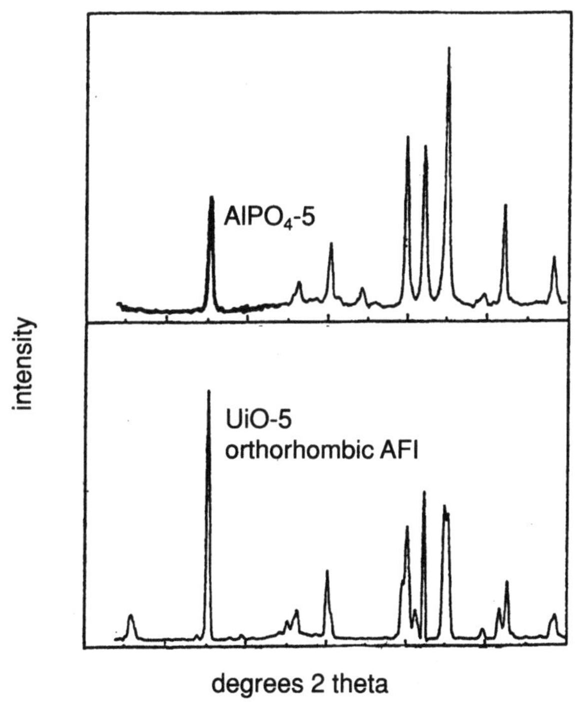

Fig. 6. A comparison of the X-ray powder diffraction pattern of $AlPO_4$-5 (AFI) (*top*) prepared by conventional synthesis methods and (*bottom*) prepared using fluoride and potassium ions. (Reprinted with permission of E. N. Halvorsen and K. P. Lillerud, Oslo, Norway)

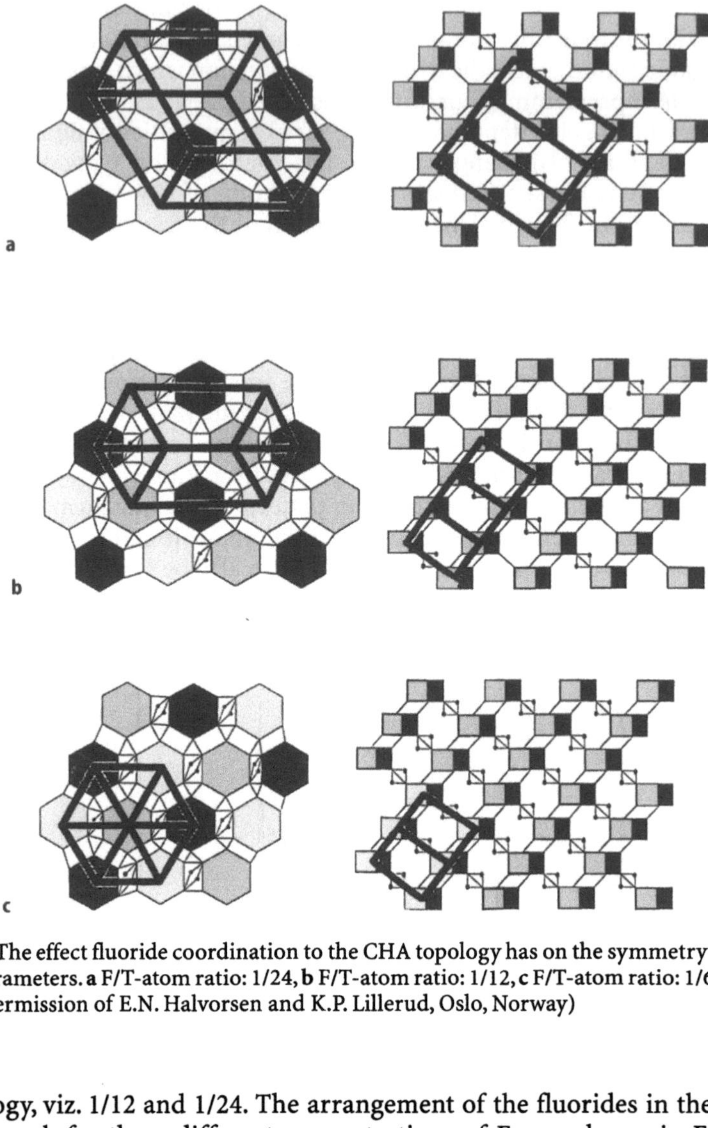

Fig. 7. The effect fluoride coordination to the CHA topology has on the symmetry and the unit cell parameters. **a** F/T-atom ratio: 1/24, **b** F/T-atom ratio: 1/12, **c** F/T-atom ratio: 1/6. (Reprinted with permission of E.N. Halvorsen and K.P. Lillerud, Oslo, Norway)

topology, viz. 1/12 and 1/24. The arrangement of the fluorides in the chabazite framework for these different concentrations of F^- are shown in Fig. 7 along with the associated changes in the unit cell parameters. The cluster building unit containing the octahedrally coordinated aluminum bridged by two fluoride ions and two phosphate is shown in Fig. 8 where a distinct foreshortening of the distance between the Al ions across the 4-ring is evident [85]. Even from non-aqueous synthesis systems, fluoride can be incorporated into this structure [86]. Upon heating to 400 °C a release of HF from the structure is observed with the generation of the neutral $AlPO_4$-chabazite.

Further sensitivity of the gel towards fluoride as a structure director has also been observed in another non-aqueous synthesis system. In the presence of dimethylformamide in triethyleneglycol the gismondine topology is formed in

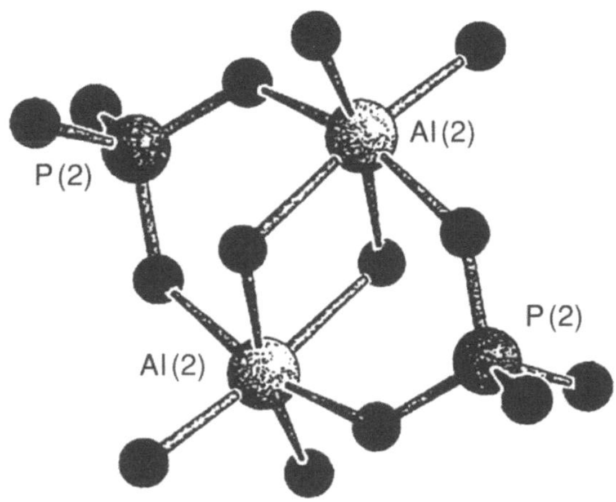

Fig. 8. Arrangement of the interconnected $[AlO_4F_2]$ octahedra

which four of the aluminum ions exist in tetrahedral sites and four in an octahedral arrangement containing bridging fluoride ions distorting the lattice as shown in Fig. 8. In the absence of fluoride the $AlPO_4$-21 structure is obtained, an indication of the structure directing role the fluoride ion can potentially have in many of these syntheses [87].

2.6 Mesoporous Aluminophosphates

The discovery of the mesoporous silicas by researchers at Mobil prompted the examination of the interaction of aluminophosphate gels with surfactants [88]. In the silica-based systems, three distinct mesophases were observed: hexagonal, cubic, and lamellar. The type of phase obtained depends on the conditions of synthesis [89–91]. The use of similar linear alkylenediamines and cyclic diamines in the aluminophosphate system has given rise to the formation of lamellar phases [92–99]. The addition of primary and ternary amines containing long alkyl chains can also result in the formation of lamellar $AlPO_4$ phases with X-ray d-spacings indicative of nanometer range layer spacing [100–102]. Using non-aqueous routes produced similar results [103–105]. Two unique layer stoichiometries have been identified and appear to be independent of the nature of the organic surfactant added to induce their formation. The basic unit consists of $[Al_3P_4O_{16}]^{-3}$ containing a six-member ring of alternating Al and P that is capped with a PO_4 unit.

Mesoporous phases containing coaxial cylindrical surfactant/$AlPO_4$ bilayers have also been observed from hydrothermal systems containing $C_{12}H_{25}NH_2$ under similar conditions which produce lamellar phases [106]. This unusual type of arrangement of the inorganic phase and surfactant is illustrated in Fig. 9. Similar arrangements have been reported in the silicate system [107]. Mixed organic containing systems containing surfactants and quaternary

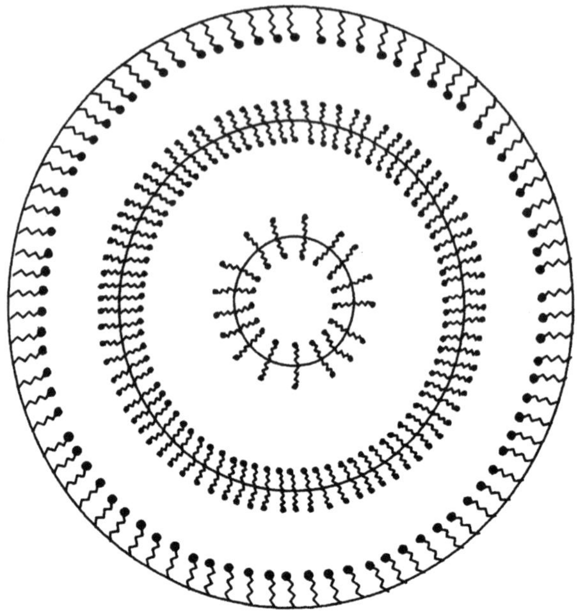

Fig. 9. Schematic illustration of the inorganic-organic bilayers coaxially wrapped around a rod-like micelle and viewed down the axis

amines, such as tetramethylammonium cations, result in the generation of a pure hexagonal aluminophosphate material [108, 109].

The inorganic reagents for these lamellar syntheses were generally phosphoric acid and catapal B alumina which are initially mixed to form an aluminophosphate gel. The hexagonal phase was obtained using aluminum hydroxide as the aluminum source. The surfactant is added to this mixture. Crystallization temperatures of 100 °C have been reported with synthesis times generally around 24 hours (or overnight).

A detailed examination of the lamellar aluminophosphate phases prepared with differing Al : P ratio in the gel reveals a strong preference for phosphate to form these layered structures. In the absence of aluminum in the gel, an organophosphoric acid layered structure was obtained. Conversely, the absence of phosphoric acid does not produce any crystalline phase. The preparation of layered phases with a 1 : 1 Al : P ratio under such acidic conditions requires Al : P ratios in the gel to be adjusted to 2 : 1. The deficiency in aluminum in these structures results in the formation of sites containing phosphate-organic linkages [102]. The aluminum in the layers exists as tetrahedral aluminum or hydrated (octahedral) aluminum. Because of the strong association between the surfactant and the phosphate, removal of the surfactant generally results in the collapse of the structure though some reversibility in the framework octahedral aluminum to tetrahedral aluminum indicates the ability of this structure type to be somewhat flexible in the waters of hydration. The hexagonal mesoporous aluminophosphate exhibits better stability than the lamellar forms withstanding calcination temperatures to 500 °C with surface areas between 740 and 930 $m^2 g^{-1}$.

The addition of fluoride ion also results in the preparation of layered fluoroaluminophosphates. These materials are shown to contain sheets having building units in which three AlO_4F_2 octahedra and one AlO_3F tetrahedron are linked through tetrahedral phosphate giving rise to eight-member rings within the sheet [110, 111].

2.7 New Directions in Porous $AlPO_4$ Synthesis

The enormous number of new aluminophosphate materials first reported by Union Carbide in the early 1980s prepared by adding organic amines to d'Yvoire's simple Al:P:H_2O crystallization mixture initiated a very active research effort in the search for new and large pore-size materials. Major advances in new materials occurred in this area, many in parallel with activities in the synthesis of new zeolites and silicates. Key areas are highlighted in Fig. 10. New phases are still being found from the organic aluminophosphate mixtures through the addition of different and more complex amines [112] but other pure $AlPO_4$ phases have been prepared through the addition of fluoride ions such as UiO-7 which is related to ZAPO-M1, a novel zinc containing aluminophosphate [113, 114]. The addition of other inorganic cations such as magnesium have been shown to produce the new DFO topology using standard synthesis con-

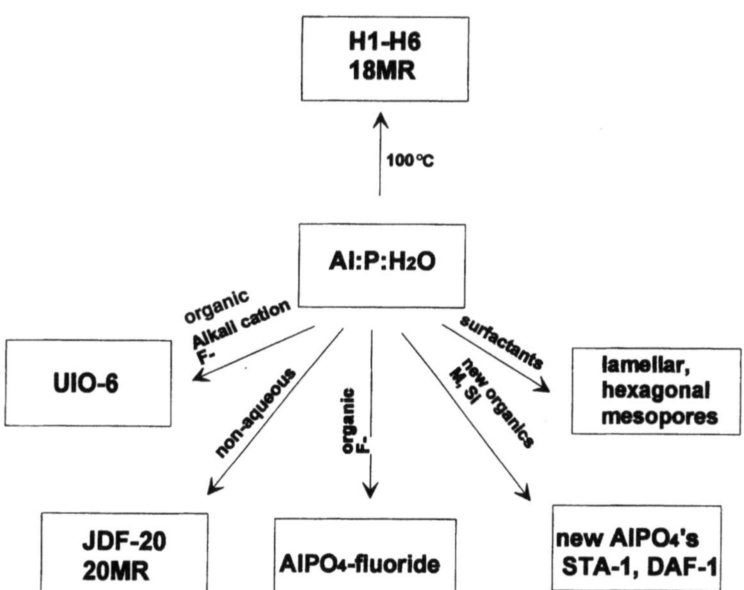

Fig. 10. A schematic diagram of the effect various different synthesis methods have on the production of aluminophosphate phases

ditions [115]. The use of non-aqueous media for synthesis has produced some known phases as well as a new larger pore material, the 20-member ring phase JDF-20 [116–118]. Surfactant addition does lead to mesoporous materials, both lamellar and hexagonal with framework arrangements still under investigation. Alkali fluoride addition to organoaluminophosphate synthesis produces a pure microporous structure UiO-6 with a novel building unit [119]. In the absence of the alkali only a mixture of this phase and $AlPO_4$-5 is observed. The use of inorganic salts such as KF adds a new direction to the area of aluminophosphate synthesis.

The generation of microporous phosphates containing other framework ions in the absence of aluminum represents yet another pathway to porous crystalline solids. Ions such as gallium, cobalt, zinc, and iron which may prefer octahedral coordination give rise to a new set of bonding rules beyond those Löwenstein formulated for tetrahedral aluminum coordination. Though these ions have resulted in a plethora of new frameworks, their synthesis method has not varied significantly from those of the aluminophosphates. The use of organic amines to control the pH and possibly to contribute to the pore architecture, the use of fluoride which becomes an integral part of the structure, and the use of non-aqueous solvents all have led researchers to isolate new phases.

3 Gallophosphate Analogs of the $AlPO_4$s

The synthesis of gallosilicate analogs of the zeolites were reported almost simultaneously with reports of the aluminosilicates [120]. Thus it would be expected that gallium oxide/hydroxide would react with phosphoric acid under similar conditions which give rise to the $AlPO_4$ molecular sieves to generate analogous $GaPO_4$ phases [121–131]. Gallium tends to prefer octahedral coordination rather than occupying tetrahedral sites and it is not surprising that gallophosphates containing a predominance of octahedral gallium have been discovered. Preference for association with hydroxides and fluorides over water is also observed limiting the $GaPO_4$ materials to $GaPO_4$ hydroxides and $GaPO_4$ fluorides.

Successful synthesis of many of the gallophosphates has been accomplished using gels containing fluoride anions. The cation that is associated with the fluoride ions in the synthesis of the $GaPO_4$s plays a critical role in the nature of the resulting phase. HF generally produces microporous structures, amines tend to form cages. The ammonium cation has a very strong templating effect in these systems which overrides any effects the presence of an added amine might have [132, 133].

The pH also dictates the resulting $GaPO_4$ phase that can form generally by encouraging different environments around the gallium. At very low pH octahedral coordination is preferred, as the pH is increased, trigonal bipyramidal coordination begins to be favored. Changing the pH can be accomplished (at fixed HF concentration) by increasing the H_3PO_4 concentration to increase the acidity or increasing the organic amine to provide a basic environment. In the system: $Ga_2O_3/P_2O_5/HF/1,3\text{-diaminopropane}/H_2O$, three novel $GaPO_4$ phases have been isolated. At pH between 5 and 10 ULM-3 ($Ga_3(PO_4)_3F_2 \cdot H_3N$

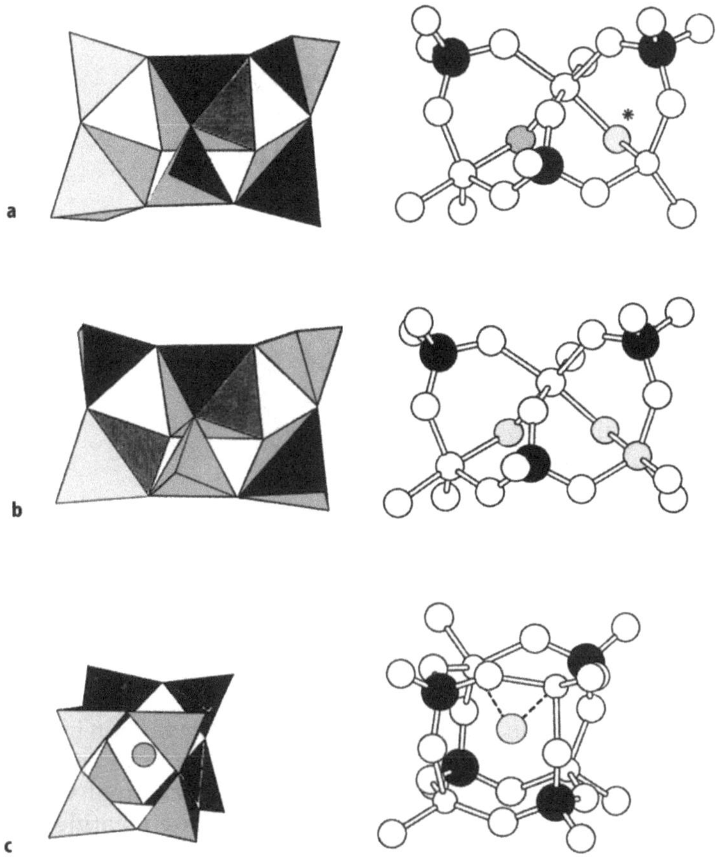

Fig. 11. Perspective view of **a** sbu A, **b** sbu B, and **c** sbu C in both polyhedral and ball and stick representation. *Black spheres*, P: *large open circles*, O; *small open circles*, Ga: *small gray circles*, F except in **a** where the *asterisk* on the gallium tetrahedron corresponds to OH. (Reproduced with permission of Gauthier-Villars Publishers)

$(CH_2)_3NH_3 \cdot H_2O$) is formed. Between pH 2 and 4 ULM-4 ($Ga_3(PO_4)_3F_2 \cdot H_3N(CH_2)_3NH_3$) crystallizes. At pH less than one, ULM-6 ($Ga_4(PO_4)_4F_2 \cdot H_3N(CH_2)_3NH_3$) is isolated. At very high pH Ga_2O_3 (pH > 10) precipitates. All three $GaPO_4$ phases isolated are microporous. ULM-3 and ULM-4 contain gallium that is in a trigonal bipyramidal arrangement with four oxygens and one fluoride as well as octahedral coordination with two fluorides. These two polyhedra are linked via the fluoride. The different arrangements of the building units observed in the gallophosphate structures are shown in Fig. 11, and their arrangement to form the various porous phases is shown in Fig. 12. ULM-6, formed at the very low pH contains gallium only in octahedral coordination with shared edges between the gallium ions [134–137].

In the HF systems, the concentration of the amine dictates the type of species which form. However, excess amine tends to suppress microporous crystal

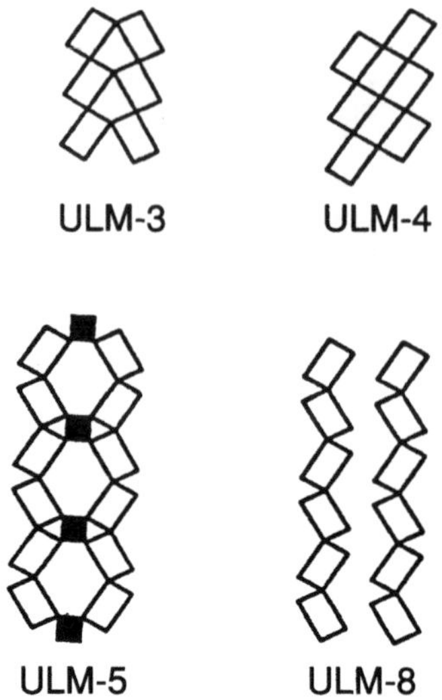

Fig. 12. Connectivity of the SBU-6 (represented by *rectangles*) in ULM-3, ULM-4, ULM-5 and ULM-8. In ULM-5, another type of SBU (SBU4-4) formed by four PO_4 tetrahedra and four GaO_4 tetrahedra ensures the connection of the SBU-6. (Reproduced with permission of Academic Press)

formation as the amine can act as a strong ligand for the gallium ion. A series of diamines of differing chain lengths have been studied by Ferey and co-workers. ULM-3 can be prepared with diamines having a chainlength of 4 and 5 [138]. When the carbon chain is increased to 6–8, a new phase identified as ULM-5 ($Ga_{16}(PO_4)_{14}(HPO_4)_{14}F_7(OH)_2 \cdot (H_3N(CH_2)_3NH_3)_4 \cdot 6H_2O$) is produced [139]. This material contains three different building units, one related to that of ULM-3 and -4, a second that is very similar to the first where the trigonal bipyramidal gallium has become tetrahedral $GaO_3(OH)$ and the third building unit consisting of an octamer similar but not identical to that which is found in the cloverite structure [140]. The octamer is illustrated in Fig. 11. This gives rise to a 16-member ring pore opening with a free aperture of 12.20 × 8.34 Å. A similar 16-member ring material was prepared from a batch composition: 1 GaO(OH) : 1 H_3PO_4 : 1 HF : 0.65 cyclohexylamine: 0.75 tripropylamine: 60 H_2O. The tripropylamine is added to adjust the pH of the system and is not found in the final crystalline product. Without it only $GaPO_4$ quartz is formed [141]. This material exhibits significant thermal stability loosing X-ray crystallinity at temperatures in excess of 800 °C. Above this temperature it is the fluoride that is lost as $GaPO_4$ cristobalite is formed. Spherical amines produce other structure types where cages are preferred rather than channels. DABCO (1,4-diazabicyclo[2.2.2]octane) produces ULM-1 and ULM-2 which contains cavities accessible through 8-member rings [142, 143].

The stability of the organic additive under the conditions of synthesis does have a critical role in determining the outcome of a synthesis. The polyamine, guanidinium carbonate $[C(NH_2)_3]_2CO_3$, was found to decompose under normal

$GaPO_4$ synthesis conditions to release NH_4^+ to the crystallization mixture. Since ammonium cations are very strongly structure directing in this system, the resulting phase formed was that of the $GaPO_4$ analog of leucophosphite, $AlPO_4$-15, $GaPO_4$-C7, and $GaPO_4 \cdot H_2O$ [144–148].

3.1 Chains and Layered Phases in the $GaPO_4$s

Chain structures have been observed in the aluminophosphate system [149–151]. In a mixed water-dimethyl sulfoxide system (10 Me_2SO : 13 H_2O) a $GaPO_4$ phase containing infinite chains of $[GaH(PO_4)_2]^{-2}$ has been observed [152]. The basic chain unit consists of one gallium ion and two phosphate ions, both with tetrahedral coordination. All vertices of the GaO_4 tetrahedron are occupied by PO_4 units. The cation consists of the protonated diamine used in this synthesis, $[H_3N(CH_2)_3NH_3]^{+2}$. Fluoride does not appear to be associated with this structure even though crystallization was observed from an HF containing gel.

Two layered $GaPO_4$ phases have been isolated, one from a gel containing tris(2-aminoethyl)amine, the other from a piperazine containing system [153, 154]. These have been identified as ULM-8 and -9, respectively. The ULM-8 layers are composed of the same building unit found in ULM-3,-4, and -5. The geometry of the amine in this synthesis is considered to be key to the formation of a layered structure.

3.2 Cobalt Substitution in $GaPO_4$ Synthesis

The addition of cobalt to gallophosphate gels alters the chemistry of the system significantly producing several phases which are isostructural with known zeolite topologies as well as a new phase. To date the CoGAPO-laumontite and gismondine structures and a CoGAPO-LTA could be isolated from $GaPO_4$ gels containing small amounts of cobalt. The latter phase was obtained from an ethyleneglycol solvent system. In addition, a new structure was obtained and identified as the new topology CGF also prepared using ethylene glycol as a solvent [155–158].

4 Beryllophosphate Analogs of the Zeolites

Though beryllium is a very toxic metal with which to work, its electronic structure will result in the generation of a negatively charged framework when associated with phosphate and thus makes these materials intriguing as true zeolite analogs. Union Carbide first reported the incorporation of beryllium in the SAPO synthesis in 1988 producing many phases that were related to the known aluminophosphates previously synthesized [159]. Beryllophosphate gels, like the aluminophosphates are formed at low pH. Because of the lack of neutrality in the beryllophosphate gel, the synthesis chemistry is significantly different from that of the neutral aluminophosphates.

Harvey and Meier were the first to show that synthetic beryllophosphates could be prepared, some with zeolite-like topologies, others with novel structures [160]. The synthesis procedure was similar to that of the early aluminosilicate zeolite syntheses and differed significantly from the aluminophosphates, as inorganic cations were employed and no organic amine was necessary. Inorganic cations proved to be strongly structure directing in these systems. A different structure was produced by changing the alkali cation. Mixed cation systems, also examined, produced yet another phase though crystallization times were significantly longer, viz. 21 days instead of 7 days. Conventional temperatures between 100 and 200 °C were employed.

The similarity between the beryllophosphates and the aluminosilicates was further substantiated with the crystallization of the faujasite structure from the sodium beryllophosphate gel at lower temperatures and shorter crystallization time (4 °C, 30 min) [161–164]. In the lithium beryllophosphate system, lowering the temperature results in the production of the analog to Li A(BW) [162]. Higher pressures and the presence of extra-framework ions such as Cl anions yielded feldspathoid-like structures including SOD and LOS topologies [165, 166].

The addition of organic cations resulted in the preparation of an interrupted framework structure which contains beryllium ions coordinated to water and/or P-OH groups [167]. In a mixed NH_4/di-isopropylamine system the beryllophosphate obtained did not incorporate the organic moiety but did take up the ammonium cation [168]. Thermal treatment of this material showed a decomposition of the ammonium cation around 600 °C.

It is interesting to point out that the crystallization of zeolite phases produces very small crystals, while the aluminophosphate crystals tend to be somewhat bigger. The beryllophosphate phases appear to readily form at very low temperatures and short crystallization times with substantially larger crystals.

5
Other Metal Phosphate Phases

With $CoPO_4$ being isoelectronic with $BePO_4$, it is not surprising that several cobalt phosphate molecular sieves have been reported using organic and inorganic cations as structure directing agents [169]. An early report of a synthesis from cobalt phosphate gels using organic diamines did produce a microporous cobalt phosphate [170]. In the presence of the inorganic cations the structures produced contained cobalt in tetrahedral coordination and topologies related to tridymite and ABW, the $CoPO_4$-ABW having a chiral arrangement. The synthesis of vanadium phosphates and arsenates have also been considered, and novel structures have been reported though none with zeolite structures [171–173]. An ethylene diamine templated oxyfluorovanadato phosphate has also been described [174]. It consists of an asymmetric V_2PO_8F unit where the fluoride bridges the two vanadium cations in the structure. The amines are located along the tunnels within the structure. Synthesis of this phase followed closely the synthesis of the microporous gallophosphate materials. A molybdenum phosphate is also known [175], and layered and chain iron phosphates have also been prepared using fluoride ions and organic amines in their synthesis [176, 177].

Zinc phosphate gels have resulted in a wealth of new phases containing porous structures. These phases have been prepared from aqueous gels containing zinc salts, phosphate and alkali cations. With pH adjusted to 5 using hydroxide, several phases have been reported. The zinc phosphate analog to the zeolite ABW ($LiZnPO_4 \cdot H_2O$) undergoes a structural rearrangement upon dehydration at 100 °C to produce a semicondensed phase which contains the lithium cations encased in (what is descriptively referred to as) squashed 6-ring channels [178]. At higher pH (ca. 11.5) in the presence of sodium hydroxide, zinc oxide and phosphoric acid, a very new chiral topology results with the composition $Na_{12}[Zn_{12}P_{12}O_{46}] \cdot 12\,H_2O$ [179]. The structure consists of unusual 4-ring squares and triple 4-ring chains. A view of the resulting helix is shown in Fig. 13. The pores are defined by highly distorted 12-member rings along with 6- and 8-member rings. Synthesis in the presence of other alkali cations results in the generation of more microporous phases [180]. This synthesis also proceeds under even stronger basic conditions and crystallization temperatures of 70 °C. An unusual feature of some of these phases is an OZn_4 unit which gives rise to a spiro-type three-member ring. An example of one of these structures is shown in Fig. 14. The structure/synthesis relationship of the zinc phosphate materials is strongly dependent on both the pH of the system as well as the ratio of cation to zinc to phosphorus. A summary of materials identified from these syntheses over the entire range of pH is given in Table 3.

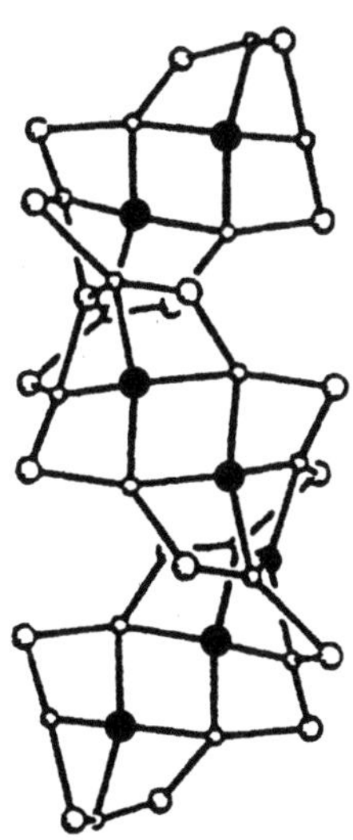

Fig. 13. The tetrahedral helix in $NaZnPO_4 \cdot H_2O$. *Large circles* represent Zn ions, *small circles* P ions. Framework oxygen is omitted for clarity

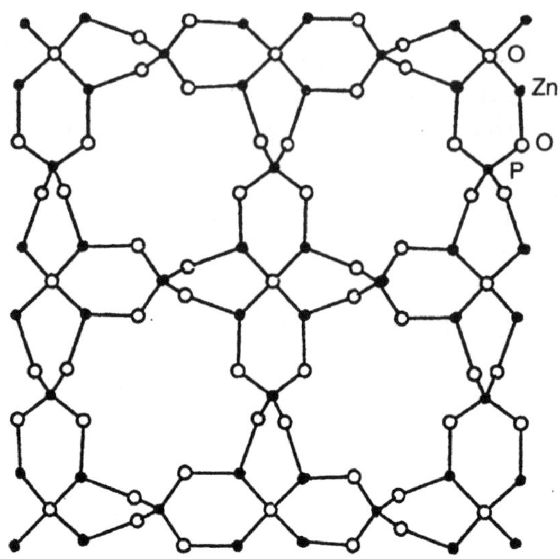

Fig. 14. [001] projection of a layer ($-0.15<z<0.15$) of the $Rb_3Zn_4O(PO_4)_3 \cdot 3.5H_2O$ structure, showing the connectivity of four spiro-5 units into 8-ring windows

Table 3. Summary of sodium zinc phosphate phases

Formula	pH[a]	Ratio	Structure
$Zn_3(PO_4)_2 \cdot 4H_2O$	1–2	0:3:2	hopeite
$Na_2Zn(HPO_4)_2 \cdot 4H_2O$	2[c]	2:1:2	layered; 4-and 12-rings
$NaZn_2(PO_4)(HPO_4)$	2–4	1:2:2	corrugated layers; 3-, 4-rings
$Na_3(ZnPO_4)_3 \cdot 4H_2O$	6–8	1:1:1	sodalite; 4-, 6-rings
$Na_6Zn_3(PO_4)_4 \cdot 3H_2O$	10	6:3:4	semicondensed 3-dimensional
$NaZnPO_4 \cdot H_2O$	11	1:1:1	new chiral frameworks, 4-, 6-, 12-rings
$Na_3Zn_4O(PO_4)_3 \cdot 6H_2O$	12.5	3:4:3	novel 3-, 8- ring framework
$Za_2Zn(OH)PO_4 \cdot 7H_2O$	13	2:1:1	novel 3-ring chain structure
ZnO	13+	0:1:0	
$NaZn(OH)_3$	14[d]	2:1:0	

[a] Starting solution pH.
[b] Na : Zn : P molar ratio of solid product.
[c] At 4 °C.
[d] Concentrated NaOH solution.

6 Evaluating the Periodic Table

Synthesis of zeolite analogs based on other framework compositions have come quite far afield from the original goal. Though the synthesis of all of these phases has utilized methods that have been initially developed to produce aluminosilicate zeolites and silicate molecular sieves, the chemistry of the individual ions has led to the crystallization of tunnel and porous structures which have adapted to the coordination needs of the individual framework ions. The ability to produce novel open frameworks has been shown to not be completely bound to the use of specific organic amines as structure directors, but to a whole host of factors most notably the attraction of these various metal anions to accommodating hydroxide and fluoride into their coordination spheres. There will always be a joy in building new structures from novel building units for many of us in the field of microporous materials. The hope is that some of these phases will result in a critical breakthrough in the fields of catalysis, adsorption, and solid state materials. It is a hope that is presently still only a wistful vision conceived in the mind.

Acknowledgements. The author wishes to thank Ms. Keera Klear and Ms. Crystal Oliver for their assistance in the literature search for this manuscript and Prof. Larry Kevan for sharing some of his recent publications so that they may be included. Thanks go to Clark Atlanta University for allowing the time and resources to prepare this paper. Partial support came from NSF Grant CHE-9510484.

7
References

1. Haseman JF, Lehr JR, Smith JP (1950) Soil Science Society Proceedings 76
2. Sauchelli V (1965) The Chemistry of Phosphate Fertilizers, Reinhold, New York
3. Wilson ST, Lok BM, Messina CA, Cannan TR, Flanigen EM (1982) J Am Chem Soc 104:1146
4. Lok BM, Messina CA, Patton LR, Gajek RT, Cannan TR, Flanigen EM. (1984) J Am Chem Soc 106:6092
5. Richardson JW, Pluth JJ, Smith JV (1989) Naturwissenschaften 76:467
6. Meier WM, Groner M (1981) J Solid State Chem 37:204
7. Cartlidge S, Meier WM (1984) Zeolites 4:218
8. Szostak R (1997) unpublished work
9. Kniep R, Mootz D, Vegas A (1972) Acta Crystallog B29:2292
10. Kniep R, Mootz D, Vegas A (1977) Acta Crystallog. B33:263
11. Smith JV, Ribbe PH (1961) J Geol 74:197
12. Simpson DR (1977) Am Mineral 62:351
13. McConnell D, Verhoek FH (1963) J Chem Educ 40:512
14. Peacor DR, Rouse RC, Ahn J-H (1987) Am Mineral 72:816
15. Rouse RC, Peacor DR, Merlino S (1989) Am Mineral 74:1195
16. Moore PB, Shen J (1983) Nature 306:356
17. Szostak R, Kuvadia R, Brown JL, Thomas TL (1989) In: Jacobs PA, van Santen RA (eds) Zeolites: Facts, Figures, Future. Elsevier, Amsterdam, Stud Surf Sci Catal 49:439
18. d'Yvoire F (1961) Bull Soc Chim Fr 1762
19. d'Yvoire F (1961) Compt rend 252:414
20. d'Yvoire F (1958) Compt rend 247:297
21. Duncan B, Stöcker M, Gwinup D, Vinje K, Szostak R (1992) Bull Chem Soc Fr 129:98
22. Duncan B, Sorby K, Ulan JG, Szostak R (1990) Catal Lett 7:367
23. Duncan B, Szostak R (1994) unpublished results
24. Davis ME, Saldarriaga C, Montes C, Garces J, Crowder C (1988) Nature 129:98
25. McCusker LB, Baerlocher Ch, Jahn E, Bülow M. (1991) Zeolites 11:308
26. Davis ME, Montes C, Garces JM (1989) In: Occelli ML, Robson HE (eds) Zeolite Synthesis. ACS Symp Series, 398, Am. Chem. Soc Washington D.C, p 291
27. Davis ME, Montes C, Hathaway PE, Garces JM (1989) In: Jacobs PA, van Santen RA (eds) Zeolites: Facts, Figures, Future. Elsevier, Amsterdam, Stud Surf Sci Catal 49:199

28. Schmidt W, Schüth F, Reichert H (1992) Zeolites 12:2
29. Li H-X, Davis ME, Higgins JB, Dessau RM (1993) J Chem Soc Chem Commun 403
30. Wilson ST, Lok BM, Messina C.A, Flanigen E.M (1984) In: Olson DH, Bisio A (eds) Proc. 6th Int. Conference on Zeolites, Butterworths, Guildford, p 97
31. Stöcker M, Akporiaye D (1991) Appl Catal 69:L7
32. Maistriau L, Gabelica Z, Derouane E.G, Vogt ETC, van Oene J (1991) Zeolites 11:583
33. Kenny MB, Sing KSW, Theocharis CR (1992) J Chem Soc Faraday Trans 88:3349
34. Schmidt W, Schüth F, Reichert H, Unger K, Zibrowius B (1992) Zeolites 12:2
35. Prasad S, Gunjikar VG, Balakrishnan I (1991) Thermochemica Acta 191:265
36. Akporiaye DE, Fjellvag H, Halvorsen EN, Huag T, Karlsson A, Lillerud KP. (1996) J Chem Soc, Chem Commun 1553
37. Barrer RM, Marshall DJ (1965) J Chem Soc 6616
38. Barrer RM, Marshall DJ (1965) J Chem Soc 6621
39. Barrer RM, Liquornik M (1974) J Chem Soc Dalton Trans 2126
40. Flanigen EM, Grose RW (1971) In: Molecular Sieve Zeolites I. American Chemical Society, Washington D.C, Adv Chem Series 101:76
41. Wilson ST (1991) In: van Bekkum H, Flanigen EM, Jansen JC (eds) Introduction to Zeolite Science and Practice. Elsevier, Amsterdam, Stud Surf Sci Catal 58:137 (1991)
42. Newalkar BL, Jasra RV, Kamath V, Bhat TSG (1994) J Chem Soc Chem Commun 1041
43. Szostak R. (1997) Molecular Sieves: Principles of Synthesis and Identification, 2nd edn. Thompson, London
44. Wilson ST, Lok BM, Messina CA, Cannan TR, Flanigen EM (1983) In: Stucky GD, Dwyer FG (eds) Intrazeolite Chemistry. ACS Symp. Series, 218 American Chemical Society, Washington D.C, p 79
45. Batista J, Kaucic V, Hocevar S (1991) Croatia chemica acta 64:139
46. Wilson ST, Flanigen EM (1989) In: Occelli ML, Robson HE (eds) Zeolite Synthesis. ACS Symp Ser 398. American Chemical Society, Washington D.C., p 329
47. Zahedi-Niaki MH, Joshi PN, Kaliaguine S (1996) J Chem Soc Chem Commun 47
48. Zahedi-Niaki MH, Joshi PN, Kaliaguine S,(1996) J Chem Soc Chem Commun 1373
49. Testa R. Szostak R, unpublished results
50. Choudhary VR, Akolekar DB, Singh AP, Sansare SD (1988) J Catal. 111:254
51. Hampson B, Leach HF, Lowe BM, Williams DC (1989) Zeolites 9:521
52. Gelsthorpe MR, Theocharis CR (1986) J Chem Soc Chem Commun 781
53. Jansen JC, Arafat A, Barakat A.K, van Bekkum H (1992) In: Occelli ML, Robson H (eds) Synthesis of Microporous Materials. Van Nostrand Reinhold, New York, p 507
54. Afafat A, Jansen JC, Ebaid AR, van Bekkum H (1993) Zeolites 13:162
55. Girnus I, Jancke K, Vetter R, Richter-Mendau J, Caro J (1995) Zeolites 15:33
56. Du H, Fang M, Xu W, Meng X, Pang W (1997) J Mater Chem 7:551
57. Girnus I, Hoffmann K,, Marlow F, Caro J, Doring G (1994) Microporous Materials 2: 537
58. Bennet JM, Cohen JP, Flanigen EM, Pluth JJ, Smith JV (1983) In: Intrazeolite Chemistry American Chemical Society, Washington D.C, ACS Symp Series 218:109
59. Pluth JJ, Smith JV, Bennett JM (1986) Acta Crystallogr C42:283
60. Parise JB (1984) J Chem Soc Chem Commun 1449
61. Pluth JJ, Smith JV (1987) Acta Crystallogr C43:866
62. Goepper M, Guth JL (1991) Zeolites 11:477
63. Parise JB, Day CS (1985) Acta Crystallogr C41:515
64. Bennett JM, Cohen JM, Artioli G, Pluth JJ, Smith JV (1985) Inorg Chem 24:188
65. Simmens A, McCusker LB, Baerlocher Ch (1991) Zeolites 11:654
66. Bennett JM, Kirchner RM (1992) Zeolites 12:338
67. Parise JB (1985) In: Drzaj B, Hocevar S, Pejovnik S (eds) Zeolites, Synthesis, Structure, Technology and Application. Elsevier, Amsterdam, Stud Surf Sci Catal 24:271
68. Pluth JJ, Smith JV, Bennett JM, Cohen JP (1984) Acta Crystallogr C40:2008
69. Meinhold RH, Tapp NJ. (1990) J Chem Soc, Chem Commun 219
70. Barrie PJ, Smith ME, Klinowski J (1991) J Chem Phys Lett 180:6

71. Peeters MPJ, de Haan JW, van de Ven LJM, van Hooff JHC (1992) J Chem Soc Chem Commun 1560
72. Prasad S, Balakrishnan I, Vetrivel R (1992) J Phys Chem 96:3096
73. Peeters MPJ, de Haan JW, van de Ven LJM, van Hooff JHC (1993) J Phys Chem 97:5363
74. Keller EB, Meier WM, Kirchner RM (1990) Solid State Ionics 43:93
75. Martens JA, Feijen E, Lievens JL, Grobet PJ, Jacobs PA (1991) J Phys Chem 95:10025
76. Vogt ETC, Richardson JW (1990) J Solid Sate Chem 87:469
77. Akporiaye DE, Stöcker M (1993) In: Von Ballmoos R, Higgins JB, Treacy MMJ (eds) Proceedings 9th Int. Zeolite Conf, Butterworth-Heinemann, Stoneham, p 563
78. Martens JA, Geerts H, Grobet PJ, Jacobs PA (1992) In: Derouane EG, Lemos F, Naccache C, Ribeiro FR (eds) Zeolite Microporous Solids: Synthesis, Structure, and Reactivity. NATO ASI Ser. C, Kluwer, Dordrecht, 352:477
79. Vinje K, Ulan J, Szostak R, Gronsky R (1991) Appl Catal 72:361
80. Sorby K, Szostak R, Ulan JG, Gronsky R (1990) Catalysis Lett 6:209
81. Jelinek R, Chmelka BF, Wu Y, Grandinetti PJ, Pines A, Barrie PJ, Klinowski J (1991) J Am Chem Soc 113:4097
82. Kessler H, Patarin J, Schott-Darie C (1994) In: Jansen C, Stöcker M, Karge HG, Weitkamp J (eds) Advanced Zeolite Science and Applications, Jin Studies in Surface Science and Catalysis 85:75
83. Szostak R, Duncan B, Aiello R, Nastro A, Vinje K, Lillerud K-P (1992) In: Occelli ML, Robson H (eds) Synthesis of Microporous Materials, V. 1. Van Nostrand Reinhold, New York, p 241
84. Qui S, Pang W, Kessler H, Guth JL (1989) Zeolites 9:440
85. Halvorsen EN (1996) PhD Thesis, University of Oslo, Norway
86. Oliver S, Kuperman A, Lough A, Ozin G.A (1997) J Mater Chem 7:807
87. Paillaud JL, Marler B, Kessler H (1996) J Chem Soc, Chem Commun 1293
88. Kresge CT, Leonowicz ME, Vartuli JC, Beck JS (1992) Nature 359:710
89. Huo Q, Margolese DI, Ciesla U, Feng P, Gier TE, Sieger P, Leon R, Petroff PM, Schuth F, Stucky GD (1994) Nature 368: 317
90. Monnier A, Schüth F, Huo Q, Kumar D, Margolese D, Maxwell RS, Stucky GD, Krishnamurthy M, Petroff P, Firouzi A, Janicke M, Chmelka BF (1993) Science 261:1299
91. Vartuli JC, Schmitt KD, Kresge CT, Roth W. , Leonowicz ME, McCullen SB, Hellring SD, Beck JS, Schlenker JL, Olson DH, Sheppard EW (1994) Chem Mater 6:2317
92. Jones RH, Thomas JM, Xu R, Huo Q, Cheetham A.K, Powell AV (1991) J Chem Soc, Chem Commun 1266
93. Morgan K, Gainsford G, Milestone N (1995) J Chem Soc Chem Commun 425
94. Bruce DA, Wilkinson AP. White MG, Bertrand JA (1995) J Chem Soc, Chem Commun 2059
95. Chippindale AM, Natarajan S, Thomas JM, Jones RH (1994) J Solid State Chem 11:18
96. Thomas JM, Johes RH, Xu R, Chen J, Chippindale AM, Natarajan S, Cheetham AK (1992) J Chem Soc Chem Commun 929
97. Jones RH, Chippindale AM, Natarajan S, Thomas JM (1994) J Chem Soc Chem Commun 565
98. Kraushaar-Czarnetzki B, Stork WHJ, Dogterom RJ (1993) Inorg Chem 32:5029
99. Barrett PA, Jones RH (1995) J Chem Soc, Chem Commun 1979
100. Chenite A, Le Page Y, Karra VR, Sayari A (1996) J Chem Soc, Chem Commun 413
101. Sayari A, Karra VR, Reddy JS, Moudrakovski IL (1996) J Chem Soc, Chem Commun 411
102. Sayari A, Moudrakovski I, Preston KF (1996) Chemistry of Materials 8:2080
103. Oliver S, Kuperman A, Coombs N, Lough A, Ozin GA (1995) Nature 378:47
104. Oliver S, Kuperman A, Ozin A (1996) Chem Commun 1761
105. Oliver S, Kuperman A, Lough A, Ozin GA (1996) Inorg Chem 35:6373
106. Chenite A, Le Page Y, Karra VR. Sayari A (1996) J Chem Soc Chem Commun 413
107. Huo Q, Leon R, Petroff PM, Stucky GD (1995) Science 268:1324
108. Zhao D, Luan Z, Kevan L (1997) J Chem Soc, Chem Commun 1009
109. Zhao D, Luan Z, Kevan L (1997) J Phys Chem B 101: (in press)

110. Riou D, Loiseau T, Ferey G (1993) J Solid State Chem 102:4
111. Renaudin J, Ferey G (1995) J Solid State Chem 120:197
112. Noble GW, Wright PA, Lightfoot P, Morris RE, Hudson KJ, Kvick A, Graafsma H (1997) Angew Chem Int Ed 36:81
113. Akporiaye DE, Fjellvag H, Halvorsen EN, Hustveit J, Karlsson A, Lillerud KP (1996) J Chem Soc Chem Commun 601
114. Marler B, Patarin J, Sierra L (1995) Microporous Mater 5:151
115. Wright PA, Jones RH, Natarajan S, Bell RG, Chen JS, Hursthouse MB, Thomas JM (1993) J Chem Soc Chem Commun 633
116. Gao Q, Li S, Xu R (1994) J Chem Soc Chem Commun 1465
117. Huo Q, Xu R. (1990) J Chem Soc Chem Commun 783
118. Huo Q, Xu R, Li S, Ma Z, Thomas JM, Jones RH, Chippindale AM (1992) J Chem Soc Chem Commun 875
119. Akporiaye DE, Fjellvag H, Halvorsen EN, Huag T, Karlsson A, Lillerud KP (1996) J Chem Soc Chem Commun 1553
120. Barrer RM, Baynham JW, Bultitude FW, Meier WM (1959) J Chem Soc 195
121. Xu R, Chen J, Feng S (1991) In: Inui I, Namba S, Tatsumi T (eds) Chemistry of Microporous Crystals. Elsevier, Amsterdam, Studies in Surface Science and Catalysis 60:63
122. Yang G, Feng S, Xu R, Huaxue (1988) J Struct Chem 7:235
123. Yang G, Feng S, Xu R (1987) J Chem Soc, Chem Commun 1254
124. Parwise JB (1986) Acta Crystallogr C42:144
125. Wang T, Yang G, Feng S, Shang C, Xu R (1989) J Chem Soc, Chem Commun 948
126. Parise JB (1986) Acta Crystallogr C42:670
127. Parise JB (1985) Inorg Chem 24:4312
128. Wilson ST, Woodward NA, Flanigen EM, Eggert HG (1987) Eur Patent Appl 226, 219
129. Merrouche A, Patarin J, Soulard M, Kessler H, Anglerot D (1992) In: Occelli ML, Robson H (eds) Zeolite and Pillared Clay Synthesis. Van Nostrand Reinhold, New York 365:645
130. Loiseu T, Ferey G (1992) J Chem Soc, Chem Commun 1197
131. Loiseau T, Ferey G (1993) Eur J Solid State Inorg Chem 30:369
132. Ferey G, Loiseau T, Lacorre P, Taulellè F (1993) J Solid State Chem 105: 179
133. Taulelle F, Loiseau T, Maquet J, Livage J, Ferey G (1993) J Solid State Chem 105:191
134. Loiseau T, Retoux R, Lacorre P, Ferey G (1994) J Solid State Chem 111:427
135. Ferey G, Loiseau T, Riou D (1994) Mater Science Forum 152/153:125
136. Cavellec M, Riou D, Ferey G (1994) Eur J Solid State Inorg Chem 31:583
137. Ferey G (1995) J Fluorine Chem 72:187
138. Loiseau T, Taulelle F, Ferey G (1996) Microporous Materials 5:365
139. Loiseau T, Ferey G (1994) J Solid State Chem 111:403
140. Estermann M, McCusker LB, Baerlocher C, Merroche A, Kessler H (1991) Nature 352:320
141. Ferey G, Loiseau T (1996) J Mater Chem 6:1073
142. Louiseau T, Ferey G (1992) J Chem Soc Chem Commun 1197
143. Louiseau T, Ferey G (1993) Eur J Solid State Inorg Chem 30:369
144. Loiseau T, Ferey G (1994) Eur J Solid State Inorg Chem 31:575
145. Moore PB (1972) Amer Mineral 57:397
146. Pluth JJ, Smith JV, Bennett JM, Cohen JP (1984) Acta Crystallogr C40:2008
147. Wang T, Yang G, Feng S, Shang C, Xu R (1989) J Chem Soc, Chem Commun 948
148. Mooney-Slater RCL (1966) Acta Crystallogr 20:526
149. Jones RH, Thomas JM, Xu R, Huo Q, Xu Y, Cheetha A, Bieber D (1990) J Chem Soc Chem Commun 1170
150. Tieli W, Long Y, Wengin P (1990) J Solid State Chem 89:392
151. Oliver S, Kuperman A, Lough A, Ozin GA (1996) Chem Mater 8:2391
152. Loiseau T, Serpaggi F, Ferey G (1997) J Chem Soc Chem Commun 1093
153. Serpaggi F, Loiseau T, Riou D, Ferey G (1994) Eur J Solid State Inorg Chem 31:595
154. Riou D, Ferey G (1994) Eur J Solid State Inorg Chem 31:605
155. Chippindale AM, Walton RI (1994) J Chem Soc Chem Commun 2453
156. Yu J, Chen J, Xu R (1996) Microporous Materials 5:333

157. Cowley AR, Chippindale AM (1996) J Chem Soc Chem Commun 673
158. Chippindale AM, Cowley AR (1997) Zeolites 18:176
159. Flanigen EM, Lok BMT, Patton RL, Wilson ST (1988) U.S. Patent 4,737,353
160. Harvey G, Meier WM (1989) In: Jacobs PA, Van Santen RA (eds) Zeolites: Facts, Figures, Future. Elsevier, Amsterdam, Studies in Surface Science and Catalysis 49:411
161. The faujasite zeolite X crystallizes at low temperatures (100 °C) from sodium aluminosilicate gels. Higher temperatures or longer crystallization times result in the formation of the gismondine structure
162. Gier TE, Stucky GD (1991) Nature 349:508
163. Harrison WTA, Gier TE, Moran KL, Nicol JM, Eckert H, Stucky GD (1991) Chem Mater 3:27
164. Nenoff TM, Harrison WTA, Nicol JM, Stucky GD (1992) Zeolites 12:770
165. Gier TE, Harrison WTA, Stucky GD (1991) Angew Chem 103:1191
166. Harrison WTA, Gier TE, Stucky GD (1993) Zeolites 13:242
167. Harrison WTA, Gier TE, Stucky GD (1991) J Mater Chem 1:153
168. Long Y, Wenqin P (1990) J Chem Soc, Chem Commun 787
169. Feng P, Bu X, Stucky GD (1997) J Am Chem Soc 119:2497
170. Chen J, Jones RH, Natarajan S, Hursthouse MB, Thomas JM (1994) Angew Chem Int Ed Engl 33:639
171. Soghomonian V, Chen Q, Haushalter RC, Zubieta J, O'Conner CJ (1993) Science 259:1596
172. Soghomonian V, Chen Q, Haushalter RC, Zubieta J (1993) Angew Chem Int Ed Eng 32:610
173. Nenoff TM, Stucky GD, Harrison WTA (1994) Zeit Kristallogr 209:892
174. Riou D, Ferey G (1994) J Solid State Chem 111:422
175. Haushalter RC, Mundi LA (1992) Chem Mater 4:31
176. Ferey G, Cavellec M, Riou D (1997) Inorg Chem 36:2187
177. Cavellec M, Riou D, Ferey G (1994) J Solid State Chem 112:441
178. Harrison WTA, Gier TE, Nicol JM, Stucky GD (1995) J Solid State Chem 114:249
179. Harrison WTA, Gier TE, Stucky GD, Broach RW, Bedard RA (1996) Chem Mater 8:145
180. Harrison WTA, Broach RW, Bedard RA, Gier TE, Bu X, Stucky GD (1996) Chem Mater 8:691

Synthesis and Characterization of Molecular Sieves Containing Transition Metals in the Framework

Giovanni Perego · Roberto Millini · Giuseppe Bellussi

Eniricerche S.p.A., Via F. Maritano 26, I-20097 San Donato Milanese, Italy

Molecular Sieves, Vol. 1

1 Introduction

Apart from applications as molecular sieves, zeolites were considered for many years only as acid catalysts and used for reactions performed at relatively high temperature, in the gas phase. Only after isomorphous replacement of transition elements for silicon and aluminum had been demonstrated, were red-ox properties of zeolites seriously considered. This achievement not only allows the opening of new perspectives for selective oxidations, but it also draws the attention toward the capabilities of zeolites to generate free radicals in the presence of even trace amounts of oxygen, affecting the catalyst performances also in acid catalyzed reactions [1].

In the early 1950s, Goldsmith reported the successful replacement of Si by Ge in Thomsonite [2]. A few years later, Barrer et al. reported the synthesis of Thomsonite, zeolite A, Faujasite and Harmotome containing Ga and/or Ge in the framework [3].

For catalytic purposes, transition metals are particularly attractive. However, up to the end of the seventies, many doubts existed about the possibility to incorporate transition metal ions in the framework. Probably the first attempt to synthesize molecular sieves with transition metal ions in the framework (formally, they are not zeolites since this term indicates microporous crystalline aluminosilicates only) was performed by Young in 1967, who claimed the hydrothermal synthesis of titano- and zirconosilicates having zeolitic properties [4]. However, the lack of crystallographic information made the real microporous nature of these materials questionable. An extensive approach to the problem of isomorphous substitution in zeolites began only in the early 1980s.

In the present paper, we review the main results obtained in the preparation of transition metal-containing molecular sieves. In particular, the attention will be focused on the synthesis procedures and the physico-chemical evidence supporting the effective framework incorporation of the heteroatom.

2 General Aspects of Transition Metal-Containing Molecular Sieves

The capability of a given ion M^{n+} to be incorporated into a molecular sieve framework is generally considered to be dependent on the ionic radii and on the tendency of the element to assume the tetrahedral coordination in its oxide. Ione et al. [5] systematically approached this problem by using the Pauling criterion, $\rho = r_c/r_o$ (where r_c and r_o are the cation and anion radii, respectively) [6]. Based on the values of ρ, the tetrahedral coordination should be stable only for cations with ρ in the range 0.414–0.225 (Table 1).

On the contrary, up to now incorporation has been well assessed for elements which do not satisfy the above rule (e.g., B, Ga, Ti, Fe, see Table 2). Probably, at low concentration of the metal (concerning most cases), the geometrical parameters become less important than other factors related to the crystallization conditions. For these reasons it is not easy to identify simple criteria for predicting the ability of a given element to be incorporated in the silica framework.

Table 1. Values of ρ_{cr} for various coordination polyhedra [6]

Coordination number	Polyhedron	Critical values (ρ_{cr})
3		$0.225 < \rho < 0.147$
4	Tetrahedron	$0.414 < \rho < 0.225$
6	Octahedron	$0.592 < \rho < 0.414$
7		$0.645 < \rho < 0.592$
8	Square antiprism	$0.732 < \rho < 0.645$
8	Cube	$0.732 < \rho < 0.645$
9		$0.732 < \rho < 0.645$
12	Cubo-octahedron	$1.000 < \rho < 0.732$

Table 2. Ions that can attain the tetrahedral coordination and corresponding value of ρ, calculated with the ionic radii reported in [7] and by considering a $r_{0=} = 1.35$ Å. (The ions for which the framework incorporation is well assessed are reported in bold)

Ion	ρ	Ion	ρ
B^{3+}	0.185	**Ge^{4+}**	0.393
Si^{4+}	0.296	Co^{4+}	0.400
Be^{2+}	0.304	Co^{2+}	0.533
Mn^{5+}	0.348	**Ti^{4+}**	0.415
Mn^{4+}	0.393	Mo^{5+}	0.444
Cr^{5+}	0.359	**Ga^{3+}**	0.452
Cr^{4+}	0.407	**Fe^{3+}**	0.467
V^{5+}	0.367	Fe^{2+}	0.570
V^{4+}	0.496	Zr^{4+}	0.541
Al^{3+}	0.393	**Zn^{2+}**	0.548

3
Titanium-Containing Molecular Sieves

Incorporation of titanium into high-silica frameworks has been the object of many studies, especially after the discovery of titanium-silicalite-1 (also known as TS-1) and of its exceptional catalytic properties in oxidation reactions involving hydrogen peroxide. Unique among the transition metal-substituted molecular sieves, TS-1 has become an industrial catalyst.

Other microporous titanosilicates with the MFI/MEL, ZSM-48, BETA and ZSM-12 framework structures have been synthesized but little evidence has been shown for these materials which demonstrates real incorporation of the transition metal into the framework.

Finally, we have to mention the existence of a small family of natural titanosilicate molecular sieves including Zorite [8] and Vinogradovite [9], in which Ti assumes octahedral coordination.

In the following, the attention will be focused mainly on the synthesis and characterization of TS-1.

3.1
Titanium-Silicalite-1 (TS-1)

3.1.1
Synthesis

The crucial problem in the synthesis of titanium-containing molecular sieves is to find proper conditions for obtaining isolated tetrahedrally coordinated titanium species. In fact, in the presence of even small amounts of water, the titanium compounds used in the reaction mixture undergo hydrolysis very quickly and then polymerize and precipitate in the form of hydrated TiO_2. The large amount of work performed on TS-1 allows some of the critical parameters of the synthesis procedure to be explained, though the overall mechanism is not yet fully understood.

The preparation method, firstly reported by Taramasso et al. [10] in 1983, is based on the controlled hydrolysis of an aqueous solution containing tetraethylorthosilicate (TEOS), tetraethylorthotitanate (TEOT) and tetrapropylammonium hydroxide (TPAOH) as template. This procedure is known as the "mixed alkoxide" method. Another procedure has been proposed, known as the "dissolved titanium" method, which makes use of colloidal silica instead of TEOS [11]. The mixed alkoxide method allows the best control of the synthesis parameters, assuring the crystallization of high quality products.

3.1.1.1
The "Mixed Alkoxide" Method

The procedure, described in Fig. 1, involves the hydrolysis of a TEOS/TEOT mixture by addition of aqueous TPAOH. Once the hydrolysis is completed, the gel is charged into a stainless steel autoclave and heated at 175 °C for 10 days.

Bellussi and Fattore identified the SiO_2/TiO_2 ratio in the reaction mixture, the alkali content and the crystallization temperature as the main parameters affecting the characteristics of TS-1 [12].

Titanium-silicalite-1 crystallizes with a titanium content lower than that present in the precursor mixture (Fig. 2). The maximum Ti content achievable corresponds to a SiO_2/TiO_2 molar ratio of ca. 40; any attempts to overcome this limit failed, the excess of titanium having been found in the form of extra-framework TiO_2 (anatase). Crystallization of pure TS-1 occurs between 100 and 200 °C; above 200 °C, formation of anatase occurs (Fig. 3).

Though some authors claim the possibility to synthesize TS-1 in the presence of sodium ions [13–15], it has been demonstrated that even trace amounts of alkali metal ions in the reaction mixture (sodium and/or potassium impurities of commercial TPAOH solutions), prevents or limits to some extent the incorporation of titanium into the framework [16–19]. Adding NaOH to the solution of alkali-free tetrapropylammonium hydroxide results in increasing overall Ti content but decreasing Ti incorporation in the solid recovered. At high sodium concentration, anatase was detected by X-ray diffraction analysis together with TS-1. This indicates that alkali cations may promote the formation of insoluble

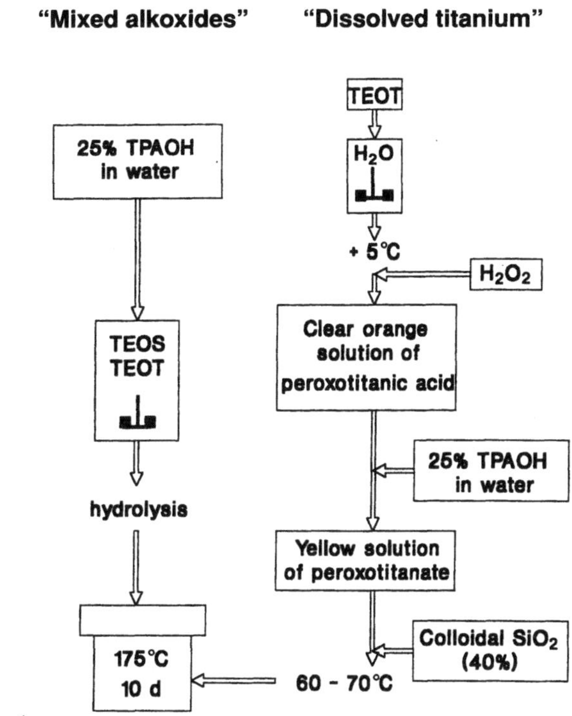

Fig. 1. Synthesis procedures for TS-1

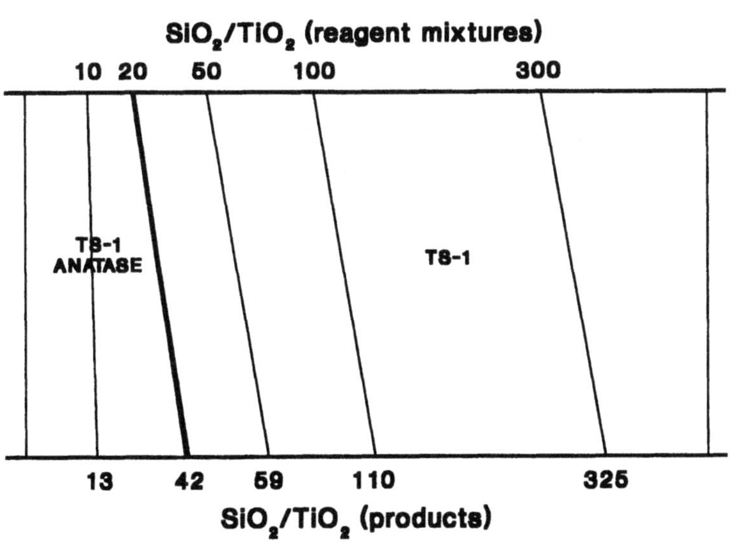

Fig. 2. Influence of the SiO_2/TiO_2 molar ratio in the reagent mixture on the composition of TS-1 [12]

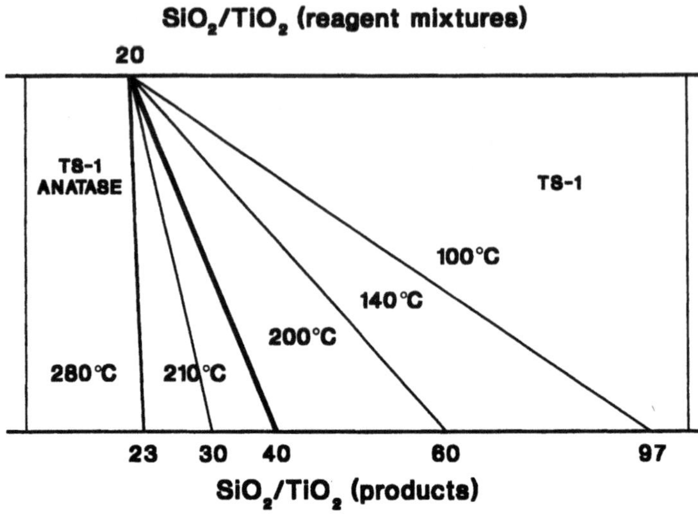

Fig. 3. Influence of the crystallization temperature on the composition of TS-1 [12]

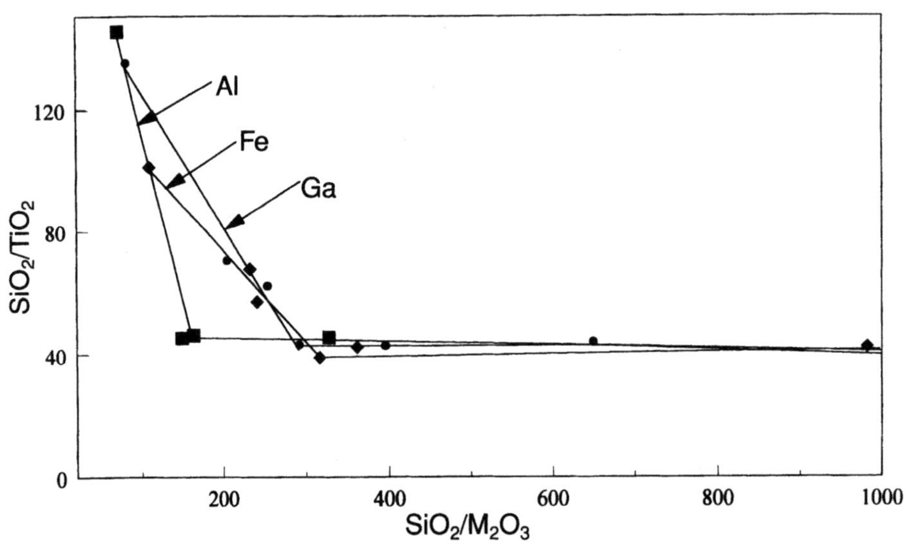

Fig. 4. Influence of trivalent elements on the composition of (Al, Ga, Fe)-TS-1 [12]

titanium species [12]. Alkali metal ions may also alter the equilibrium of silicate species formed during hydrolysis and condensation, through the complexation of terminal Si-O-groups [20].

A rather different behavior was observed when trivalent metal cations such as B^{3+}, Al^{3+}, Ga^{3+} and Fe^{3+} were added to the reaction mixture. These ions partially suppress Ti incorporation (Fig. 4) but do not favor any segregation of extra-framework titanium species [12]. This made it possible to synthesize bifunctional catalysts such as [Ti,Al]-MFI, containing both acid [AlO_4] and oxidation [TiO_4] sites [19,21].

As stated above, a maximum of about 2.5 atoms per unit cell (x=[Ti]/([Ti]+[Si])=0.025) has been achieved. Starting from the assumption that this limit is determined by kinetic rather than thermodynamic reasons, much effort has been made attempting the synthesis of TS-1 with higher Ti content. Following the original recipe, van der Pol and van Hooff studied the influence of synthesis parameters (silica and TPAOH sources, OH^-/Si and Si/Ti ratios, sol concentration, crystallization conditions and time) on morphological and compositional characteristics of TS-1 [22]. They claim to have obtained TS-1 samples with x up to ca. 0.04, supporting this conclusion with XRD and IR evidence. According to Kraushaar, the addition of aqueous TPAOH represents the most critical step, because it gives rise to instantaneous hydrolysis of TEOT with possible formation of insoluble and stable titanium species. Precipitation of these species can be avoided if the organic template solution is added slowly and there are enough silanol groups for the condensation reaction to occur with the monomeric hydrolyzed titanium species (in other words, the hydrolysis rates of silica and titania sources are comparable) [23]. Following Thangaraj et al. the limit of x=0.025 arises from the fact that TEOT is hydrolyzed much faster than TEOS. For this reason they proposed the replacement of TEOT by the less hydrolizable tetrabutylorthotitanate (TBOT), claiming the possibility to synthesize pure TS-1 with x up to 0.09 [24–27]. For avoiding instantaneous hydrolysis of the titania source, TBOT is dissolved in dry isopropyl alcohol prior to addition to the TEOS/TPAOH solution; the complete hydrolysis of the mixture is carried out in an open vessel and the gel crystallized under autogeneous conditions [24–27]. More recently, Tuel and Ben Taarit performed a systematic study using different couples of tetraalkylorthosilicates and orthotitanates, reaching the conclusion that the extent of Ti incorporation in TS-1 effectively depends on both titania and silica source used; the higher extent (x=0.05) is obtained using tetramethylorthosilicate and tetrapropyl- or tetrabutylorthotitanate [28]. In order to avoid precipitation of titanium, the use of aqueous $TiCl_3$ as titania source has been recently proposed [29]. The material crystallized was found to have catalytic activity similar to that displayed by TS-1 synthesized by classical methods, but no information is reported concerning the maximum extent of titanium incorporation achievable. These results originated a wide debate (still open today) about the real Ti incorporation, leading to the hypothesis of different titanium sites in the framework (see Section 3.1.2).

Little information is reported on the use of organic template agents different from TPAOH. A TBAOH/TEAOH mixture [30] and hexapropyl-1,6-hexanediammonium hydroxide (di-TPAOH) [31] were used; the characteristics of the products are reported to be the same as those obtained with TPAOH. In preparation runs with mixtures of TPAOH/TEAOH and TPAOH/NH_4OH the second template agent is claimed to influence the crystallization kinetics without being occluded in the pores [32].

3.1.1.2
The "Dissolved Titanium" Method

This method is more complicated than the mixed alkoxide technique (Fig. 1). Hydrogen peroxide is added to the titanium-containing solution in order to form stable peroxo-complexes able to release slowly the metal ion during the hydrothermal treatment. The gel composition, the hydrothermal conditions and the characteristics of the products are similar to those described for the mixed alkoxide method.

3.1.1.3
Other Methods

Guth et al. [33, 34] and Qiu Shinlun et al. [35] proposed starting from a reaction gel containing fluoride ions at low pH (6.5–7.5). No evidence is reported about Ti incorporation in the materials obtained by this method. Kooyman et al. observed anatase impurity in TS-1 samples obtained by this synthesis route [36] while the formation of octahedral extra-framework titanium was confirmed by EXAFS experiments [37].

Padovan et al. [38, 39] developed a method based on the impregnation of dried microspherical porous silica up to incipient wetness with an aqueous solution obtained by hydrolyzing the titania source (tetraisopropylorthotitanate) with TPAOH, followed by calcination at 448 K in a sealed glass tube. Pure and well crystallized TS-1 is obtained after 10 hours, while extra-framework titanium species start to appear after longer crystallization times.

An improvement of this method was recently proposed, which involves the wetness impregnation of a SiO_2-TiO_2 cogel, prepared by an acid-base two-step process, with aqueous TPAOH, followed by crystallization at 170 °C for 24 hours [40–42]. The Ti content of TS-1 is controlled by the SiO_2-TiO_2 cogel composition; the upper Ti concentration limit, x, in the final product was found to be 0.023.

3.1.2
Structure Characterization

The assessment of true incorporation of the heteroatom in a zeolite framework assumes paramount importance. In fact, the physico-chemical and catalytic properties of the materials strictly depend on the location of heteroatoms in framework sites. This is particularly true for titanium: it is well known that a dramatic difference in catalytic performance exists between a well synthesized TS-1 sample and one containing extra-framework titanium species.

The accurate structure characterization of transition metal-containing molecular sieves is made difficult by the generally low heteroatom concentration. Due to the practical impossibility of determining the site geometry by direct structural methods, spectroscopic studies, together with the application of theoretical methods, have been widely used (Table 3). Actually, a lot of data exists, mainly concerning TS-1. In the following, we will try to summarize these results,

Table 3. Structural and spectroscopic techniques used for characterizing Ti-containing molecular sieves

Technique	Results	Ref.
XRD	Quantification of framework and extra-framework Ti	[45, 46]
FT-IR	Detection and coordination of framework Ti	[45, 47]
RAMAN	Detection of extra-framework TiO_2	[47]
DRS UV-Vis	Quantification of extra-framework Ti; coordination of framework Ti	[47]
EXAFS/XANES	Coordination of Ti	[59, 61]
EPR	Coordination of Ti; qualitative estimation of framework and extra-framework Ti species	[52, 55]

bearing in mind that the characterization methods and the experimental evidence achieved for TS-1 are applicable to the other Ti-containing molecular sieves as well.

Titanium silicalite-1 is characterized by an X-ray powder diffraction (XRD) pattern typical of orthorhombic MFI-type zeolite (Pnma space group). Only at low Ti content ($x < 0.01$) is a monoclinic $P2_1/n$ symmetry observed, a phenomenon already observed for the aluminum (ZSM-5 [43]) and boron parent structures (H-BOR-C [44]).

Independent of the lattice symmetry, a linear dependence of the lattice parameters (determined by the least-squares fit to the interplanar spacing of selected reflections in the XRD pattern [45] or by the more accurate full-profile fitting analysis [46]) on the Ti content has been found (Fig. 5). The equation relating the unit cell volume to the Ti content (Table 4) is particularly useful for determining the real framework composition directly from XRD analysis; by comparing this with the Ti content resulting from elemental analysis, the amount of possible extra-framework Ti species can be estimated [46].

Because the linear dependence of lattice parameters on framework composition indicates that TS-1 follows a Vegard's type law, the lattice expansion is expected to derive from the larger Ti-O bond length, d_{Ti-O}, with respect to the Si-O bond length, d_{Si-O}. According to that which was proposed for the boron-containing MFI (H-BOR-C [44]), the unit cell volume, V_x, of TS-1 containing a molar fraction, x, of Ti can be related to the unit cell volume of the pure silica end-member, silicalite-1, V_{Si}, through the equation:

$$V_x = V_{Si} - V_{Si}[1 - (d_{Ti-O}/d_{Si-O})^3]\,x \tag{1}$$

By applying Eq. (1) to the results of Fig. 5D and by attributing to d_{Si-O} the value typically observed in zeolites, 1.61 Å, a d_{Ti-O} of 1.80 Å is obtained [46].

A linear expansion of lattice parameters was reported also for TS-1 samples with $x \gg 0.025$, but with a lower slope (dV/dx) (Fig. 6) [22, 24–28]. In fact, the unit cell volumes reported for $x = 0.039$ (5389.4 Å^3) [22], for $x = 0.091$ (5396.5 Å^3) [24] and for $x = 0.056$ (5389 Å^3) [28] are very close to the value observed by Millini et al. for a Ti sample with $x = 0.025$ (5392.3 Å^3) [46]. Furthermore, by applying Eq. (1) to the data reported in Refs. [22, 24–28], a d_{Ti-O} of 1.67–1.73 Å is obtained.

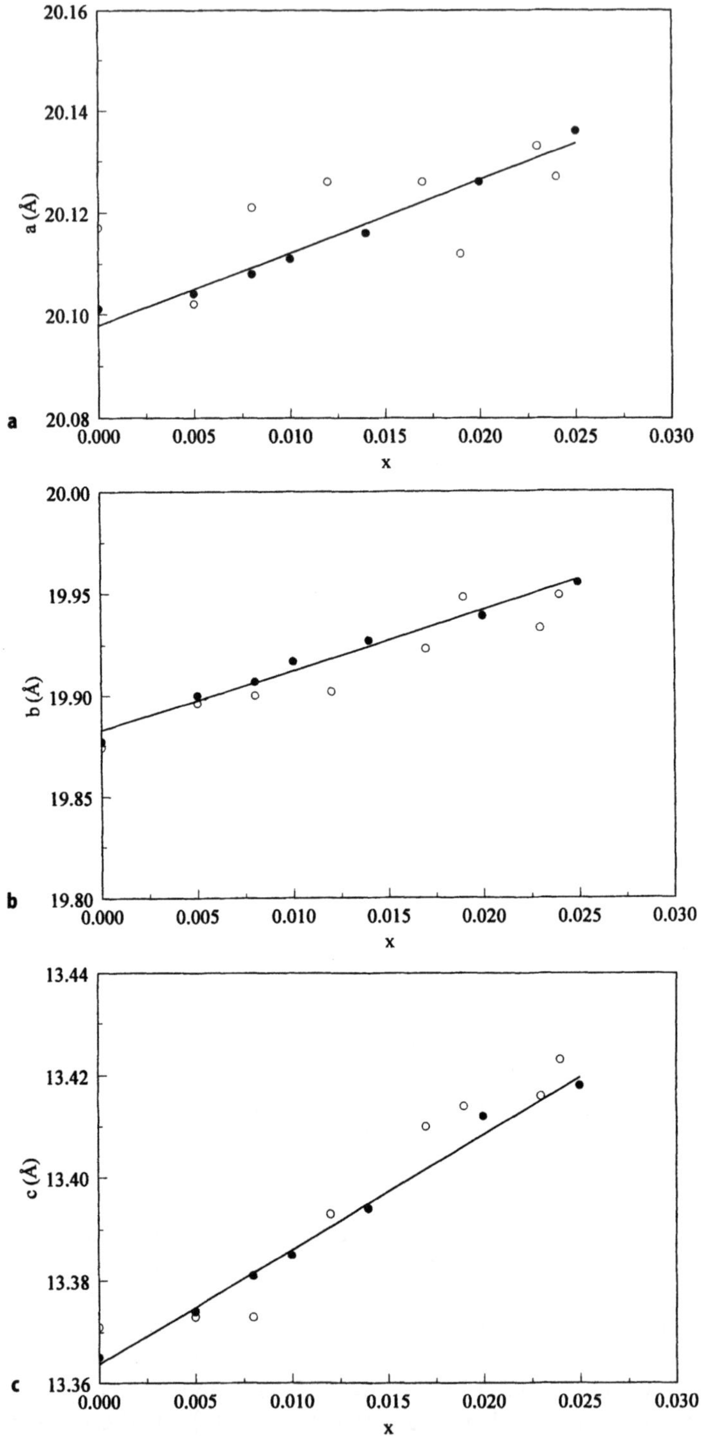

Fig. 5 a–d. Unit cell parameters and volume variation as a function of Ti content, x = [Ti]/([Ti] + [Si]). (○) data from [45], (●) from [46]

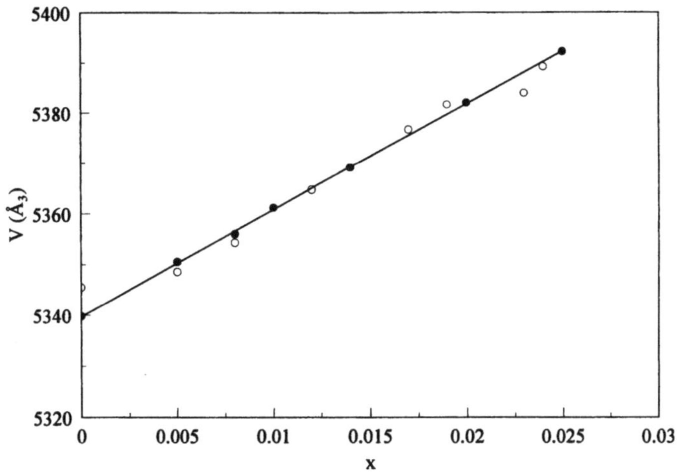

Fig. 5 d. (continued)

Table 4. Linear regression analysis of unit cell parameters in TS-1 [a]

	Ref. [45]			Ref. [46]		
	u	v	r	u	v	r
a	0.665	20.112	0.5854	1.429	20.098	0.9868
b	2.935	19.876	0.9487	2.975	19.882	0.9924
c	2.425	13.364	0.9665	2.281	13.363	0.9957
V	1946	5341.8	0.9874	2210.4	5339.4	0.9993

[a] According to the equation: $y = ux + v$; r, correlation coefficient.

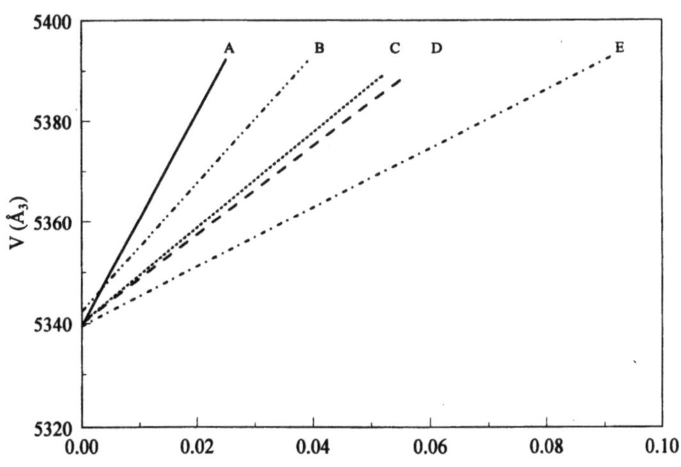

Fig. 6. Unit cell volume variation as a function of Ti content reported by different authors: (A) [46], (B) [22], (C) and (D) [28], (E) [24–27]

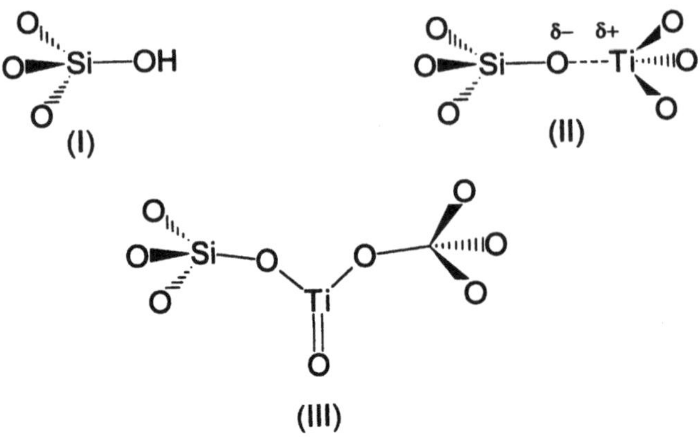

Fig. 7. Proposed assignments for the 960 cm^{-1} IR absorption [47]

On the basis of these results, Tuel and Ben Taarit suggested the existence of different Ti-sites in TS-1 [28]. This topic will be discussed in the following.

A characteristic absorption is observed in the IR spectra of TS-1, located at 960–970 cm^{-1}. This band, not observed in pure silica and aluminum-substituted zeolites, is usually regarded as a "fingerprint" of TS-1 and, more generally, of the Ti-containing molecular sieves. The absorption intensity has been found to be proportional to the Ti molar fraction, x [45], though dependent to some extent on the degree of hydration [47] and even on crystal size [48]. For this reason, it is not suitable for a careful determination of framework titanium. The assignments proposed for this absorption [47] are represented in Fig. 7. The assignment to silanol groups (I), known to absorb at 985 cm^{-1} in amorphous silica [49], may be excluded because the absorption frequency is not sensitive to deuterium exchange [48]. The assignment to the Si-O stretching of the polarized Si-$O^{\delta-}$-$Ti^{\delta+}$ bond (with Ti in tetrahedral coordination) (II) was proposed by Boccuti et al. [47]. The 960 cm^{-1} band is present also in the Raman spectrum of pure TS-1 as well as in Ti-containing cristobalite and silica glass [50], the latter being known to contain tetrahedral titanium [51]. The assignment (II) accounts well for the sensitivity of the band frequency to the treatment of TS-1 with ^{17}O and ^{18}O labeled water, according to the variation of reduced mass of the Ti-O-Si harmonic oscillator, after substitution of ^{16}O with heavier isotopes [48].

A more detailed interpretation of this absorption was performed by Zecchina et al. [52]. The authors depicted two different situations, which depend on the equivalence or non-equivalence of the force constants of the Si-O and Ti-O bonds. In the first case, the replacement of Ti for Si induces a shift of the B mode of vibration of the $[TO_4]$ units from 1120 (T=Si) to 960 cm^{-1} (T=Ti), the shift being simply due to mass effects. In the second case a higher polarity is considered for the Ti-O bond, leading to the conclusion that the 960 cm^{-1} absorption is due to the local mode of a $[SiO_4]$ unit (large difference in polarity between the two bonds) or to the stretching modes of the $[Ti(OSiO_3)_4]$ cluster (small difference in the bond polarity) [52].

The assignment to titanyl groups (III) would be consistent with the above evidence coming from treatment with labeled water but not with the observed features of the DRS UV-Vis spectrum of TS-1. In fact, a charge transfer (C.T.) phenomenon involving a Ti=O bond is expected to occur at 30000–35000 cm^{-1} [47]. Really, the only absorption observed in the electronic spectrum of pure and anhydrous TS-1 (vacuum dried sample) is located at 48000–50000 cm^{-1}, which is consistent with a C.T. transition involving [TiO_4] or [O_3Ti-OH] moieties [47]. Extra-framework Ti species may be quantified by measuring the absorption bands located at 42000 cm^{-1} (isolated hexa-coordinated Ti) and 30000–35000 cm^{-1} (crystalline TiO_2 phases), with a limiting concentration of about 0.03 wt%.

Tetrahedral coordination of Ti is peculiar of anhydrous TS-1. When TS-1 is exposed to H_2O, NH_3 or CH_3OH vapor [47] or immersed in H_2O [52, 53], Ti assumes hexa-coordination. The evidence comes from the electronic spectra, showing the C.T. band shifted to 43000 cm^{-1}, as a consequence of the more ionic character of hexa-coordinated Ti compared to the tetrahedrally coordinated Ti.

Both IR and DRS UV-Vis spectra of TS-1 are sensitive to H_2O_2. The 960 cm^{-1} band disappears in the spectrum of the yellow colored adduct formed by treating TS-1 with aqueous H_2O_2 while being replaced by a weak shoulder in the 900–1000 cm^{-1} region [54, 55]. Upon thermal decomposition of the adduct, the original absorption is almost completely recovered. An absorption at 26000 cm^{-1} appears in the DRS UV-Vis spectrum for the H_2O_2 adduct, while the intensity of the original C.T. band decreases [53, 54]. The same band was observed for $(NH_4)_2[Ti(O_2)F_5]$ adsorbed on silicalite-1, confirming that the reaction of H_2O_2 with TS-1 leads to the formation of hydroperoxo species bonded to framework Ti [53].

When the H_2O_2 treatment is made in basic media, a light yellow (or white) stable material is recovered, which was found to be an anionic peroxo derivative stabilized by the presence of the alkali metal ion. The adduct does not display oxidation properties, unless the pH of the suspension is brought to neutrality [55].

Tuel et al. performed an EPR investigation of TS-1, after reduction of Ti(IV) to the paramagnetic Ti(III) [56], compared to a Ti(III)-exchanged H-ZSM-5. The authors concluded that Ti exists in tetrahedral coordination in TS-1, completely different from that in the exchanged sample, typical of Ti(III) in distorted octahedral coordination. The same approach has been proposed as a qualitative tool for discriminating between framework and extra-framework Ti species [52].

Important contributions come from the application of X-ray absorption spectroscopy (EXAFS/XANES). Comparing the XANES spectra of dehydrated TS-1 with those of several model compounds, Behrens et al. concluded that TS-1 contains tetrahedral, square pyramidal and octahedral Ti, in the approximate ratio 10:15:75 [57, 58]; the same authors did not exclude that at ambient conditions all the Ti atoms assume the octahedral coordination [57]. However, these results are affected by the poor quality of the TS-1 sample used, probably containing the majority of the Ti in extra-framework positions, as indicated by the reported value of the unit cell volume (lower than that of the pure silica parent structure) [57, 58].

The quality of the material is fundamental for a reliable EXAFS analysis to be performed. This was the main reason for the difficulties encountered by Schultz et al. in trying to fit the spectra of TS-1 samples synthesized both by direct and secondary synthesis; as a matter of fact, the Ti-O bond length could be determined ($d_{Ti\text{-}O}$ = 1.85 Å) but not the coordination number [59].

More accurate EXAFS/XANES characterization on pure and well characterized TS-1 samples was recently performed independently by two research groups [60–62].

The comparison between the XANES spectra of TS-1 and anatase allowed the presence of octahedrally coordinated Ti to be excluded, at least in anhydrous samples.

From the EXAFS spectra of dehydrated samples, indications were achieved for a first coordination shell of Ti consisting of O atoms only, a Ti-O bond distance (R_{Ti-O}) of 1.80–1.81 Å and a coordination number (N_{Ti-O}) close to 4. Both R_{Ti-O} and N_{Ti-O} were found to increase when the product was in the presence of H_2O or NH_3 as a consequence of the coordination of the polar molecules to Ti, in agreement with the evidence from DRS UV-Vis data [61, 62].

An important result concerns the value of R_{Ti-O}, which is in very good agreement with the average Ti-O bond length observed in several four-coordinated Ti compounds [60] and with that derived from the analysis of the XRD data [46] (see above).

Based on a detailed analysis of the EXAFS spectra of TS-2 (see Section 3.2), Trong On et al. proposed novel Ti site structures, involving tetrahedrally coordinated Ti species sharing [SiO_4] tetrahedra and forming bridges across the zeolite channels (Fig. 8) [63, 64]. This should be considered a structural defect rather than a real isomorphous substitution, since it involves a disruption of the tectosilicate framework. According to the authors, such a site structure would impose less strain to the framework while justifying the lower lattice expansion observed for the Ti-rich TS-1 samples prepared according the procedure of Thangaraj et al. [24–27] and Tuel and Ben Taarit [28] with respect to that observed for the TS-1 samples prepared by the original recipe. However, the existence of these Ti site structures was confuted by Pei et al. [60], the hypothesis of Trong On et al. being based on an erroneous interpretation of EXAFS spectra.

3.1.3
Theoretical Studies

In spite of the large interest in Ti-containing molecular sieves, the application of theoretical methods to the problem of Ti siting (i.e., Ti site geometry and preferential substitution sites) has been made only recently. In fact, most of the studies concerning MFI-type molecular sieves have concerned the aluminum-containing parent structure (ZSM-5).

Jentys and Catlow applied defect energy minimization techniques to the problem of Ti substitution in a monoclinic MFI-type framework [65]. The conclusion was that no evidence exists for preferential substitution sites while the minimized Ti-O bond distance, d_{Ti-O}, was found to be 1.77–1.78 Å. In the same paper, the Ti site proposed by Trong On et al. [63, 64] (site I, Fig. 8) was taken into account, optimizing the geometry by "ab initio" quantum mechani-

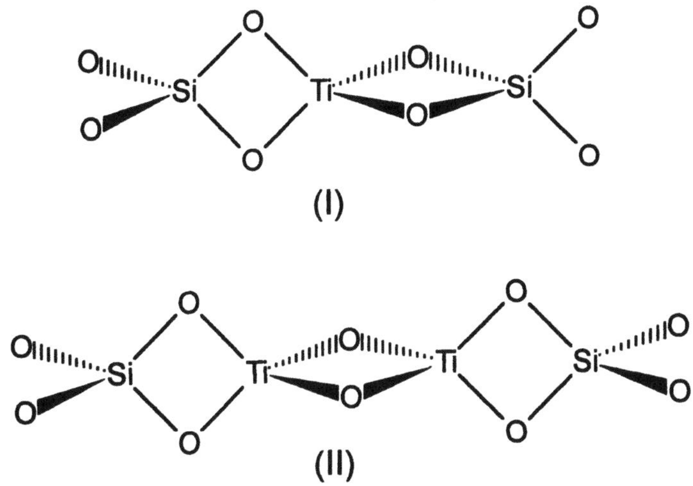

Fig. 8. Ti site structures proposed by Trong On et al. [63, 64]

cal methods. The conclusion drawn by the authors is that framework incorporation is preferred over the formation of species with Ti in edge-sharing or bridging position.

In another paper, an ab initio quantum mechanical approach was followed [66]. The Local Density Functional (LDF) calculations, performed on open pentameric $[T(OSiO_3H_3)_4]$ clusters (T=Si or Ti) with the geometry derived from the crystal structure of orthorhombic MFI zeolite [67], gave the following results:

1. The minimum energy of the Ti-clusters corresponds to a d_{Ti-O} of 1.80 Å, in agreement with the corresponding value derived experimentally (see Section 3.2)
2. No preferential substitution sites exist for Ti, in agreement with the conclusions of Jentys and Catlow [65].

Finally, molecular dynamics calculations were performed by Oumi et al. in order to obtain information about the anisotropic lattice expansion experimentally observed for TS-1 [68]. Using a simple two-body interatomic potential, they predicted an equilibrium d_{Ti-O} of 1.85 Å, slightly higher than that observed both experimentally and theoretically. Comparing the effect of Ti substitution on the lattice expansion over the twelve independent T-sites, the authors concluded that T8 is the most probable site for Ti incorporation.

3.2 Synthesis and Characterization of Other Titanium-Containing Molecular Sieves

The early reports on this topic appeared in the patent literature for titanium-containing MOR, SOD and MTW. However, the claims were not supported by clear evidence about the structural features of titanium [69].

Incorporation of titanium in the framework of silicalite-2 (all-silica ZSM-11, MEL) received much attention, due to the close similarity between the structures of silicalite-1 and -2 [70–72]. In fact, Perego et al. demonstrated that the framework of a boron-substituted ZSM-11 (BOR-D [73]) does not have the MEL topology but is really formed by a stacking sequence of inversion-center-related layers (typical of the MFI topology) interrupted by stacking faults corresponding to mirror-related layers (typical of the MEL topology) [71]. The structural model adopted, based on a random occurrence of stacking faults with a 25% probability, accounts well for the features present in the XRD pattern of BOR-D, silicalite-2 and ZSM-11.

The synthesis of titanium-silicalite-2 (TS-2) was firstly claimed by Bellussi et al. by using tetrabutylammonium hydroxide (TBAOH) as a structure-directing agent [74]. Subsequently, Reddy et al. [75, 76] adopted the procedure developed by Thangaraj et al. for the synthesis of TS-1 [24–26]; the authors claim the possibility to synthesize TS-2 with a wide range of SiO_2/TiO_2 molar ratio (17–∞) with a maximum of 4 Ti atoms per unit cell in the solid [76]. As for TS-1, the presence of alkali metal ions in the synthesis mixture favors the formation of extra-framework titanium species.

Tetrabutylammonium hydroxide seems to be the only template agent able to form TS-2. Other organic molecules, such as 1,8-diaminooctane and trimethylbenzylammonium hydroxide, which are known to behave as structure-directing agents for the aluminum-analog (ZSM-11), are unable to promote the formation of TS-2 [76].

A family of titanium-containing molecular sieves with pentasil-type framework structures is represented by titanium-silicalite-3 (TS-3) [74], characterized by a framework topology similar to that of TS-2 but with a different degree of stacking faults. In fact, as found by Perego et al. for the boron-substituted MFI/MEL molecular sieves, the frequency of stacking faults may be controlled by choosing the appropriate pair of tetraalkylammonium ions (e.g., TMAOH/TPAOH, TMAOH/TBAOH, TEAOH/TBAOH) [72].

In 1992, Serrano et al. [77] reported the synthesis of the titanium derivative of ZSM-48 starting from a gel containing diaminooctane as a structure directing agent and peroxytitanates and fumed silica as titania and silica sources, respectively. The minimum Si/Ti molar ratio reported was 26.1 ($x=0.037$), without formation of extra-framework titanium species [77]. A recent contribution concerning Ti-ZSM-48 confutes the above data, indicating the upper limit of $x=0.015$ [78]. This limit was tentatively attributed to the inefficient reaction between fumed silica and peroxotitanate, as the latter undergoes hydrolysis upon addition of the diaminooctane solution. Hexamethonium hydroxide was also used as a structure-directing agent for the synthesis of Ti-ZSM-48, with TBOT and TEOS as titania and silica sources, respectively. The Ti incorporation seems to be more difficult compared to the former method since in a sample containing 1 wt.-% titanium ($x=0.007$) extra-framework TiO_2 was detected [79].

The incorporation of 1.5 titanium atoms per unit cell ($x=0.031$, the highest value ever reported for Ti-ZSM-48), without formation of extra-framework species, was obtained in a synthesis procedure recently developed, involving the use of trimethylpropylammonium hydroxide (TMPAOH) [80]. The materials synthesized according to this procedure showed remarkable catalytic activity in the oxidation of ethanol with H_2O_2, while those prepared according the original recipe did not show any activity in the hydroxylation of phenol [78].

The paragraphs above describe the synthesis of medium pore molecular sieves. Many attempts have been made also to synthesize large pore molecular sieves (ring openings with twelve or more members).

Titanium-Beta was synthesized by Camblor et al. using the "dissolved titanium" method without addition of H_2O_2 [81–83]. It is worthy to note that such a molecular sieve can be crystallized only in the presence of aluminum so that it is really a ternary [Si,Al,Ti] compound. Lowering the concentration of aluminum in the reaction gel results in an increasing titanium content in the solid, the upper limit corresponding to 2.4 Ti per unit cell (x=0.038) with a minimum aluminum content of 0.74 Al per unit cell (x=0.011) [82]. Again, the presence of alkali metal ions in the synthesis gel favors the precipitation of amorphous titanosilicate species, hampering Ti incorporation. Ti-Beta is an efficient catalyst in the oxidation of large organic molecules with H_2O_2 [82, 83].

The large pore Ti-ZSM-12 (or TS-12) has been recently prepared by Tuel by adopting a procedure based on the mixed alkoxide method [84]. TEOS and TBOT were used as silica and titania sources, respectively, while the template agent was hexamethylene bis(diethylmethylammonium hydroxide). The incorporation of Ti was limited to 0.45 Ti per unit cell (x=0.016) [84].

The hydrothermal synthesis of the naturally occurring minerals Zorite and Vinogradovite was firstly reported by Chapman and Roe [85]. Pure Zorite-like materials can be prepared by heating a hydroalcoholic gel containing TEOT, Ludox, NaOH and, eventually, $NaH_2PO_4 \cdot H_2O$ to 200 °C. The framework topology of this material originates from vertex-sharing [SiO_4] tetrahedra and [TiO_6] octahedra. Eight-membered ring channels, formed by tetrahedra only, run along [010], while twelve-membered ones, formed by eight tetrahedra and four octahedra, run along [001]. Vinogradovite was obtained together with ANA by using $Na_2Al_2O_4 \cdot 3H_2O$ instead of $NaH_2PO_4 \cdot H_2O$. The structure is composed of pyroxene chains joined to edge-sharing Ti octahedra that form brookite columns. These polyhedra define one-dimensional, 4 Å channels that contain zeolitic water.

3.3
Titanium-Containing Molecular Sieves by Secondary Synthesis

This route of synthesis consists of treating a properly selected molecular sieve with a volatile or soluble compound of the element to be inserted into framework sites. A well known application of this method concerns the partial replacement of aluminum by silicon in FAU-type zeolites [86, 87].

In the case of titanium, this route has been attempted for the framework types listed in Table 5, using aqueous solution of $(NH_4)_2TiF_6$ [88] or $TiCl_4$ vapor [89–91] as titania source. When the microporous precursor is contacted with aqueous $(NH_4)_2TiF_6$ at 75–95 °C, a partial dealumination of the framework occurs accompanied by deposition of a large amount of titanium. However, no clear evidence supporting true incorporation of Ti is given [54, 88].

Table 5. Ti-containing molecular sieves prepared by secondary synthesis

Precursor	SiO_2/Al_2O_3 [a]	Pretreatment	Ti salt	SiO_2/Al_2O_3 [b]	SiO_2/TiO_2 [b]	Ref.
FAU	5.08	NH_4^+ form	$(NH_4)_2TiF_6$	7.55	7.76	[88]
MAZ	6.62	NH_4^+ form	$(NH_4)_2TiF_6$	8.82	8.04	[88]
LTL	5.8	NH_4^+ form	$(NH_4)_2TiF_6$	7.36	12.74	[88]
PHI	3.77	NH_4^+ form	$(NH_4)_2TiF_6$	6	4.88	[88]
MFI	30.82	NH_4^+ form	$(NH_4)_2TiF_6$	58.34	13.22	[88]
MFI	46.49	–	$(NH_4)_2TiF_6$	62.2	90	[88]
MFI	50	dealum.	$TiCl_4$	2000	ca. 80	[89]
MFI	25	–	$TiCl_4$	25	36	[90]
FAU	not given	H^+ form	$TiCl_4$	not given	not given	[90]
Beta	not given	H^+ form	$TiCl_4$	not given	not given	[90]
MFI	73 [c]	–	$TiCl_4$	86 [c]	40	[91]
MFI	73 [c]	–	$TiCl_4$	99 [c]	27	[91]
MFI	50	dealum.	$TiCl_4$	not given	not given	[54]
MFI	50	–	$(NH_4)_2TiF_6$	not given	not given	[54]
silicalite-1	∞	–	$TiCl_4$ (gas)	∞	200–660	[92]
silicalite-1	∞	–	$TiCl_4$ (liq.)	∞	120–150	[92]

[a] Composition of zeolite precursor, molar ratio.
[b] Molar ratio after treatment with the titanium salt.
[c] SiO_2/B_2O_3 molar ratio.

The use of $TiCl_4$ vapor at 200–500 °C seems to be more efficient [54, 89–92]. When dealumination pretreatment is performed on the precursor, usually by acid leaching, structural defects are created in which Ti incorporation presumably takes place. When no pretreatment is made, the reaction with $TiCl_4$ does not lead to reduction of the aluminum content [90]; if any substitution occurs, extra-framework aluminum oxide species are presumably formed. Evidence of partial incorporation of titanium come from the presence of the typical 960 cm^{-1} IR absorption.

The synthesis of Ti-containing MFI was attempted starting from the borosilicate analog (BOR-C [91]). The reaction was followed at different contact times and the product analyzed by IR and XPS. Boron is partially replaced by titanium in the early stage of reaction, as monitored by the 960 cm^{-1} IR band and the Ti(2p) signal typically observed in TS-1 samples prepared by the direct procedure. Increasing the contact time, a second Ti(2p) signal, typical of titanium oxide species, starts to appear. The explanation given is that only boron atoms located near the crystal surface are substituted, due to the steric hindrance of $TiCl_4$; the excess $TiCl_4$ reacts with the surface silanol groups producing extra-framework titanium oxide [91].

More recently, Kooyman et al. tried to react pure silica MFI (silicalite-1) with $TiCl_4$ both in the gas and in liquid phase [92]. Incorporation of Ti was expected to occur in the structural defects formed during the fast crystallization method used for synthesizing the precursor. Apparently, the attempt was unsuccessful, the titanium having been found to segregate as TiO_2 on the crystal surface.

4
Vanadium-Containing Molecular Sieves

In the previous section, we have examined in detail the large number of papers dealing with the synthesis and characterization of Ti-containing molecular sieves, which give an idea of the importance of such materials. However, from the practical point of view, Ti is only one of the elements which promote oxidative catalytic properties. Among them, vanadium is surely one of the most important: vanadium oxide supported on several oxidic matrices (e.g., SiO_2, TiO_2, Al_2O_3, P_2O_5) is in fact an active catalyst for selective oxidations and ammoxidation of organic compounds [93–96]. Furthermore, V displays notable catalytic properties in oxidation reactions involving H_2O_2, making it similar to Ti. This prompted a number of studies on the possibility to incorporate V into the framework of microporous materials such as MFI, MFI/MEL and more recently ZSM-48 and NCL-1. In the following, the synthesis and characterization of these materials will be discussed.

4.1
Synthesis

4.1.1
Pentasil-Type Vanadosilicates

As for Ti-containing molecular sieves, the incorporation of V has been mainly attempted in pentasil-type frameworks (MFI and MFI/MEL), while little information is available about the possibility to synthesize vanadosilicates with the framework topology of other zeolites. Table 6 summarizes the data reported in the literature. Hydrothermal synthesis is the route used in most cases.

Inui et al. used a "rapid crystallization method", which produces crystals of a uniform size [98]. This method makes use of a gel precipitated from a solution containing the vanadium and silica (water glass) sources, the organic template agent and NaCl. After centrifugation and milling, the gel is mixed with a solution containing the same starting materials (decant-solution) and the mixture is heated in an autoclave to 210 °C, in a N_2 atmosphere [99, 100]. The aluminum found in the crystallized products (Si/Al = 106–132, i.e., 0.7–0.9 Al atoms per unit cell) presumably comes from water glass [100]. The type of V source affects the extent of heteroatom incorporation as well as the morphological and catalytic properties of the products [99]. No incorporation occurs when NH_4VO_3 is used, probably because of the low interaction between vanadate ions and the silica gel. Starting from $V(acac)_3$, V-MFI with high crystallinity, uniform crystal size, high BET surface area and high catalytic activity was obtained [99].

Following a more simple hydrothermal procedure, Kornatowski et al. were able to incorporate up to ca. 1 wt.-% of V into the MFI framework starting from vanadates of alkali metals [101]. EPR analysis indicated that most of the vanadium is highly dispersed in the framework as distorted fourfold coordinated V^{4+}, the remainder forming clustered extra-framework V^{4+} species [101]. After calcination in air, the signal attributed to isolated V^{4+} species vanished due to

Table 6. Synthesis of V-containing molecular sieves

Entry	V source	Reactant template agent[a]	Gel		Products			Ref.
			Si/V[b]	Na^+/V[b]	Si/V[b]	Si/Al[b]	Cryst. phase	
1	VCl_3	HMDA	53	0.5	63	120	MFI	[97]
2	VCl_3	TPA-Br	n.g.[c]	n.g.[c]	178	132	MFI	[99]
3	$V(acac)_3$	TPA-Br	90	n.g.[c]	176	119	MFI	[99]
4	$VOSO_4$	TPA-Br	n.g.[c]	n.g.[c]	174	123	MFI	[99]
5	$VO(C_2O_4)$	TPA-Br	n.g.[c]	n.g.[c]	246	107	MFI	[99]
6	NH_4VO_3	TPA-Br	90	n.g.[c]	–	106	MFI	[99]
7	M^+VO_3	TPA-OH	n.g.[c]	n.g.[c]	85	–	MFI	[101]
8	NH_4VO_3	TPA-Br	40	1.0[f]	152	–	MFI	[102]
9	Various[g]	TPA-OH	15	4	n.g.[h]	–	MFI	[103]
10	VCl_3	HMDA	n.g.[c]	n.g.[c]	21.7	198	MFI[d]	[104]
11	VCl_3	TPA-OH	31	–	50	–	MFI	[105]
12	VCl_3	TPA-OH	31	0.3	28	–	MFI[d]	[105]
13	$VOSO_4$	TPA-OH	50	–[e]	98	770	MFI	[106]
14	$VO(C_2O_4)$	TPA-Br	5	0.5	42	–	MFI	[107]
15	VCl_3	TPA-OH	15	–	58	–	MFI	[108]
16	VCl_3	TPA-OH	15	–	118	–	MFI	[108]
17	VCl_3	TBA-OH	n.g.[c]	n.g.[c]	30.8	669	MFI/MEL[d]	[104]
18	$VOSO_4$	TBA-OH	20–160	–	41–300	–	MFI/MEL	[109]
19	VCl_3	TBA-OH	40	–	68	–	MFI/MEL	[109]
20	NH_4VO_3	TBA-OH	40	–	182	–	MFI/MEL	[109]
21	VCl_3	HMDA	n.g.[c]	n.g.[c]	21.6	455	ZSM-48[d]	[104]
22	$VOSO_4$	DAO	30–100	–	59–191	–	ZSM-48	[110]
23	$VOSO_4$	HMTA	48–175	7–30	150–400	–	NCL-1	[111]

[a] HDMA = hexamethylenediamine; TPA = tetrapropylammonium; TBA = tetrabutylammonium; DAO = 1,8-diaminooctane; HMTA = hexamethylenebis (triethylammonium bromide).
[b] Molar ratios.
[c] Not given.
[d] Collapse to crystobalite upon calcination at 550 °C in air.
[e] NH_3 (0.11 eq) was used.
[f] $M^+=NH_4^+$; synthesis run in the presence of fluoride ions, pH = 7.9.
[g] V sources: $NaVO_3$, V_2O_5, NH_4VO_3, $VO(isopropylate)_3$, $VO(acetylacetonate)_2$, $VOSO_4$.
[h] The V content is given in wt% relative to the total mass of the as-prepared samples.

oxidation to non-paramagnetic V^{5+}. On the contrary, the clustered V^{4+} species remain unchanged.

Similar results were obtained by Hong et al. for the hydrothermal synthesis of V-MFI at pH below 8 and in the presence of NH_4F [102]. In contrast to the results reported in [99], NH_4VO_3 allows a greater extent of V incorporation than $VOSO_4 \cdot 5H_2O$ (Si/V = 152 and 326, respectively). Also in this case, the hyperfine structure observed in the EPR spectra of as-synthesized materials agrees with atomically dispersed V^{4+}. Furthermore, when the sample synthesized with ammonium vanadate is oxidized in air, only a broad signal attributed to undispersed V^{4+} species appeared, while the original features were restored after reduction in H_2 at 450 °C [102]. Higher V content was observed in V-MFI prepared with VO(isopropylate)$_3$ [103].

Generally, V-containing molecular sieves are thermally stable. However, V-containing MFI, MFI/MEL and ZSM-48 prepared with VCl_3 are reported to collapse to cristobalite upon calcination in air at 550 °C, but not in an atmosphere of H_2 [104]. The explanation that the structural collapse is due to oxidation of the atomically dispersed V^{3+} to V^{4+} is not convincing since the presence of air during the synthesis probably induces the oxidation of V^{3+} even before the hydrothermal treatment. More convincing is the hypothesis made by Bellussi et al. that the thermal stability of these materials in oxidative conditions is strongly affected by the presence of sodium [105]. In fact, thermally stable V-MFI was crystallized from alkali-free gels, though the V atoms were found to be easily removable from the framework by simple mild treatments with aqueous ammonium acetate solutions. When running the synthesis in the presence of NaOH, MFI was formed together with an amorphous or an unidentified crystalline phase and the calcination in air caused the structure to collapse to cristobalite [105].

Rigutto and van Bekkum demonstrated that in the absence of alkali and in a N_2 atmosphere, $VOSO_4 \cdot 5H_2O$ is an effective source for obtaining materials which are stable upon thermal and chemical treatment (V atoms are only partially removed from the framework upon ion exchange) [106]. The latter behavior is in contrast with that reported by Bellussi et al., who observed an increase of the Si/V ratio from 110 to 760 after six exchange treatments [105].

Few successful results were claimed for other V-containing molecular sieves. The synthesis of the V-analog of ZSM-11 (V-MFI/MEL) was firstly reported by Habersberger et al., by using TBAOH, silica sol, VCl_3 and NaOH [104]. Similarly to V-MFI, the presence of Na^+ and the use of VCl_3 led to thermally labile materials collapsing to cristobalite upon calcination in air. Thermally stable V-MFI/MEL samples (named VS-2) were prepared in the absence of alkali starting from $VOSO_4 \cdot 5H_2O$ [109]. Syntheses were performed with a wide range of Si/V ratio in the reaction gel (20–160), reaching a 50% incorporation efficiency ($41 < Si/V < 300$) [109b]. The Si/V ratio was affected only marginally by treatments with ammonium acetate [109b]. On the contrary, extensive removal of V was observed after steam treatment at 873 K in a flow of 100% H_2O [109a].

4.1.2
Other Vanadium-Containing Molecular Sieves

Habersberger et al. reported for the first time the synthesis of V-ZSM-48 with hexamethylene diamine as a structure directing agent, VCl_3 and NaOH. The product was found to collapse to cristobalite upon calcination in air at 550 °C [104].

More recently, Tuel and Ben Taarit claimed the synthesis of V-ZSM-48, by heating a gel containing fumed silica, $VOSO_4 \cdot 5H_2O$ and 1,8-diaminooctane [110] to 180 °C. Starting from a Si/V ratio in the range 30–100, ca. 50% of V was retained in the crystallized products, with a maximum of 0.8 V atoms per unit cell ($Si/V = 59$).

Reddy et al. reported the synthesis of V-containing NCL-1 from a gel containing fumed silica, NaOH, vanadyl sulfate and hexamethylene bis (triethylammonium bromide). Incorporation of V is rather limited, the lowest Si/V ratio obtained being 150 [111]. Though the structure of NCL-1 is still unknown, adsorption and diffusivity experiments indicate a pore size close to that of mordenite (ca. 7 Å) [112]. Therefore, V-NCL-1 would represent the unique example of V incorporation in a large pore molecular sieve.

By adopting a hydrothermal digestion of vanadium-alumino-silica glasses in basic solutions at 120 °C, Dubanska claimed the possibility to synthesize a number of V-molecular sieves including NU-15, ANA, LTA, KF, SOD [113]. However, no conclusive evidence supporting true incorporation of the heteroatoms was given.

4.2 Vanadium-Containing Molecular Sieves by Secondary Synthesis

Three different routes were attempted:

1. Ion-exchange of FAU-type zeolites with aqueous solutions of VO^{2+} salts, followed by partial dehydration under vacuum [114, 115]
2. High temperature (> 300 °C) solid-state reaction of zeolites with V_2O_5 in air or N_2 [116–120]
3. Reaction of zeolites with $VOCl_3$ in gas phase at 520–570 °C [121, 122]

No incorporation of V was achieved by using method 1, with the VO^{2+} ions located in extra-framework sites coordinated to framework oxygen atoms.

A cation migration from V_2O_5 to strong acid sites to give isolated V^{4+} species was observed in H-ZSM-5 [116–119] and in H-Y zeolites [120] by high temperature solid-state reaction with V_2O_5. However, negligible framework incorporation was detected.

When method 3 was used on MFI-type zeolites, isolated V^{4+} species were formed in both framework and extra-framework positions [121, 122]. Incorporation occurred at the defective sites formed upon partial dealumination of the framework, while the extra-framework V^{4+} ions replaced H^+ as counterions of Al sites.

4.3 Structure Characterization

The chemistry of V is rather complex because of the existence of three stable oxidation states, depending on the synthesis procedure adopted as well as on the post-synthesis treatments. Most of the V-containing molecular sieves described in the literature contain V^{4+} and/or V^{5+} species, and only few V^{3+} ions. Several authors pointed out that the $V^{4+} \rightarrow V^{5+}$ transition, taking place upon calcination in air, is reversible when the sample is calcined in a reducing atmosphere (H_2) [99, 102, 106–108, 110].

Similarly to Ti-containing molecular sieves (Sect. 3.2), incorporation of V is supported by spectroscopic and structural evidence. In calcined samples, a band located at ca. 960 cm^{-1} was observed in the IR region, not present in the silica parent structures [101, 105, 109–111]. This absorption, present also in Ti-containing analogs, was attributed to the asymmetric stretching vibration of V-O-Si bonds. Its intensity was found to be linearly correlated with the V content both in VS-2 [109] and in V-ZSM-48 [111].

Again, a linear dependence of the unit cell volume with the V-content was observed for both VS-2 [109] and V-ZSM-48 [111], while no data are available on V-MFI.

Electron spin resonance and ^{51}V-NMR are generally used for deriving information about the dispersion and coordination of V^{4+} and V^{5+} species, respectively. The ESR spectra of V^{4+}-containing molecular sieves generally show the anisotropic hyperfine splitting originating from the nuclear spin of ^{51}V (I=7/2, natural abundance 99.8%). Such a feature is assumed to be typical of atomically dispersed V^{4+} species within the framework [123]. In some cases, this pattern is superimposed to a broad singlet, attributed to agglomerated extra-framework V^{4+} species. Information about geometry and local symmetry of atomically dispersed V^{4+} atoms can be obtained from the values of g tensors and hyperfine coupling constants (Table 7). Though the reported values of g tensors and A parameters indicate that both local symmetry and covalent character slightly change from one sample to another, the authors generally agree that isolated V^{4+} species exist in more or less distorted C_{4v} or D_{4h} symmetry, the T_d symmetry having been observed in few cases [103, 108]. Kornatowski et al. reported that in as-prepared V-MFI framework V is present as V^{4+} and V^{5+} with

Table 7. ESR parameters of V^{4+}-containing molecular sieves

Entry[a]	Treatment[b]	$g_{\parallel}$	$g_{\perp}$	$A_{\parallel}$	$A_{\perp}$	Local symmetry[c]	Ref.
2	H_2	1.927	1.994	191	89	-	[99]
4	H_2	1.957	2.015	196	91	-	[99]
7	a.s.	1.939	1.961	192.5	87.8	dist. s. pl.	[101]
8	a.s.	1.939	1.998	184	77	-	[102]
8	H_2	1.925	1.997	189	84	-	[102]
11	a.s.	1.95	2.00	191	55	dist. s. pl.	[105]
12	a.s.	1.93	1.97	191	89	dist. s. pl.	[105]
13	a.s.	1.935	1.994	183	69	s.py.	[106]
13	H_2	1.935	1.996	179	65	s.py.	[106]
14	a.s.	1.949	1.99	185	72.5	dist. s.pl.	[107]
15	air			203	80	octahedral	[108]
16[d]	H_2	1.9117	1.9625	154	68	tetrahedral	[108]
16[d]	air	1.9355	1.9827	203	80	octahedral	[108]
18	a.s.	1.932	1.981	185	72	dist. oct.	[109]
18	H_2	1.933	1.972	185	71	dist. s.py.	[109]
22	a.s.	1.936	1.99	188	70	dist. oct.	[110]
23	a.s.	1.929	1.973	197	72	-	[111]

[a] See Table 6.
[b] a.s. = as synthesized; air = calcined in air; H_2 = calcined in H_2.
[c] Proposed coordination of the V species.
[d] Data collected at 77 K.

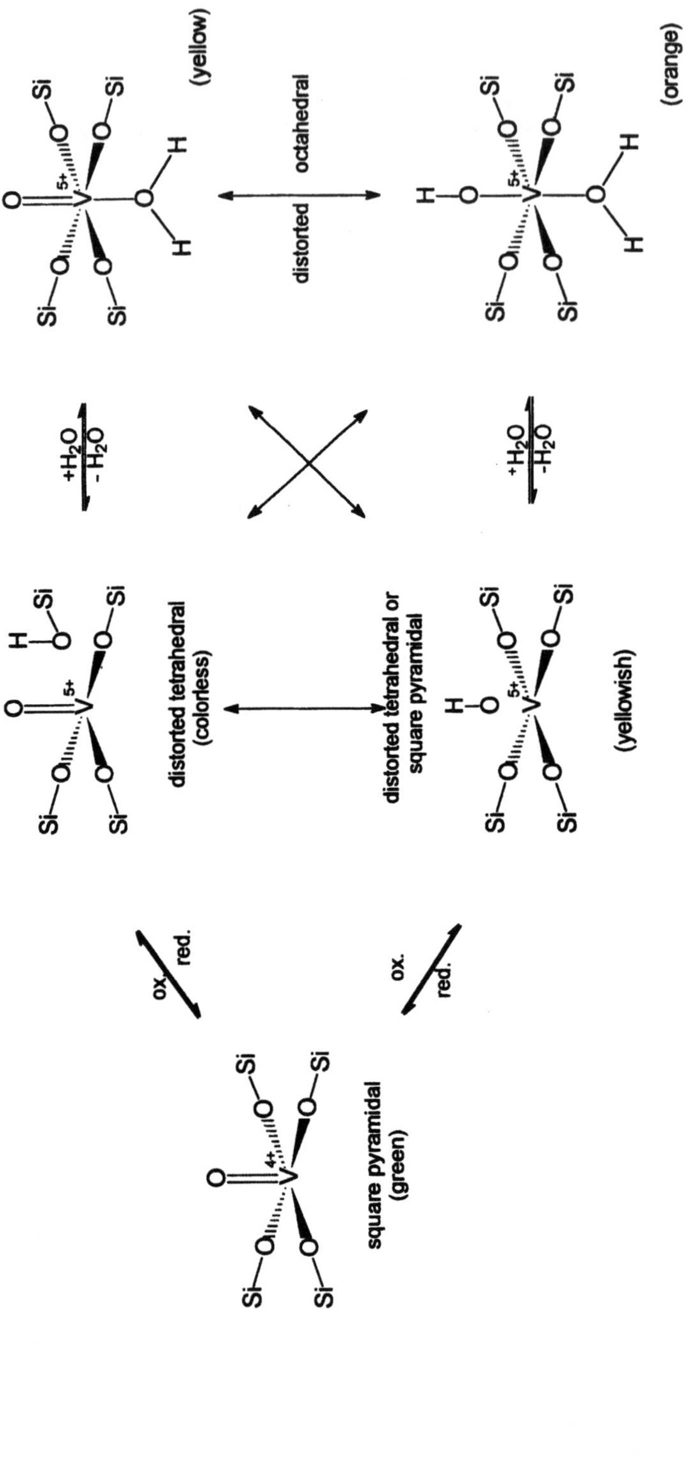

Fig. 9. V site structures in as-synthesized and calcined V-MFI (after Kornatowski et al. [103])

different coordinations, the former assuming the tetrahedral or square pyramidal one, depending on the arrangement of OH groups within the coordination polyhedra (Fig. 9) [103]. After calcination in air, V^{4+} is fully oxidized to V^{5+} preserving the tetrahedral and square pyramidal coordination. Upon adsorption of water, the formation of strongly distorted octahedral coordination has been postulated.

The local symmetry of V^{5+} was investigated by ^{51}V NMR, taking advantage of the large amount of data available for model compounds (e.g., Na_3VO_4 and V_2O_5 containing V in distorted tetrahedral and octahedral coordination, respectively). There is a general consensus about the tetrahedral coordination of V^{5+} (Table 8), though differences exist in peak symmetry, chemical shift and side bands, probably related to slight variations of the local symmetry.

Based on experimental evidence, the structures drawn in Fig. 10 were proposed for V framework sites ("framework satellite"). None of these structures represents a real isomorphous substitution of Si. Though the mechanism of formation has not yet been clarified, it is probable that these species are related to framework defects, which are known to exist in the crystallized molecular sieves [124–127]. Internal silanol groups are supposed to form through a reversible hydrolysis of Si-O-Si bonds [124] or to originate from a Si vacancy with a nest of four Si-OH clustered around it [126–128]. The latter situation accounts well for Al incorporation by secondary synthesis [128]. According to Zecchina et al. defects in silicalite-1 arise from one or more missing $[SiO_4]$ tetrahedra, producing nanodefects or microcavities within the structure [129, 130]. The presence of single missing $[SiO_4]$ seems to be questionable because of the high steric hindrance of four adjacent H atoms, while the hypothesis of a few adjacent missing $[SiO_4]$ units accounts well for the sorption and IR data [131].

Table 8. ^{51}V NMR data for V-containing molecular sieves

Entry[a]	Treatment[b]	δ (ppm)	δ_{iso} (ppm)	Local symmetry[c]	Ref.
7	air	–	-557	tetrahedral	[101]
11	a.s.	–	-576	tetrahedral	[105]
12	air	–	-385	octahedral	[105]
13	air	-500	–	tetrahedral	[106]
14	a.s.	–	-672/-680	tetrahedral	[107]
15	air	-480	–	tetrahedral	[108]
18	air	–	-573	tetrahedral	[109]
22	air	-530	-575	tetrahedral	[110]

[a] See Table 6.
[b] a.s. = as-synthesized; air = calcined in air.
[c] Proposed coordination of the V species.

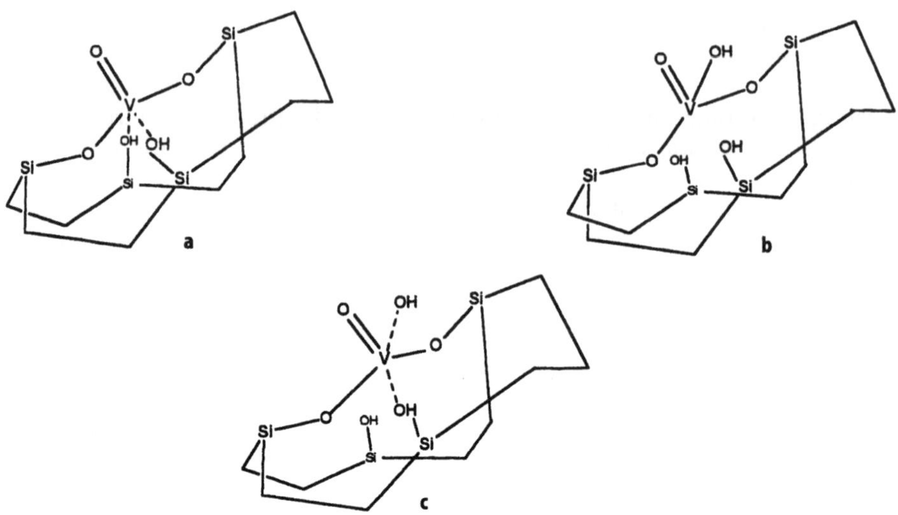

Fig. 10. V site structures proposed by Rigutto et al. [106] (**a**, **b**) and by Centi et al. [108] (**c**)

5 Iron-Containing Molecular Sieves

Incorporation of Fe into zeolite frameworks has been investigated both for better understanding the status of the trace amounts of ferric ions always present in natural zeolites and for synthesizing materials potentially useful for novel catalytic applications. Based on ionic radii, one should predict the incorporation to be rather difficult, Fe^{3+} (r=0.63 Å) being favored with respect to Fe^{2+} (r=0.77 Å). Nevertheless, isomorphous substitution of Al by Fe is known to occur both in natural [132] and in synthetic molecular sieves [133]. The first ferrisilicate reported in the literature ($MFeSi_2O_6$, M=K, Rb, Cs), isostructural with leucite, was prepared by solid-state reaction at 1300–1650 °C starting from a mixture of M_2CO_3, Fe_2O_3 and SiO_2 [134]. Subsequently, the preparation of Fe-containing molecular sieves was attempted adopting the classical hydrothermal procedure, with successful results achieved for a large number of medium- and large-pore frameworks [133].

5.1 Synthesis

As for Ti- and V-containing molecular sieves, the synthesis conditions must assure the suppression of the polymerization/precipitation reactions of iron oxide/hydroxide species. Though the aquo-chemistry of Fe^{3+} is well documented (see [133] and references therein), little information is available on the behavior of Fe sources in the presence of silica, particularly under the hydrothermal conditions necessary for the crystallization of molecular sieves. Monomeric octahedral aquo-hydroxocomplexes of iron are stable only at $pH < 2$. At higher pH values, these species tend to undergo polymerization and finally to precipitate; to avoid this, Fe^{3+} must be complexed with chelating agents (e.g., oxalic, citric, tartaric acids, EDTA). Strong evidence exists supporting the interaction of Fe^{3+} with silicic acid in dilute aqueous solutions [135, 136]. The monomeric species formed, $FeSiO(OH)_3^{2+}$, are relatively stable in dilute solutions but they may undergo polymerization when the concentration of Fe^{3+} exceeds 5×10^{-5} mol l^{-1}. Iron was also found to strongly interact with partially polymerized silica at pH>2 [137].

Though the above observations refer to conditions very far from those typically used in zeolite syntheses, they were useful for driving successful crystallization of Fe^{3+}-containing silica frameworks, the most representative being reported in Table 9. All the synthesis procedures involve the addition of a monomeric or oligomeric silica source to an acidic solution of Fe^{3+} (sulfate, nitrate or chloride). After the ferrisilicate complexes are formed, the organic template is added and the pH adjusted to the desired value before the hydrothermal treatment. When a template agent able to form a strong complex with Fe^{3+} (e.g., amines, pyridine) is required, it must be added after the ferrisilicate complexes are formed, otherwise the Fe incorporation is severely limited [139]. The best incorporation efficiency was obtained working with oligomeric silica sources (sodium silicate, N-brand-silicate from The PQ Corp.), while highly polymerized silicas (e.g., Ludox, Cab-O-Sil) were unable to complex all the Fe^{3+} ions, with the consequent precipitation of iron hydroxides [139]. Use of the latter silica sources requires their previous prolonged digestion in NaOH to decrease their degree of polymerization. In any case, the use of TEOS is preferred since the other silica sources generally contain non-negligible amounts of aluminum impurities. The source of Fe^{3+} is commonly chosen from the soluble salts (nitrate, sulfate or chloride, Table 9), without any influence of the anion type on the characteristics of the final products.

5.1.1
Pentasil-Type Ferrisilicates

Most of the literature on this topic deals with the MFI structure [138–151]. Ball et al. [140] compared the synthesis of Al- and Fe-MFI making use of colloidal silica (Ludox): crystallization of Al-MFI is faster while both crystallinity and crystal size decrease with increasing iron content. By following the EPR signal at g=4.3 (corresponding to Fe^{3+} in tetrahedral coordination, see Sect. 5.2) the authors observed an increase of Fe incorporation within the first 24 hours, followed by a reduction due to a secondary nucleation of silicalite-1. Kotasthane et al. compared the crystallization kinetics of Al- and Fe-MFI starting from sodium silicate and triethylbutylammonium bromide (TEBABr) as structure-directing agent [144]. The apparent activation energies for nucleation (E_n) and crystallization (E_c) were evaluated, and the crystallization data fitted to the Avrami-Erofeev equation, which allowed them to conclude that the rates of nucleation and crystal growth are lower when Al^{3+} is replaced by Fe^{3+}. When the synthesis of Fe-MFI was run with TEOS [151], the amount of Fe in the solid, the degree of crystallinity and the Fe^{3+} ESR signal intensity were found to increase

Table 9. Typical syntheses of ferrisilicate molecular sieves

Entry	Template agent[a]	SiO_2 source	Fe^{3+} source	M	Al	Cryst. phase	Ref.
1	TPABr	Na-silicate	sulfate	–	traces	MFI	[138]
2	TPABr	N-brand silicate	nitrate	–	–	MFI	[139]
3	TPAOH/Br	Ludox/Na-silicate	nitrate	NaOH	traces	MFI	[140]
4	TPABr	Na-silicate	sulfate	–	traces	MFI	[141]
5	TPABr	Aerosil	nitrate	K_2CO_3	traces	MFI	[142]
6	HMDA	waterglass	sulfate	–	traces	MFI	[143]
7	TEBABr	Na-silicate	sulfate	–	traces	MFI	[144]
8	$N(C_2H_5)_3$	silica sol	chloride	NaOH	yes	MFI	[145]
9	n-$C_3H_7NH_2$	Na-silicate	sulfate	–	–	MFI	[146]
10	TPABr/NH_4F	TEOS/Aerosil	chloride	–	–	MFI	[147]
11	TPABr	waterglass/Cataloid	chloride	NaOH	traces	MFI	[149]
12	TEBABr	Na-silicate	oxalate	–	–	MFI	[150]
13	TPAOH	TEOS	nitrate	–	–	MFI	[151]
14	DAO/TBAOH	TEOS	nitrate	NaOH	–	MFI/MEL	[152]
15	Pyrrolidine	fumed silica	nitrate	NaOH	–	MTT	[157]
16	TEAOH	TEOS	sulfate	NaOH	–	BETA	[160]
17	TEAOH	fumed silica	oxalate	Na/KOH	–	BETA	[150]
18	EP-Br	Na-silicate	nitrate	–	traces	TON	[155]
19	EP-Br	silica sol	sulfate	KOH	traces	TON	[156]
20	TEABr	fumed silica	nitrate	NaOH	–	MOR	[162]
21	TMACl	Na-silicate	nitrate	–	–	SOD	[153]
22	Diquat-6	Na-silicate	nitrate	–	traces	EUO	[158]
23	Diquat-6	TEOS	nitrate	NaOH	–	ZSM-48	[133]
24	EN-TEAm	Na-Al-silicate	chloride	–	yes	ZSM-48	[159]
25	MTEABr	Na-silicate	nitrate	–	–	MTW	[133]
26	TMAOH	TEOS	nitrate	NaOH	–	NU-1	[154]

[a] TMA = trimethylammonium; TEA = tetraethylammonium; TPA = tetrapropylammonium; TBA = tetrabutylammonium; HMDA = hexamethylendiamine; TEBA = triethylbutylammonium; DAO = 1,8-diaminooctane; Diquat-6 = 1,6-hexamethonium dibromide; EN = ethylendiamine; TEAm = triethylamine; MTEA = methyltriethylammonium; EP = 1-ethyl-piridinium.

as a function of time in the same manner up to about 60 hours of crystallization. No secondary nucleation was observed, but only part of the Fe^{3+} was effectively incorporated into the silica framework.

Iron-MFI can be synthesized also in the presence of NH_4F (Si/F=2) by using a ferrisilicate xerogel prepared by refluxing TEOS with a ferric chloride solution or Aerosil and $FeCl_3 \cdot 6H_2O$, obtaining large and often twinned crystals [147, 148]. A homogeneous distribution of Fe was found by EDS analysis in samples prepared from ferrisilicate xerogel, while those prepared from Aerosil showed an iron-rich core (Si/Fe ca. 30) and a silicon-rich outer shell (Si/Fe>10,000). These characteristics probably derive from the different iron distribution in the starting materials: homogeneous in the ferrisilicate xerogel and concentrated on the silica surface in Aerosil.

Using the so-called rapid crystallization method [98], Fe-MFI was prepared, with almost all the Fe^{3+} in tetrahedral coordination (i.e., incorporated in the MFI framework) and an iron content up to 3.2 wt.-% (as Fe_2O_3) [149].

An interesting method was recently proposed by Kumar et al., useful for promoting the incorporation of those metal ions which form insoluble hydroxide/ oxides in basic media (e.g., Ti, Fe) [150]. The method is based on the use of suitable complexing agent (such as acetylacetone or oxalic acid) forming fairly stable complexes which slowly release the metal ions during the hydrothermal treatment. This method was successfully applied in the synthesis of Fe-MFI and Fe-Beta (with Fe-oxalate) as well as of TS-1 (with Ti-acetylacetonate).

The successful incorporation of Fe^{3+} into the MFI/MEL intergrowth structure, to give the ferrisilicate analogs of ZSM-11, was reported by Reddy et al. [152]. Samples with a Si/Fe ratio of ca. 20 were synthesized by using 1,8-diaminooctane (DAO) or TBAOH, TEOS and ferric nitrate.

5.1.2
Other Ferrisilicate Molecular Sieves

The synthesis of several other ferrisilicate molecular sieves have been reported (Table 9). The procedures followed by the various authors were generally derived from those reported for the aluminosilicate analogs.

Iron-SOD was synthesized with a wide range of iron concentrations (SiO_2/Fe_2O_3 as low as 7) by Szostak and Thomas, starting from a gel containing sodium silicate, iron nitrate and TMACl as structure directing agent [153]. It was shown that the rate of crystallization of Fe-SOD decreases with increasing iron concentration.

In the synthesis of Fe-NU-1, the precipitation of iron hydroxide was prevented by dissolving ferric nitrate in ethyl alcohol and by adding this solution to TEOS [154]. However, the reproducibility of the synthesis was poor, since SOD and MTN were often obtained as by-products.

Borade et al. were the first to report the synthesis of Fe-ZSM-22 (TON) with Si/Fe=43.4, by using 1-ethyl-pyridinium bromide (EP-Br) as organic template [155]. Most of the iron was incorporated into the lattice, while small amounts of extra-framework iron oxide were detected by XPS. The acidic strength of the framework hydroxyls was found to be in the order: Al-ZSM-22 > Fe-ZSM-22 > Al-ZSM-5. More recently, Singh reported that Fe-ZSM-22 can be prepared with a Si/Fe ratio as low as 39 [156].

A ferrisilicate analog of the medium pore zeolite ZSM-23 (MTT) was synthesized by using fumed silica, ferric nitrate, sodium hydroxide and pyrrolidine [157]. The ferrisilicate with $SiO_2/Fe_2O_3=118$ and substantially aluminum-free, displayed significant ion exchange capacity (K/Fe=0.9) and catalytic activity in the m-xylene isomerization reaction, though the low strength of acid sites was responsible of the lower activity with respect to the aluminosilicate analog.

Two different ferrisilicates were synthesized by using 1,6-hexamethonium dibromide (Diquat-6): Fe-EUO [158] and Fe-ZSM-48 [133]. The former is particularly interesting in view of a possible use as a catalyst, because, due to its peculiar pore structure, it could exhibit product shape selectivity typical of a medium pore zeolite (10-membered ring openings) and restricted transition state shape selectivity typical of a large pore zeolite (large side pockets along the channels). Fe-Al-ZSM-48 has been synthesized also in non aqueous media (mixture of ethylendiamine and triethylamine), in the form of alumino-ferrisilicate [159].

5.1.3
Fe^{3+} Incorporation in "Large Pore" Molecular Sieves

The preparation of Beta [150, 160, 161], MOR [162], MTW [133] has been reported. The experimental evidence achieved for these materials reasonably support Fe^{3+} incorporation.

Several papers deal with the isomorphous substitution of Fe for Al in low silica molecular sieves such as FAU and LTL. This possibility was unambiguously demonstrated in 1972 when McNicol and Pott obtained strong evidence for the framework location of the Fe^{3+} impurities present in NH_4-FAU and in MOR [163]. A few years later, Derouane et al. found that the iron impurity (ca. 2000 ppm) present in NH_4-FAU obtained by exchanging a commercial NaY zeolite can be simultaneously present in three forms: Fe^{3+} species in the aluminosilicate framework, Fe^{3+} ions acting as counterions and Fe_3O_4 or other Fe^{3+} compound with strong exchange spin-spin interactions precipitated on the zeolite [164]. Traces of Fe^{3+} may be incorporated into the NH_4-Y zeolite during the ion-exchange with an aqueous $FeCl_3$ solution [165].

Pure ferrisilicates with the FAU and LTL structure have never been synthesized. The first attempt at synthesizing Fe-containing FAU was performed by Ratnasamy et al. [166] by adding a homogeneous aluminum-ferrisilicate gel (seeds) to an aqueous gel mixture containing sodium silicate, sodium aluminate, aluminum sulfate and ferric sulfate. The minimum Si/Fe ratio obtained, estimated by NMR spectroscopy, was 17.2 with a Si/Al ratio of 2.98. Zi et al. found that the presence of iron in the reaction mixture is detrimental to the quality of Y zeolite crystals. As the iron concentration in the starting gel increases, the crystallinity of the products decreases with possible co-crystallization of other phases, such as herschelite, noselite, and zeolites P and A [167].

The partial replacement of Al by Fe (up to 0.3 atomic fraction of (Al+Fe)) in the LTL framework was achieved by adding a ferric nitrate solution to a reaction gel containing aluminum trihydrate, potassium hydroxide and fumed silica [168]. The upper limit of 0.3 was recently overcome by using potassium ferrate(VI) as source of iron [169]. The tetrahedral coordination of the FeO_4^{2-} ion is moderately stable in strong alkaline media which is expected to favor the nucleation of zeolite species. Using this iron source, the authors were able to synthesize LTL zeolite in which a mole fraction of Al_2O_3 up to 0.825 was replaced by Fe_2O_3 in the framework; attempts to increase the iron content (0.90–1.00 mole fraction) led to amorphous products. Once again, the increasing amount of iron in the reaction mixture causes a decrease of the crystallization rate: 100% crystalline LTL with a 0.825 mole fraction of Fe required ca. 300 h crystallization, while that with 0.25 was obtained after 100 h.

5.2 Preparation of Ferrisilicate Molecular Sieves by Secondary Synthesis

Skeels and Flanigen were the first to propose a method for the incorporation of Fe^{3+} by treating the ammonium or the hydronium form of FAU, MOR and LTL zeolites with a solution of $(NH_4)_3FeF_6$ at 75–95 °C for between 30 min and 48 h [170]. The solution of $(NH_4)_3FeF_6$ was prepared either by dissolving the required amount of salt or by adding NH_4F or $(NH_4)HF_2$ to a solution of FeF_3. The amount of iron found in the final products was generally equivalent to the amount of aluminum extracted from the zeolite precursors. Since no other crystalline phases were detected and the crystals were substantially free of defects, the authors concluded that such a preparation effectively leads to the incorporation of Fe into the zeolite framework.

The solid-state reaction between HZSM-5 and iron compounds (FeO, Fe_3O_4, $FeCl_3$) was investigated by Kucherov and Slinkin [171]. The formation of isolated extra-framework Fe^{3+} species was observed only in samples treated with $FeCl_3$, while no migration of the trivalent element within the porous structure was observed after high temperature treatment with iron oxides. Similar results were obtained by Wichterlova et al. [172].

5.3 Structure Characterization

Several techniques were used for assessing the incorporation and for determining the status of Fe^{3+} in ferrisilicate molecular sieves. However, the first information indicating successful incorporation of iron is generally given by the color of the crystalline product; a white color, in fact, indicates the lack of segregated iron species.

More accurate information may be obtained from XRD analysis. As already stated in the previous sections, the substitution of Si (or Al) by another element induces a variation of the lattice parameters. Based on ionic radii, the incorporation of Fe^{3+} is expected to produce a significant increase of the lattice dimensions with respect to the pure silica or the aluminosilicate parent structures. Furthermore, XRD may be useful for determining the iron content in a given structure, as reported by Szostak et al., who found a linear increase of the unit cell volume of as-synthesized Fe-MFI with the decrease of the SiO_2/Fe_2O_3 molar ratio in the range 20–100 [139b]. A similar trend was observed for Fe-ZSM-11 [152], TON [156], FAU [166] and LTL [168, 169].

Direct proofs of the location and environment of Fe^{3+} in ferrialuminosilicates with the SOD [173] and LTL [174] structures were achieved by EXAFS analysis. The EXAFS spectrum of Fe-Al-SOD with Fe/Al=0.05 (i.e., the sample with the maximum concentration of framework Fe^{3+}) was fitted with a first shell of four oxygen atoms at 1.87 Å and four silicon atoms at 3.20 Å from Fe [173]. In the case of LTL, EXAFS analysis performed on a sample with one third of the framework aluminum replaced by iron, revealed that Fe^{3+} was substituted into the framework, with a first coordination shell of four oxygen atoms located at 1.85 Å and an outer shell of silicon. Anomalous dispersion powder diffraction data combined with Rietveld profile analysis agree with Fe atoms randomly distributed on the two crystallographically independent framework sites [174].

In as-synthesized Fe-MFI samples prepared in the presence of fluoride ions [147, 148], optical and EXAFS analysis agree with Fe^{3+} in a four-fold coordination with an average Fe-O distance of 1.86–1.87 Å [148]. Upon calcination and subsequent hydration, the framework Fe^{3+} cations remained tetrahedrally coordinated though EXAFS analysis would indicate a two-shell model involving three oxygen atoms at 1.87–1.89 Å and a fourth one at 2.18 Å. This was explained by the presence of adsorbed water molecules interacting with Fe^{3+} Lewis acid sites formed during calcination.

Since Fe^{3+} is paramagnetic, ferrisilicates can be usefully investigated by ESR spectroscopy. According to the earlier reports of McNicol and Pott [163] and Derouane et al. [164], three signals at g=2.0, 2.3 and 4.3 can be observed in the X-band ESR spectra of ferrisilicate molecular sieves. The commonly accepted attribution of these signals is: Fe^{3+} in exchangeable positions (i.e., acting as a counterion with respect to the framework charge), Fe^{3+} segregated in oxide/hydroxide species, and Fe^{3+} incorporated into the silica framework in distorted tetrahedral symmetry, respectively. The appearance of the signal at g=4.3 is often considered a proof of the incorporation of ferric ions into the silica framework. However, the conclusions reached from ESR analysis often contrast with the evidence achieved from other techniques. In fact, even in the absence of the signal at g=2.3 (corresponding to a lack of segregated ferric oxide/hydroxide species) and with the major part of Fe^{3+} in framework positions, the signal at g=2.0 is always present, more intense than that at g=4.3. Lin et al. assigned the ESR signals to three different Fe^{3+} framework sites in Fe-ZSM-5 [175], while Park and Chon attributed the signal at g=2.0 to tetrahedral framework Fe^{3+} (in analogy with the ESR spectrum of $FePO_4$, in which Fe^{3+} is in tetrahedral coordination) and that at g=4.3 to framework pseudo-octahedral ions [176]. More recently, this problem was faced by Goldfarb et al. who performed the characterization of several Fe-containing molecular sieves (SOD, LTL, MFI, MAZ, FAU) by X-band and Q-band ESR, Pulsed ESR and UV-Visible spectroscopy [177]. It was observed that Fe^{3+} can be incorporated in significant amounts in the framework of SOD, LTL, MFI and MAZ, like in FAU. The ESR signal at g=2 was attributed both to framework and extra-framework Fe^{3+} species and its presence does not provide evidence for framework substitution unless combined with other techniques. Also, the g=4.3 signal is not indicative of framework Fe^{3+}. In fact, the ESR spectra of Fe-SOD with low iron content, in which Fe^{3+} is almost completely incorporated into the framework [173], displays a unique narrow signal at g=2, and hence it represents a good model for further studies of Fe substitution in zeolites [178].

Another technique widely used for characterizing ferrisilicates is ^{57}Fe Mössbauer spectroscopy, a very sensitive tool able to differentiate the nuclei in slightly different electronic environments. The electron density of the nuclei, determined by the oxidation state, the coordination number and the type of ligands produce a Mössbauer isomer shift (IS), while nuclear and magnetic hyperfine interactions are responsible for the form of the spectrum.

In the case of ferrisilicate molecular sieves, it is important to distinguish between tetrahedrally and octahedrally coordinated Fe^{3+}. Following Garten et al., a tetrahedrally coordinated ferric ion has an IS < 0.3 mm s^{-1}, whereas an octahedrally coordinated one gives an IS > 0.3 mm s^{-1}, both at room temperature and referred to the metallic element [179]. At low temperature, the IS increases somewhat. However, the IS value alone is not sufficient to determine whether Td coordinated iron is in framework positions or not. Detailed ^{57}Fe Mössbauer studies were reported for Fe-MFI [142, 146, 180]. Meagher et al. [180] reached the conclusion that in as-synthesized materials essentially all the ferric iron is in tetrahedral framework sites up to a SiO_2/Fe_2O_3 ratio of 50; at higher concentrations of Fe, the heteroatom, if incorporated, occupies crystal defect sites with octahedral coordination. The authors propose the white color and an IS value of ca. 0.25 mm s^{-1} at room temperature as criteria for a good Fe-MFI sample. Application of ^{57}Fe Mössbauer spectroscopy to as-synthesized Fe-EUO [133], Fe-Beta [181] and Fe-MOR [182] demonstrated the presence of well dispersed ferric ions in the tetrahedral framework of the ferrisilicates.

As for Al^{3+}, incorporation of Fe^{3+} in tetrahedral coordination implies the formation of negative charges in the framework, hence ion-exchange properties are expected for ferrisilicate molecular sieves. Indeed, such behavior was first demonstrated by Szostak et al., with the proposal of a semiquantitative method for estimating the extent of iron incorporation in MFI, based on the exchange of the H-form with KOH, followed by the analysis of K and Fe [139b]. The highest K/Fe ratio (0.97) was obtained in a sample with a bulk SiO_2/Fe_2O_3 ratio of 171, while for a sample with $SiO_2/Fe_2O_3=51$, K/Fe was 0.40, indicating that only part of the overall iron is in framework positions. The acidic properties of ferrisilicates are responsible for their catalytic activity [133], and many efforts were made to study these properties in detail, using mainly IR spectroscopy and TPD analysis. The IR spectrum of a ferrisilicate displays notable differences with respect to the aluminosilicate analog, especially in the range 400–1200 cm^{-1}. In fact the presence of iron changes the reduced mass of the harmonic oscillator Si-O-Fe with respect to the Si-O-Si one, producing a shift to lower wavenumbers of both the symmetric and asymmetric stretching vibrations. This was really observed in the IR spectra of ferrisilicate molecular sieves. For instance, in Fe-MFI the symmetric stretching Si-O-Fe was observed at 656 cm^{-1}, compared to 800 cm^{-1} for Si-O-Si [139b]. Furthermore, the replacement of Si by Fe causes the appearance of an IR-active mode at 1015 cm^{-1} and a Raman-active one at 1020 cm^{-1}, assigned to vibration modes involving $[O_3Si\text{-}O]^{\delta-}$-tetrahedra polarized by the presence of Fe^{3+} in the structure [183].

The most interesting spectral region for evaluating the acidity of ferrisilicates is that including the stretching of hydroxyl groups (ν(OH)). Iron-ZSM-5 displays two IR bands at 3740 and 3630 cm^{-1}, attributed to terminal SiOH and bridging Si(OH)Fe groups, respectively [184]. The frequency of the former is substantially insensitive to the type of heteroatom present in the MFI structure (B, Al, Ga, Fe). On the contrary, the frequency of the latter depends on the acid strength of the Si(OH)M moiety, which varies in the following order: SiOH < B(OH)Si < Fe(OH)Si < Ga(OH)Si < Al(OH)Si. This trend was confirmed by Temperature Programmed Ammonia Desorption (TPAD) experiments, the temperature of maximum ammonia desorption being linearly correlated to the ν(OH) frequency of the latter band.

Kustov et al. investigated H-Fe-MFI samples by diffuse reflectance IR, ESR and luminescence spectroscopies, reaching the conclusion that ca. 30 % of Fe is effectively incorporated into the framework of a sample having the stoichiometry $Na_{0.04}H_{2.25}(SiO_2)_{93.41}(FeO_2)_{2.59}$ [185]. From the frequency shifts of the ν_{0-1} band (fundamental stretching vibration of OH) after weak base absorption (CO, C_2H_4, C_6H_6), the authors derived the following sequence of increasing Brønsted acidic site strength: H-(Al)-Y<H-(Fe)-ZSM-5<H-(Al)-ZSM-5. The 3630 cm^{-1} IR band was not observed in samples prepared via the acidic route in the presence of fluoride ions, indicating the lack of Brønsted acid sites [175]. Only Lewis acid sites were observed by IR analysis after pyridine absorption. This behavior was explained with the substitution of OH^- anions by F^- [175].

The acidity of Fe-ZSM-22 was compared to that of Al-ZSM-22 and Al-ZSM-5, based on IR, TPD and XPS data [155]. The IR spectra of Al-ZSM-22 and Fe-ZSM-22 are characterized by absorption at 3742 cm^{-1} (assigned to terminal silanol groups), 3725 cm^{-1} (attributed to extra-framework material), 3628 (for Fe-compound) and 3606 cm^{-1} (for the Al-derivative). The latter band disappears completely upon adsorption of pyridine, confirming the assignment to bridging, acidic OH group. According to previous observations, the higher frequency observed for the ν(OH) of ferrisilicate is in agreement with a lower acid strength compared to Al-ZSM-22. The ratio between Brønsted and Lewis acid sites, evaluated by both IR and N_{1s} XPS analysis, was about 3.5, compared with 4.6–4.7 determined in acid treated Al-ZSM-22 and in Al-ZSM-5. The NH_3 TPD analysis confirmed that the strength of acid sites of Fe-ZSM-22 is comparable to that of Al-ZSM-5 and significantly lower than that of Al-ZSM-22.

As Fe is incorporated in the FAU-type structure, the asymmetric stretching vibration of T-O-T (T=Si, Al or Fe) located at 1020 and 1150 cm^{-1}, shifts to lower frequencies while the intensity of the absorption at 775 cm^{-1} is enhanced [166]. In the hydroxyls region, new bands appear at 3570 and 3643 cm^{-1} in samples with high Fe content, assigned to (OH) vibrations of Si(OH)Fe groups.

As the Fe content increases, the thermal stability of Fe-FAU decreases: the temperature of framework collapse decreases from 1133 K (pure NaY) to 1033 K for a sample containing 5.3 wt.-% Fe_2O_3 [166].

The same behavior was observed in the case of Fe-MFI, for which a partial loss of framework Fe^{3+} occurs during the calcination necessary to eliminate the organic templates [139b]. The loss of Fe is greatly enhanced under steaming conditions (e.g., 550 °C, 100% steam) [139b, 160], displaying a behavior similar to alumino- and gallosilicates.

Evidence for incorporation of Fe^{3+} come also from magnetic moment measurements (μ_{eff} in the range 5.6–5.9 B.M. for isolated ions), the decomposition pattern of the organic template (TGA, DTA), the UV-Vis and ^{29}Si MAS NMR spectra (see [133] and references therein cited).

5.4 Theoretical Studies

The first quantum mechanical approach to the problem of siting of Fe in FAU was performed by Beran et al. [186]. They have modeled a FAU with Fe^{2+}, Fe^{3+} and $Fe(OH)^+$ ions localized in S_{II} and $S_{I'}$ extra-framework positions or with Fe^{3+} in the framework sites, using the CNDO/2 method on a cluster representing the six-membered ring opening. No difference was found for the framework properties when replacing Al with Fe and the framework Fe^{3+} was predicted to be quite stable also in the presence of a reduction to Fe^{2+}.

The siting of Fe in the MFI structure was investigated by Vetrivel et al. (orthorhombic polymorph) [187] and by Lewis et al. (monoclinic polymorph) [188]. Semi-empirical Extended Hückel Molecular Orbital (EHMO) calculations were carried out on dimeric clusters $(OH)_3$-T-O-T-$(OH)_3$ (T = Si^{4+} or Fe^{3+}) with the geometry derived from the structural model determined by Olson et al. [189]. The relative substitution energy in the twelve independent T-sites, calculated without geometrical optimization, varies in the range 18–55 kJ mol^{-1}; the authors indicate site T10, located on the 10-membered ring of the sinusoidal channel, as the most favorable for Fe incorporation.

Defect energy minimization techniques were used for studying the siting of Fe and of charge compensating protons in monoclinic (Fe,Si)-ZSM-5 [188]. The defect energies of the 24 independent T-sites vary within a 82 kJ mol^{-1} range, with the majority of the sites lying within 67 kJ mol^{-1}. Following the notation of van Koningsveld et al. [190], the most stable sites are T19 and T18, while the least stable ones are T4, T12 and T14. The presence of a charge compensating proton induces the elongation of the Fe-O(H) bond from 1.82 (mean Fe-O bond distance) to 2.014–2.04 Å. This type of geometry fits the Fe K-edge EXAFS curves of Fe-ZSM-5. Another result of this approach concerns the variation of pore geometry which occurs upon incorporation of Fe, with the sinusoidal channel diameter reduced by between 0.4 and 0.8 Å and an increase of the diameter of the straight channel of about 0.2 Å. Similar results have been found in the approach of Vetrivel et al. [187]. These results account well for the higher shape selectivity properties displayed by Fe-ZSM-5 with respect to Al-ZSM-5. Evidence in favor of a lower Brønsted acidity of the ferrisilicate with respect to the aluminosilicate was also achieved [187, 188].

6
Other Transition Metal-Containing Molecular Sieves

Though several patents claim the synthesis of many transition-metal containing molecular sieves [191], there is a general lack of evidence supporting their effective incorporation.

The synthesis of chromium-containing molecular sieves has received much attention for the potential catalytic properties of these materials. In fact, it is known that Cr-based materials catalyze several reactions such as dehydrogenation of alkanes [192] and alcohols [193], oligomerization and polymerization of olefins [193]. The literature deals essentially with Cr-pentasils. The early patents claimed Cr incorporation on the basis of the absence of crystalline chromium oxide and the catalytic activity displayed in the isomerization of xylenes. More recently, Cornaro et al. reported that Cr-silicalite-1 can be obtained from a gel containing TEOS, Cr nitrate and alkali free TPAOH (in the presence of Na, green spots of Cr_2O_3 were observed) [194]. Highly crystalline products were obtained in a reproducible way only when the pH was higher than 8.5, while at lower values uncompleted or no crystallization takes place. The maximum Cr content achieved corresponds to Si/Cr = 85 (in as-synthesized form); after calcination and ammonium exchange, the Si/Cr ratio increases to 248, with Cr homogeneously distributed in the crystals, as shown by STEM-EDS analysis. The easy Cr leaching observed during the exchange treatment suggests that the heteroatom is weakly bonded to the silica framework, probably in octahedral coordination.

Sugimoto et al. prepared Cr-silicalite-1 using morpholine (tetrahydro-1,4-oxazine) as template agent [195]. Incorporation of Cr was poorly supported by IR, XRD and EPR measurements: in the H^+-exchanged form, Cr(V) species were identified together with highly dispersed octahedrally coordinated Cr(III) species. TPD experiments showed a low Brønsted acidity, responsible for the high conversion of methanol to C_2-C_4 olefins [195].

Incorporation of Cr does not occur when Na^+ is present in the reaction mixture, probably because of the formation of oligomers during the hydrothermal reaction [196]. These species influence the crystallization process, inhibiting the crystal growth rate in the c-direction. In the final products, Cr is present on the external crystal surface in the form of a stable silicate phase. Other authors claim that Cr(III) can be incorporated also in the presence of alkali [197], with Si/Cr as low as 23. After mild template burnoff (300 °C), and subsequent treatment with water or ammonium nitrate solution, more than 60% of the Cr is removed. Upon calcination at 500 °C, part of the remaining Cr aggregates in form of Cr_2O_3 crystals and the remainder is oxidized to Cr(V) strongly bound to the silica framework. These sites are able to interact with different molecules (e.g., pyridine, NH_3) and display redox properties.

Synthesis of Cr-MFI was performed also in the presence of fluoride ions at pH 6.5–7.0, with a notable amount of Al_2O_3 (up to 1.00 wt.-%) [198]. Incorporation of Cr was weakly supported by IR, XRD and XPS data.

Recently, Cr incorporation was claimed in silicalite-2 (CrS-2) prepared by heating at 170 °C an alkali-free gel containing TEOS, chromium nitrate and TBA-

OH [199]. The authors supported Cr incorporation with XRD, IR and ESR evidence. In particular, they attributed the 963 cm^{-1} IR band to Si-O-Cr stretching mode, and the ESR signal at g=1.97 to an oxochromium(V) species. CrS-2 was reported to be an efficient catalyst for the chemoselective epoxidation of olefins in the presence of tert-butyl hydroperoxide.

The synthesis of the zincosilicate VPI-7 demonstrated that also bivalent transition metal ions can be incorporated into the silica framework in tetrahedral coordination [200]. VPI-7, prepared by using TEAOH as the structure directing agent, is isostructural with the beryllosilicate Lovdarite (LOV) [201]. However, the low thermal stability renders VPI-7 of little interest for practical uses.

7 Conclusions

The synthesis of molecular sieves containing transition metals in the framework represents a reliable route for preparing materials with novel catalytic properties. In spite of the large number of papers and patents claiming the incorporation of several transition metal ions in different microporous silica frameworks, unambiguous evidence in favor of the isomorphous substitution of silicon has been achieved only for Ti and Fe, for the former limited to few framework types. Some evidence exists also for V and Cr, but the data available do not allow definite conclusions to be drawn in this regard.

For Ti-MFI (TS-1) a lot of experimental data have been collected which adequately demonstrate the existence of one type only of Ti-site (tetrahedral coordination in the anhydrous state, expanding up to five and six coordination after adsorption of H_2O, NH_3, H_2O_2, etc.).

Nevertheless, several aspects of the synthesis need to be investigated. In particular, it seems crucial to achieve more detailed information about the phenomena involved in the stabilization of the monomeric Ti species in solution and their interaction with silica. A deeper investigation of the chemistry of titania-silica gels might open the way for synthesizing other Ti-containing molecular sieves.

For Fe-containing molecular sieves, isomorphous substitution has been proved for several framework types. The crucial step in the synthesis may be identified in the stabilization of monomeric Fe^{3+} species at low pH .

Though stable to reduction, Fe^{3+} tends to be partially removed from the framework by thermal treatment especially under steaming conditions. This behavior may be explained with the low tendency of this ion to attain the tetrahedral coordination. As for Al, the framework Fe^{3+} confers acidic and hence catalytic properties to the ferrisilicate molecular sieves.

It is worth pointing out how the present knowledge of these classes of materials has required a great amount of work, involving different research groups and multidisciplinary approaches and how critical it is to assess unambiguously framework incorporation of small amounts of metal in microporous materials.

8 References

1. Cavani F, Giordano G, Pedatella M, Trifirò F (1994) Stud Surf Sci Catal 84:1425
2. Goldsmith JR (1952) Min Mag 29:952
3. Barrer RM, Baynham JW, Bultitude FW, Meier WM (1959) J Chem Soc 195
4. Young DA (1967) US Patent 3 329 480 and US Patent 3 329 481
5. Ione KG, Vostrikova LA, Mastikhin VM (1985) J Mol Catal 31:355
6. Pauling L (1960) "The Nature of Chemical Bond" 3rd edn. Cornell University Press, Ithaca
7. Shannon RD (1976) Acta Crystallogr A32:751
8. Sandomirskii PA, Belov NV (1979) Kristallografiya 24:1198
9. Rastsvetaeve RK, Adrianov VI (1984) Kristallografiya 29:681
10. Taramasso M, Perego G, Notari B (1983) US Patent 4 410 501
11. Taramasso M, Manara G, Fattore V, Notari B (1987) US Patent 4 666 692
12. Bellussi G, Fattore V (1991) Stud Surf Sci Catal 69:79
13. Ruren X, Wenquin P (1985) Stud Surf Sci Catal 24:27
14. Saleh RY (1985) European Patent Appl 132 550
15. Kornatowski J, Malinowski M (1989) Proceedings of the Eighth International Conference on Zeolites, Amsterdam, Recent Research Reports, p 49
16. Notari B (1988) Stud Surf Sci Catal 37:413
17. El Hage Al Asswad J, Nagy JB, Gabelica Z, Derouane EG (1989) Proceedings of the Eighth International Conference on Zeolites, Amsterdam, Recent Research Reports, p 475
18. Notari B (1991) Stud Surf Sci Catal 60:343
19. Bellussi G, Carati A, Clerici MG, Esposito A (1991) Stud Surf Sci Catal 63:421
20. Goepper M, Li H-X, Davis ME (1992) J Chem Soc Chem Commun 1665
21. Thangaraj A, Kumar R, Sivasanker S (1992) Zeolites 12:135
22. Pol AJHP van der, Hooff JHC van (1992) Appl Catal A: General 92:93
23. Kraushaar B (1989) PhD Thesis, University of Eindhoven, The Netherlands
24. Thangaraj A, Kumar R, Mirajkar SP, Ratnasamy P (1991) J Catal 130:1
25. Thangaraj A, Eapen MJ, Sivasanker S, Ratnasamy P (1992) Zeolites 12:943
26. Thangaraj A, Sivasanker S (1992) J Chem Soc Chem Commun 123
27. Mirajkar SP, Thangaraj A, Shiralkar VP (1992) J Phys Chem 96:3073
28. Tuel A, Ben Taarit Y (1994) Appl Catal A: General 110:137
29. Gao X, Suo J, Li S (1995) J Chem Soc Chem Commun 835
30. Tuel A, Ben Taarit Y, Naccache C (1993) Zeolites 13:454
31. Tuel A, Ben Taarit Y (1994) Zeolites 14:594
32. Tuel A, Ben Taarit Y (1993) Microporous Materials 1:179
33. Guth JL, Kessler H, Higel JM, Lamblin JM, Patarin J, Seive A, Chezeou JM, Wey R (1989) ACS Symp Ser 398:176
34. Popa JM, Guth JL, Kessler H (1988) European Patent Appl 292 363
35. Qiu Shinlun, Pang Wenqin, Yao Shangquin (1989) Stud Surf Sci Catal 49A:133
36. Kooyman PJ, Jansen JC, Bekkum H van (1992) Proceedings of the Ninth International Conference on Zeolites, Montreal, p 505
37. Lopez A, Tuilier HM, Guth JL, Delmotte L, Popa JM (1993) J Solid State Chem 102:480
38. Padovan M, Leofanti G, Roffia P (1989) European Patent Appl 311 983
39. Padovan M, Genoni F, Leofanti G, Petrini G, Trezza G, Zecchina A (1991) Stud Surf Sci Catal 63:431
40. Uguina MA, Ovejero G, Grieken R van, Serrano DP, Camacho M (1994) J Chem Soc Chem Commun 27
41. Uguina MA, Serrano DP, Ovejero G, Grieken R van, Camacho M (1995) Appl Catal A 124:391
42. Serrano DP, Uguina MA, Ovejero G, Grieken R van, Camacho M (1995) Microporous Materials 4:273
43. Wu EL, Lawton SL, Olson DH, Rohrman Jr AC, Kokotailo GT (1979) J Phys Chem 83:2777

44. Taramasso M, Perego G, Notari B (1980) Proceeding of the Fifth International Conference On Zeolites, Naples, p 40
45. Perego G, Bellussi G, Corno C, Taramasso M, Buonomo F, Esposito A (1987) Proceedings of the Seventh International Conference on Zeolites, Tokyo, p 129
46. Millini R, Previde Massara E, Perego G, Bellussi G (1992) J Catal 137:497
47. Boccuti MR, Rao KM, Zecchina A, Leofanti G, Petrini G (1989) Stud Surf Sci Catal 49:133
48. Bellussi G, Carati A, Clerici MG, Maddinelli G, Millini R (1993) J Catal 133:220
49. Boccuzzi F, Coluccia S, Ghiotti G, Morterra C, Zecchina A (1978) J Phys Chem 82:1298
50. Pilz W, Peuker Ch, Tuan VA, Fricke R, Kosslick H (1993) Ber Bunsenges Phys Chem 97:1037
51. Varshal BG, Bobrov AV, Marvin BN, Iljuchin VV, Belov NV (1974) Dock Akad Nak SSSR 216:374
52. Zecchina A, Spoto G, Bordiga S, Ferrero A, Petrini G, Leofanti G, Padovan M (1991) Stud Surf Sci Catal 69:251
53. Geobaldo F, Bordiga S, Zecchina A, Giamello E, Leofanti G, Petrini G (1992) Catal Lett 16:109
54. Huybrechts DRC, Vaesen I, Li HX, Jacobs PA (1991) Catal Lett 8:237
55. Clerici MG, Ingallina P, Millini R (1993) Proceedings of the Ninth International Conference on Zeolites, Montreal, p 445
56. Tuel A, Diab J, Gelin P, Dufaux M, Dutel J-F, Ben Taarit Y (1990) J Mol Catal 63:95
57. Behrens P, Felsche J, Vetter S, Schulz-Ekloff G, Jaeger N, Newmann W (1991) J Chem Soc Chem Commun 678
58. Behrens P, Felsche J, Newmann W (1991) Catal Today 8:479
59. Schultz E, Ferrini C, Prins R (1992) Catal Lett 14:221
60. Pei S, Zajac GW, Kaduk JA, Faber J, Boyanov BI, Duck D, Fazzini D, Morrison TI, Yang DS (1993) Catal Lett 21:333
61. Bordiga S, Boscherini F, Coluccia S, Genoni F, Lamberti C, Leofanti G, Marchese L, Petrini G, Vlaic G, Zecchina A (1994) Catal Lett 26:195
62. Bordiga S, Coluccia S, Lamberti C, Marchese L, Zecchina A, Boscherini F, Buffa F, Genoni F, Leofanti G, Petrini G, Vlaic G (1994) J Phys Chem 98:4125
63. Trong On D, Bittar A, Sayari A, Kaliaguine S, Bonneviot L (1992) Catal Lett 16:85
64. Trong On D, Bonneviot L, Bittar A, Sayari A, Kaliaguine S (1992) J Mol Catal 74:233
65. Jentys A, Catlow CRA (1993) Catal Lett 22:251
66. Millini R, Perego G, Seiti K (1994) Stud Surf Sci Catal 84C:2123
67. Koningsveld H van, Bekkum H van, Jansen JC (1987) Acta Cryst B43:127
68. Oumi Y, Matsuba K, Kubo M, Inui T, Miyamoto A (1995) Microporous Materials 4:53
69. Lok BMT, Bonita KM, Flanigen EM (1984) US Patent 4 707 345
70. Kokotailo GT, Chu P, Lawton SL, Meier WM (1978) Nature 275:119
71. Perego C, Cesari M, Allegra G (1984) J Appl Cryst 17:403
72. Perego G, Bellussi G, Carati A, Millini R, Fattore V (1989) ACS Symp Ser 398:360
73. Taramasso M, Manara G, Fattore V, Notari B (1980) GB Patent 2 024 790
74. Bellussi G, Carati A, Clerici MG, Esposito A, Millini R, Buonomo F (1989) Belgian Patent 1 001 038
75. Reddy JS, Kumar R, Ratnasamy P (1990) Appl Catal 58:L1
76. Reddy JS, Kumar R (1992) Zeolites 12:95
77. Serrano DP, Hong-Xin L, Davis ME (1992) J Chem Soc Chem Commun 745
78. Reddy KM, Kaliaguine S, Sayari A, Ramaswamy AV, Reddy VS, Bonneviot L (1994) Catal Lett 23:175
79. Reddy KM, Kaliaguine S, Sayari A (1994) Catal Lett 23:169
80. Tuel A, Ben Taarit Y (1995) Zeolites 15:164
81. Camblor MA, Corma A, Martinez A, Perez-Pariente J (1992) J Chem Soc Chem Commun 589
82. Camblor MA, Corma A, Perez-Pariente J (1993) Zeolites 13:82
83. Blasco T, Camblor MA, Corma A, Perez-Pariente J (1993) J Am Chem Soc 115:11806
84. Tuel A (1995) Zeolites 15:236
85. Chapman DM, Roe AL (1990) Zeolites 10:730

86. Beyer HK, Belenykaja I (1980) Stud Surf Sci Catal 5:203
87. Skeels GS, Breck DW (1984) Proceedings of the Sixth International Conference on Zeolites, Reno, USA, p 87
88. Skeels GW, Ramos R, Breck DW (1985) WO Patent 85/04854
89. Kraushaar B, Hooff JHC van (1988) Catal Lett 1:81; (1990) ibidem 2:43
90. Ferrini C, Kouvenhoven HW (1990) Stud Surf Sci Catal 55:53
91. Carati A, Contarini S, Millini R, Bellussi G (1990) ACS Symposium on Synthesis and Properties of New Catalysts, Boston (USA), Mat Res Soc Ext Abstracts (EA-24):47
92. Kooyman PJ, Waal P van der, Verdaasdonk AJ, Jansen KC, Bekkum H van (1992) Catal Lett 13:229
93. Bond GC, Sarkany AJ, Parfitt GD (1979) J Catal 57:476
94. Bond GC, Koenig P (1982) J Catal 77:309
95. Gellings PJ (1985) Catalysis 7:105
96. Centi G, Trifirò F, Ebner JR, Franchetti V (1989) Chem Rev 28:400
97. Marosi L, Stabenow J, Schwarzman M (1980) DE Patent 2 831 631
98. Inui T, Yamase O, Fukuda K, Itoh A, Tarumoto J, Morinaga N, Hagiwara T, Takegami Y (1984) Eighth International Congress of Catalysis – Preprint, Vol III, p 569
99. Miyamoto A, Medhanavyn D, Inui T (1986) Appl Catal 28:89
100. Inui T, Medhanavyn D, Praserthdam P, Fukuda K, Ukawa T, Sakamoto A, Miyamoto A (1985) Appl Catal 18:311
101. Kornatowski J, Sychev M, Goncharuk V, Baur WH (1991) Stud Surf Sci Catal 65:581
102. Hong SB, Kim CG, Uh YS, Park YK, Woo SI (1992) Korean J Chem Eng 9:16
103. Kornatowski J, Wichterlova B, Rozwadowski M, Baur WH (1994) Stud Surf Sci Catal 84A:117
104. Habersberger K, Jiru P, Tvaruzkova Z, Centi G, Trifirò F (1989) React Kinet Lett 39:95
105. Bellussi G, Maddinelli G, Carati A, Gervasini A, Millini R (1993) Proceedings of the Ninth International Conference on Zeolites, Montreal, p 207
106. Rigutto MS, Bekkum H van (1991) Appl Catal 68:L1
107. a) Fejes P, Marsi I, Kiricsi I, Halasz J, Hannus I, Rocjeknbauer A, Tasi Gy, Korecz L, Schoebel Gy (1991) Stud Surf Sci Catal 69:173, b) Fejes P, Halasz J, Kiricsi I, Kele Z, Hannus I, Fernandez C, Nagy JB, Rockenbauer A, Schoebel Gy (1993) Stud Surf Sci Catal 75:421
108. Centi G, Perathoner S, Trifirò F, Aboukais A, Aissi CF, Guelton M (1992) J Phys Chem 96:2617
109. a) Hari Prasad Rao PR, Ramaswamy AV, Ratnasamy P (1992) J Catal 137:225, b) Hari Prasad Rao PR, Kumar R, Ramaswamy AV, Ratnasamy P (1993) Zeolites 13:663, c) Hari Prasad Rao PR, Belhekar AA, Hegde SG, Ramaswamy AV, Ratnasamy P (1993) J Catal 141:595
110. a) Tuel A, Ben Taarit Y (1993) Appl Catal A 102:201, b) Tuel A, Ben Taarit Y (1994) Zeolites 14:18
111. Reddy KR, Ramaswamy AV, Ratnasamy P (1992) J Chem Soc Chem Commun 1613
112. Kumar R, Reddy KR, Raj A, Ratnasamy P (1993) Proceedings of the Ninth International Conference on Zeolites, Montreal, p 189
113. Dubanska V. (1992) CERAMICS-Silikaty 36:31
114. Martini G, Ottaviani MF, Seravalli GL (1975) J Phys Chem 79:1716
115. Willigen H van, Chandrashekar TK (1983) J Am Chem Soc 105:4232
116. Kucherov AV, Slinkin AA (1987) Zeolites 7:38
117. Kucherov AV, Slinkin AA (1987) Zeolites 7:583
118. Sass CE, Chen X, Kevan L (1990) J Chem Soc Faraday Trans 86:189
119. Petras M, Wichterlova (1992) J Phys Chem 96:1805
120. Huang M, Shan S, Yuan C, Li Y, Wang Q (1990) Zeolites 10:772
121. Whittington BI, Anderson JR (1991) J Phys Chem 95:3306
122. Whittington BI, Anderson JR (1993) J Phys Chem 97:1032
123. Takahashi H, Shiotani M, Kobayashi H, Sohma J (1969) J Catal 14:134
124. Dessau RM, Schmitt KD, Kerr GT, Woolery GL, Alemany LB (1987) J Catal 104:484

125. Dessau RM, Schmitt KD, Kerr GT, Woolery GL, Alemany LB (1988) J Catal 109:472
126. Kraushaar B, Haan JW de, Hooff JHC van (1988) J Catal 109:470
127. Kraushaar B, Den LJM van de, Haan JW de, Hooff JHC van (1988) Stud Surf Sci Catal 37:167
128. Yamagishi K, Namba S, Yashima T (1991) J Phys Chem 95:872
129. Zecchina A, Bordiga S, Spoto G, Marchese L, Petrini G, Leofanti G, Padovan M (1992) J Phys Chem 96:4985
130. Zecchina A, Bordiga S, Spoto G, Marchese L, Petrini G, Leofanti G, Padovan M (1992) J Phys Chem 96:4991
131. Marra GL, Tozzola G, Leofanti G, Padovan M, Petrini G, Genoni F, Venturelli B, Zecchina A, Bordiga S, Ricchiardi G (1994) Stud Surf Sci Catal 84:559
132. Roque-Malherbe R, Diaz-Aguila C, Reguera-Ruiz E, Fundora-Lliteras J, Lopez-Colado L, Hernandez-Velez M (1990) Zeolites 10:685
133. Ratnasamy P, Kumar R (1991) Catal Today 9:327
134. Hirao K, Soga N, Kunugi M (1976) J Phys Chem 80:1612
135. Hazel D, Schock Jr RU, Gorden M (1949) J Am Chem Soc 71:2256
136. Weber Jr WJ, Stumm W (1965) J Inorg Nucl Chem 27:237
137. Porter RA, Weber Jr WJ (1971) J Inorg Nucl Chem 33:2443
138. Ratnasamy P, Borade RB, Sivasanker S, Shiralkar VP, Hedge SG (1985) Acta Phys Chem 31:137
139. a) Szostak R, Thomas TL (1986) J Catal 100:555, b) Szostak R, Nair V, Thomas TL (1987) J Chem Soc Faraday Trans I 83:487
140. Ball WJ, Dwyer J, Garforth AA, Smith WJ (1987) Proceedings of the Seventh International Conference on Zeolites, Tokyo, p 137
141. Borade RB (1987) Zeolites 7:398
142. Calis G, Frenken P, Boer E de, Swolfs A, Hefni MA (1987) Zeolites 7:319
143. Doppler G, Lehnert R, Marosi L, Trautwein AX (1988) 37:215
144. Kotasthane AN, Shiralkar VP, Hedge SG, Kulkarni SB (1986) Zeolites 6:253
145. Hagen A, Roessner F, Weingart I, Spliethoff B (1995) Zeolites 15:270
146. Lazar K, Borbely G, Beyer H (1991) Zeolites 11:214
147. Patarin J, Kessler H, Guth JL (1990) Zeolites 10:674
148. Patarin J, Tuilier MH, Durr J, Kessler H (1992) Zeolites 12:70
149. a) Inui T, Nagata H, Yamase O, Matsuda H, Kuroda T, Yoshikawa M, Takeguchi T, Miyamoto A (1986) Appl Catal 24:257, b) Inui T, Matsuda H, Yamase O, Nagata H, Fukuda K, Ukawa T, Miyamoto A (1986) J Catal 98:491, c) Inui T, Nagata H, Takeguchi T, Iwamoto S, Matsuda H, Inoue M (1993) J Catal 139:482
150. Kumar R, Raj A, Kumar SB, Ratnasamy P (1994) Stud Surf Sci Catal 84:109
151. Catana G, Pelgrims J, Schoonheydt RA (1995) Zeolites 15:475
152. Reddy JS, Reddy KR, Kumar R, Ratnasamy P (1991) Zeolites 11:553
153. Szostak R, Thomas TL (1986) J Chem Soc Chem Commun 1986:113
154. Bellussi G, Millini R, Carati A, Maddinelli G, Gervasini A (1990) Zeolites 10:642
155. Borade RB, Adnot A, Kaliaguine S (1991) Zeolites 11:710
156. Singh AP (1992) Zeolites 12:858
157. Kumar R, Ratnasamy P (1990) J Catal 121:89
158. Kumar R, Thangaraj A (1989) Proceedings of the Eighth International Conference on zeolites, Amsterdam, Recent Research Reports, p 53
159. Li R, Xu W, Wang J (1992) Zeolites 12:716
160. Kumar R, Thangaraj A, Bhat RN, Ratnasamy P (1990) Zeolites 10:85
161. Borade RB, Clearfield A (1994) Microporous Materials 2:167
162. Chandwadkar AJ, Bhat RN, Ratnasamy P (1991) Zeolites 11:42
163. McNicol BD, Pott GT (1972) J Catal 25:223
164. Derouane EG, Mestdagh M, Vielvoye L (1974) J Catal 33:169
165. Navakova J, Kubelkova L, Wichterlova B, Juska T, Dolejsek Z (1982) Zeolites 2:17
166. Ratnasamy P, Kotasthane AN, Shiralkar VP, Thangaraj A, Ganapathy S (1989) ACS Symp Ser 398:405

167. Zi G, Dake T, Ruiming Z (1988) Zeolites 8:453
168. Joshi PN, Awate SV, Shiralkar VP (1993) J Phys Chem 97:9749
169. Duke CVA, Latham K, Williams CD (1995) Zeolites 15:213
170. Skeels GW, Flanigen EM (1989) ACS Symp Ser 398:421
171. Kucherov AV, Slinkin AA (1988) Zeolites 8:110
172. Wichterlova B, Beran S, Bednarova S, Nedomova K, Dudikova L, Jiru P (1988) Stud Surf Sci Catal 37:199
173. Vaughan DEW, Strohmaier KG, Pickering IJ, George GN (1992) Solid State Ionics 53–56:1282
174. Pickering IJ, Vaughan DEW, Strohmaier KG, George GN, Via H (1993) Proceedings of the Ninth International Conference on Zeolites, Montreal, p 595
175. Lin DH, Coudurier G, Vedrine JC (1989) Stud Surf Sci Catal 49:1431
176. Park JW, Chon H (1992) J Catal 133:159
177. Goldfarb D, Bernardo M, Strohmaier KG, Vaughan DEW, Thomann H (1994) J Am Chem Soc 116:6344
178. Goldfarb D, Bernardo M, Strohmaier KG, Vaughan DEW, Thomann H (1994) Stud Surf Sci Catal 84:403
179. Garten RL, Delgass WN, Boudart M (1970) J Catal 18:90
180. Meagher A, Nair V, Szostak R (1988) Zeolites 8:3
181. Kumar R, Date SK, Bill E, Trautwein A (1991) Zeolites 11:211
182. Chandwadkar AJ, Date SK, Bill E, Trautwein A (1992) Zeolites 12:180
183. Scarano D, Zecchina A, Bordiga S, Geobaldo F, Spoto G, Petrini G, Leofanti G, Padovan M, Tozzola G (1993) J Chem Soc Faraday Trans 89:4123
184. Chu CT-W, Chang CD (1985) J Phys Chem 89:1569
185. Kustov LM, Kazansky VB, Ratnasamy P (1987) Zeolites 7:79
186. Beran S, Jiru P, Wichterlova B (1982) Zeolites 2:252
187. Vetrivel R, Pal S, Krishnan S (1991) J Mol Catal 66:385
188. Lewis DW, Catlow CRA, Sankar G, Carr SW (1995) J Phys Chem 99:2377
189. Olson DH, Kokotailo GT, Lawton SL, Meier WM (1981) J Phys Chem 85:2238
190. Koningsveld H van, Jansen JC, Bekkum H van (1990) Zeolites 10:235
191. Szostak R (1989) Molecular sieves. Principle of synthesis and identification. Van Nostrand Reinhold, New York, p 209
192. Carrà S, Forni L (1972) Catal Rev Sci Eng 5:159
193. McDaniel MP, Welch MB (1983) J Catal 82:98
194. Cornaro U, Zanetti E, Stocchi B (1991) Atti 1° Convegno Nazionale di Scienza e Tecnologia delle Zeoliti, L'Aquila, p 204
195. Sugimoto M, Katsuno H, Takatsu K, Kawata N (1992) Appl Catal A 80:13
196. Puil N van der, Widyawati, Jansen JC, Bekkum H van (1994) Stud Surf Sci Catal 84:211
197. Kucherov AV, Slinkin AA, Beyer GK, Borbely G (1995) Zeolites 15:431
198. Mambrim JST, Pastore HO, Davanzo CU, Vichi EJS, Nakamura O, Vargas H (1993) Chem Mater 5:166
199. Reni Joseph, Sasidharan M, Kumar R, Sudalai R, Ravindranathan T (1995) J Chem Soc Chem Commun 1341
200. Annen MJ, David ME, Higgins JB, Schlenker JL (1991) J Chem Soc Chem Commun 1175
201. Röhrig C, Gies H, Marler B (1994) Zeolites 14:498

Synthesis of Zeolite-Like Inorganic Compounds

S. A. Schunk · F. Schüth

Institut für Anorganische Chemie, Johann-Wolfgang-Goethe-Universität, Marie-Curie-Straße 11, D-60439 Frankfurt a. M., Germany. *E-mail: ferdi@schueth.chemie.uni-frankfurt.de*

1 Introduction

Zeolite and aluminophosphate synthesis is already highly developed to produce materials with desired properties such as catalytic activity, good adsorption or ion exchange behavior, optical transparency or satisfactory guest incorporation. Although microporous lattices of zeolite systems are powerful tools for shape

Molecular Sieves, Vol. 1

and size selective catalysis or the organization of small organic molecules, sometimes even simple features like phase purity can be a big problem that has to be overcome. In addition, zeolite systems have a number of limitations that are not favorable for many applications. Framework constituents are predominantly boron, aluminum, gallium, silicon and phosphorus as T-atoms, bridged by oxygen. Thus, these materials are group III to V element oxides of the second and third period, and it can easily be seen that the resulting materials will be insulators. Another disadvantage of zeotype materials is certainly that transition metal uptake in the framework is in most cases limited to a few percent, and overloading will often lead to a collapse of the framework. Looking at many catalytic applications this is a limiting factor and there might also arise problems for possible optical applications.

These and other limitations have been at least part of the motivation of many scientists to look for new compounds that form porous frameworks and show properties that go even beyond the properties that zeotype compounds do exhibit. Their success is proved by the variety of materials that have been found and investigated during the past two decades. However, other than in molecular sieve science, determination of the sorptive properties of such materials is not always a standard technique in more conventional solid state chemistry. Therefore, for many of the examples covered in this review, the structural porosity of the crystals has not been determined. In addition, many of the structures discussed in the following could so far not be obtained in a form where the guest species, occluded in the voids of the structures, are removed. Nevertheless, we chose to include many such examples in this chapter, because the guest species often do not seem to be an integral and necessary part of the framework and thus in principle might be removed from the structure to give an accessible pore system. There is certainly no sharp boundary between dense structures and open framework structures. Many dense structures, for instance, are often discussed in terms of interpenetrating frameworks, in which each sublattice could be described as a tunnel structure. However, in most cases, there does not seem to be a prospect to obtain any of the sublattices without the remaining part of the structure. Such materials are beyond the scope of this chapter, and if mentioned, only serve for comparison.

There are differences between the scientific communities working on molecular sieves and conventional solid state chemistry, respectively, with respect to the conventions used to describe pore dimensions. While in the zeolite community usually open diameters are given, which are important for the applications in sorption or catalysis, mostly atom center distances over the channels are quoted by solid state chemists who are usually more interested in the structural details. In the following we will mostly refer to open diameters, unless specifically stated otherwise. These values are less precise than the atom center distances, which can be determined with high precision by X-ray diffraction, but are more important when applications of such materials are envisaged.

The synthetic concepts by which these compounds are synthesized are as numerous as the different materials themselves. They range from classical hydrothermal reactions, via molten flux methods to pure precipitation reactions. As diverse as the conditions employed to synthesize such compounds are the structural motifs which are present in these structures. While classical molecular sieves like zeolites or aluminophosphates are exclusively formed from tetrahedrally oxygen-coordinated framework atoms, there are many examples with octahedrally coordinated framework atoms or even more unusual coordination spheres found in the non-classical compounds. Since the structural elements encountered are so diverse, organization of this chapter along lines of structural properties of the materials did not seem to be an appropriate approach. We therefore chose the chemical composition as the guideline for this chapter, starting with oxidic frameworks with different metal centers, gradually moving to frameworks with additional electronegative framework species and finally to oxygen free materials. For most of these chemical compositions different structures, like layers, tunnel or three-dimensional framework structures are possible, as will be seen in the following sections.

In the next section we will first present various synthetic approaches to prepare such porous frameworks and then discuss the different classes of materials that we think are outstanding and representative. For every class of material we will discuss the special properties and show the different synthetic pathways that lead to the desired structure. Although such an overview can hardly be complete, we have attempted to cover the most important examples in this review.

2
Synthetic Methods

Zeolite chemists rely almost exclusively on the use of hydrothermal reactions for the preparation of molecular sieves. Since the chemistry is very diverse, if other framework constituents are used for the formation of porous structures, there is also a higher diversity in the synthetic approaches (Fig. 1).

Hydrothermal or solvothermal reactions are also an important method for the synthesis of unconventional zeolite-like compounds. The reagents are usually dissolved or dispersed in water and then heat-treated at elevated temperatures in an autoclave. For reactions at moderate temperatures up to about 250 °C the autoclaves are normally lined with Teflon. If higher reaction temperatures are desired, the reactions are, for instance, carried out in stainless steel or gold tubes. However, in favorable cases, the use of higher temperatures might not even be necessary. The products sometimes just precipitate from solution at room temperature.

One of the most regularly employed methods in the synthesis of dense solids, viz. solid state reactions, is also in some cases used for the synthesis of open framework structures. However, such reactions usually need fairly high temperatures to reach sufficiently high reaction rates. These high temperatures are in most cases not suitable to retain an open framework structure. Solid state reactions are therefore only of limited use if open framework structures shall be prepared.

Also in the group of high temperature methods fall processes where crystallization takes place from the melt, often under addition of a "flux" which mobilizes the reagents. However, the same restrictions as for solid state reactions apply here. The high temperatures usually employed are not in accordance with the formation of open framework structures.

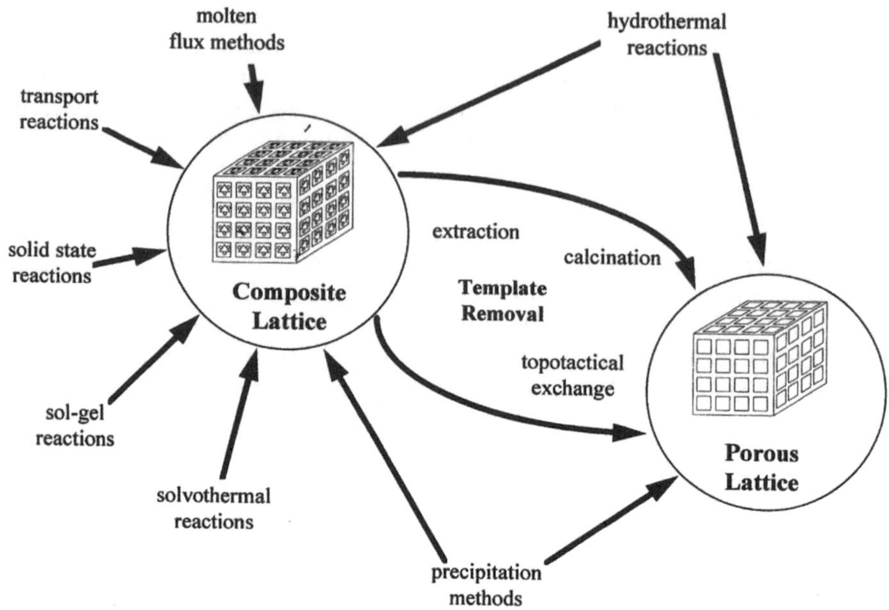

Fig. 1. Pathways for the formation of zeolite-like composite lattices and porous lattices

Other methods used in solid state chemistry are of very limited importance. Transport reactions of volatile species have in some cases been found to be a suitable pathway, like in the formation of nitridosodalites. However, these are special cases with no general applicability.

In many cases it is necessary to post-treat the prepared samples in order to obtain accessible pore systems or to exchange one guest species for another one. Here the techniques known from zeolite chemistry can be employed as well. Calcination is not so generally useful as in zeolite science, since the compounds are quite frequently susceptible to oxidation or reduction. Therefore, ion exchange or extraction methods are more important in the field of non-conventional open framework structures.

Examples of the pathways reviewed in this section will be found in the following discussion of the different groups of materials.

3 Oxidic Frameworks

3.1 Examples Occurring in Nature

3.1.1 *Hollandite and Related Materials*

Hollandite- or cryptomelane-type manganese oxides have a one-dimensional tunnel structure comprised of tunnels with an almost quadratic cross section with two MnO_6-octahedra on each side (2 × 2). The MnO_6 octahedra are linked together sharing edges and forming a wall of double chains, where each wall shares corners with neighboring walls [1]. The one-dimensional tunnels with an open diameter of 0.4–0.5 nm (Fig. 2a) generally accommodate large cations and water. A wide variety of transition metals can substitute for Mn [1], also modifying the size of the unit cell, and generally speaking it is very easy to substitute the large cations, depending on their charge and radius [2]. The synthesis of the pure manganese oxide hollandite is possible via a hydrothermal reaction of dilute solutions of $LiMnO_4$ and $Mn(NO_3)_2$ in sulfuric acid at 100 °C [3]. Evidently lithium cations stabilize voids in the lattice, balance the framework charges and prevent the formation of dense manganese oxide structures. Analogous to syn-

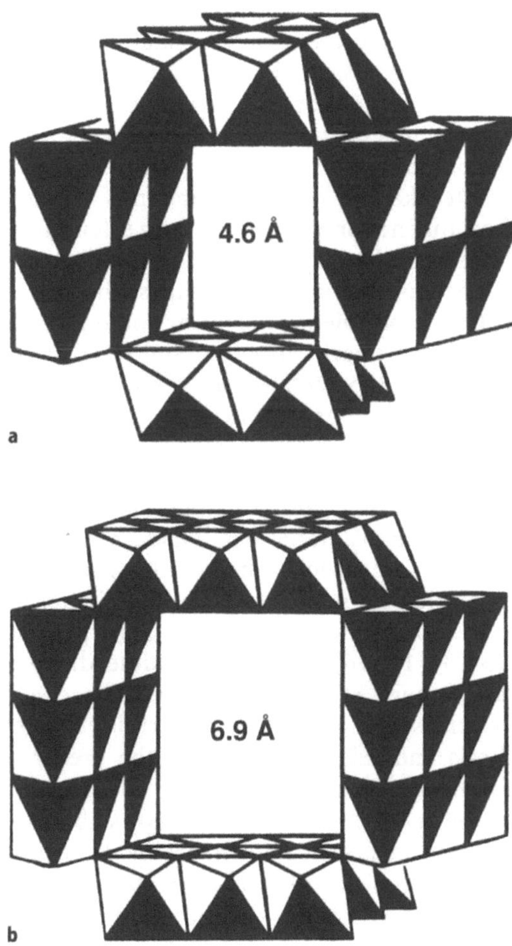

Fig. 2. a The 2 × 2 structure of hollandite is formed by edge sharing MO_6 octahedra arranged in a sheet-like manner. **b** In the todorokite structure the octahedral sheets have the 3 × 3 dimension and form bigger channels

thetic concepts developed for zeolites, more moderate conditions seem to favor the formation of the porous framework [4]. However, there are exceptions to this rule of thumb, as can be seen for the titanium-based hollandite: Totally different synthetic strategies are employed to obtain these materials [5–7]. The general synthetic method here is to heat a mixture of an alkali carbonate or metal and TiO_2 under reducing atmosphere or vacuum, if desired together with other metal oxides, to temperatures up to 1000 °C or higher. Obviously, the aim here is to decompose the carbonate at these high temperatures and form the hollandite phase. The use of a reducing atmosphere [6, 7] is important in order to favor the formation of Ti^{3+} in the hollandite framework which produces the negative framework charge. The authors report that upon heating in air the dioxide bronze even changes to non-stoichiometric titanate about 680 °C [7]. In order to prevent collapse of this framework, other metals which preferably have a valency of III are incorporated, resulting in a higher stability of the materials, which can then even be successfully employed as $DeNO_x$-catalysts in an oxidizing atmosphere at 700 °C [6]. However, it has to be emphasized here that the conversion rates for the $DeNO_x$ -reaction are not very high, one of the main reasons being the very low BET surface area of the catalyst in the order of only a few square meters per gram. The pores of this material are thus not accessible.

The list of hollandite-like and related compounds is extensive, and recent developments [8] show that not only sulfur but even selenium and tellurium can act as oxygen substituents in these systems. Such phases are prepared in similar ways to the titanium systems. Although most of these materials have no porosity detectable in adsorption experiments, they bear some resemblance to zeolitic structures. All of these frameworks have the ability of intercalating and exchanging ions via reversible topotactical redox reactions. This concept allows experiments on host-guest systems in which the guest changes the electronic state. Exchange reactions of cations can mostly be carried out under very mild conditions [2, 9], leading to a large variety of electron or ion conducting materials that are often metastable and cannot be obtained otherwise. This offers a broad field for the use of such materials: metalloxide-hydrogen bronzes are used as passive optical indicators, or especially hollandite-type manganese oxides can be used as cathode materials for non-aqueous lithium batteries [3, 10, 11].

Closely related to the hollandite structure (2 × 2) are psilomelane (2 × 3) [12, 13] and todorokite (3 × 3) [14–16] with an open tunnel diameter of 0.6–0.7 nm (Fig. 2b). All of these mineral structures have been successfully synthesized, which shows in an impressive manner how powerful the synthetic concepts of hydrothermal synthesis and selective choice of the precursor material were employed. In the following we will refer mainly to the results of Suib et al. [16–18] as they were the first to synthesize thermally stable todorokite. Looking at the synthetic conditions it has to be known that hydrothermal reactions of the polymorphs of MnO_2 are known to be very sensitive with regard to the precursor material, as well as the reaction conditions [19]. The challenge for Suib's group was to find a suitable precursor and suitable reaction conditions. The first step is the preparation of the precursor, which is precipitated as birnessite, a layered sodium type manganese oxide with an average manganese oxidation state of 3.7, which is then ion-exchanged with Mg^{2+} to yield the layered manganese oxide buserite. The conditions used to prepare the precursor are decisive with respect to the final todorokite. This precursor is converted to todorokite under hydrothermal conditions between 155 and 170 °C for 40 hours. The products obtained by that procedure were then thoroughly washed again and calcined between 200 and 500 °C for 1 hour. It is very interesting to see that the primary factor for the quality of the resulting material is the MnO_4^-/Mn^{2+} ratio that is employed to synthesize the layered precursor material, then other factors like pH, temperature, aging and autoclave treatment play an important role as well [16]. Also the Mn/Mg ratio of the layered precursors is an important factor for the crystallinity, phase purity and thermal stability of the resulting todorokite.

The best todorokites synthesized following an optimized procedure were thermally stable up to 500 °C, like natural todorokite, and their adsorption capacities were 18.2 wt.-% for cyclohexane and 20 wt.-% for carbon tetrachloride. 1,3,5-triethylbenzene was not adsorbed, probably because it is too bulky to enter the pore system. Suib and his group could already prove that synthetic todorokite is a material with a high potential for many applications which range from catalytic reactions over pure or isomorphously substituted materials [16, 18] where the materials show good catalytic activity for total and selective oxidation to their use as tunable semiconductors or sensor materials [20]. Regarding the fact that a precursor material with Mg^{2+} cations was employed for the synthesis of the (3 × 3) structure of todorokite, it does not seem improbable that, in analogy to zeolite synthesis, with bigger templating cations like amines the larger tunnel structures like (3 × 4) or (3 × 5) which are observed with TEM in natural materials might be obtained [21].

3.1.2
Pharmacosiderite and Nenadkevichite

The two examples mentioned in the previous section are not the only ones also found in nature. In addition to these purely octahedral structures, other porous materials like the structural analogues of pharmacosiderite [22, 23], an octahedral-tetrahedral potassium-iron hydroxoarsenate with 8-membered rings, and nenadkevichite [24], a titanium-niobium-silicate with 8-membered rings and octahedral-tetrahedral coordination of the framework constituents have been synthesized via gel precipitation processes combined with hydrothermal reactions. Many different framework compositions are synthetically available for these compounds, the most thoroughly investigated ones being germanium analogues [22, 25] and titaniumsilicates. In the case of the cesium-titanium-silicate analogue of pharmacosiderite it was possible to prepare single crystals, using a temperature of 750 °C and a pressure of approximately 2000 bar [26], while milder hydrothermal conditions around 200 °C resulted in the formation of a polycrystalline material with the same structure [23]. No adsorption data or catalytic results have been reported as yet to the best of our knowledge, but the ion exchange behavior of pharmacosiderite was investigated in detail. The samples studied showed high selectivities for strontium and cesium [23] which might make them useful for the clean-up of radioactive waste.

In the following we will come back to other structures that have natural analogues.

3.2
Polyoxometallates

Polyoxometallates are outstanding as multianion-cluster ligands, cryptates and clathrates, their ability to accept and store electrons is unique, and so far no other class of materials has proven to be so easily reducible as a ligand. Two different sorts of ligands can be discriminated, firstly so-called planar structures where metal cations are coordinated by terminal or bridging oxygen atoms and, secondly, defect or lacuna structures where the cations are placed in the hole of the cluster. These lacuna structures bear in their properties certain similarities to zeolite-like compounds, such as ion exchange in cavities or acidic properties. Therefore, the polyoxometallates are included in this review.

The structural chemistry of the polyoxometallates, induced by the polymerization of acid solutions of molybdenum(VI)- or tungsten(VI)-ions are certainly amongst the most complex systems that have been investigated [27–29]. The research on these systems is complicated by the fact that some of the equilibria of polyoxometallates in solutions adjust very slowly, especially for tungsten, so that it is very difficult to obtain equilibrium information. Many complementary techniques usually have to be employed to reach a realistic picture of the species present in solution.

As mentioned above, the classical polyoxoanion forming metals are molybdenum and tungsten in the oxidation state VI. It is assumed that for these metal cations the combination of ion-radius and -charge and the availability of empty d-orbitals for the formation of metal-oxygen-π-bonds is especially favorable. However, other metals can act as polyoxoanion-builders as well. Vanadium-, niobium- and tantalum-V, technetium-, rhenium-, ruthenium- and osmium-VI, chromium-, molybdenum-, tungsten-, technetium- and rhenium-V and titanium-, vanadium-, chromium-, molybdenum- and tungsten-IV can build polyoxometallate-clusters.

Three basic polyhedra, the tetrahedron, the octahedron and the icosahedron, can be used to structurally describe polyoxometallates. Very common and well known is the so-called Keggin structure ($XM_{12}O_{40}$) that many polyoxometallates tend to form in solution. The self-organization of a cluster in T_d-symmetry out of twelve units of the same symmetry even occurs if the core of the cluster is not tetrahedral. Derivatives of the Keggin structure can be isomers formed by simple symmetric operations, further condensation products or fragmented defect structures, the so-called lacuna species. The second family of ($XM_{12}O_{38}$) species with O_h-symmetry has until now not been observed for discrete polyoxometallates [28], but many derivatives are known which are basically fragments of the ideal structure like, for example, in the mineral sherwoodite. These structures are all based on MO_6-octahedra bound over edges and can be regarded as fragments of a cubic dense packing of oxygen atoms. The last of the three basic structures with I_h-symmetry is only known for a few heteropolymolybdates with lanthanoids and actinoids in the oxidation state IV. The

$(XM_{12}O_{42})$-cluster has a large cavity to embed the 4 or 5f-elements. No defect structures of this species are known so far.

Polyoxometallates are mainly synthesized in aqueous phases, classically by acidifying an alkali metal molybdate or tungstate, if desired together with a salt of a heteroelement. The underlying aqueous chemistry of the polyoxometallates is mainly coupled to the pH of the solution, and as a general tendency it can be stated for all polyoxometallates that low pH favors condensation, the formation of aggregates and finally the precipitation of the hydrous oxides, whereas under basic conditions more monomeric or smaller clusters are formed. However, one has to be aware, that the reactions taking place in solution are always equilibrium reactions where rarely just one species is present [29], and it can even happen that an isolated crystalline compound is not the main species in solution at that pH. However, chemistry with polyoxometallates cannot only be carried out in aqueous systems: Depending on the charge compensating cation, polyoxometallates are well soluble in organic solvents, from non-polar ones like benzene to polar solvents like dimethylsulfoxide. This offers many applications in the field of organometallic chemistry. Furthermore, polyoxometallates have normally a fairly high thermal stability and are not sensitive towards oxidation even with strong oxidizing agents.

Polyoxometallates are used in many applications [28], the most important one certainly being acid or oxidation catalysis [28, 30, 31]. Polyoxometallates can, thanks to their good overall solubility, be employed in homogeneous catalysis but they are equally suited for heterogeneous catalysis, since they can be easily dispersed on various solids. In solution, heteropolyacids are stronger than most mineral acids like HCl, HNO_3 or H_2SO_4, and their strength depends weakly on their composition, yet it can be stated that generally tungsten acids are markedly stronger than molybdenum ones [30]. Polyoxometallates have some advantages compared to mineral acids: Since they are stronger acids and better proton donors, they are generally more active catalysts, but they can also act as stabilizing agents for organic intermediates and enhance the reaction rate by a strong positive salt effect. In heterogeneous systems polyoxometallates are very efficient catalysts as well, compared to the conventional ones like Al_2O_3, SiO_2-Al_2O_3 or zeolites like HY. This is a clear result of their high acid strength, their high proton mobility and their ability to adsorb large amounts of polar molecules [30].

For heterogeneous catalysis it is very important to disperse the polyoxometallates on a catalytic support as the crystalline solids have a low BET surface area (1 – 5 m^2g^{-1}). The voids in the polyoxometallate structures are in most cases too small to admit substrate molecules.

To solve this problem without needing an additional support, interesting new pathways in the field of polyoxometallate synthesis are being developed. One of these approaches is the attempt of linking polyoxometallate clusters in porous frameworks [32]. Hölderich et al. brought tungsten, tungsten oxide, $H_2N(CH_2)_6NH_2$, phosphoric acid and water together in a hydrothermal reaction and obtained a framework structure built up by Dawson ions $(M_{18}X_2O_{62})$ which are linked by the two-fold protonated amine. The elliptical channels thus formed in the structure have a diameter of 0.76×0.84 nm (center to center distances which would mean an open diameter of around 0.5 nm). However, the pores are not accessible in sorption experiments with nitrogen or water after activation up to 200 °C, a fact which the authors cannot explain so far. Above this temperature the structure collapses due to decomposition of the amine. The relatively low stability of the materials is probably due to two facts: the interactions between the Dawson anions and the amine cations seem to be relatively weak and do not involve a covalent bond, and the organic molecule is, other than in zeolites, part of the framework, so that the compound collapses on removal of the organic part. If an inorganic spacer could be bonded in a similar way covalently to the Dawson ion, a stable compound with highly interesting properties should result. Somewhat similar to the approach chosen by Hölderich et al., a templated layered structure of molybdenum oxide has been synthesized hydrothermally by Guo et al. [33].

Another interesting approach to develop polyoxometallates into zeolite-like compounds are the above mentioned lacuna structures which can be combined to form bigger units embedding the central cation. In some of these complexes the central ion can be easily exchanged, but others act as clathrates or cryptates with a tight fixation of the central ion. For such clathrate or cryptate structures it is often observed that the central ion acts as a template [28] directing the properties of the resulting cluster. Some of the latest clathrate structures that have been found in polyoxometallate chemistry are the oxovanadiumorganoarsonate structures recently published by Khan and Zubieta [34]. Large cage-like clusters (Fig. 3) which are not as polar as the related organophosphonate materials, have a cavity which is sufficiently big to allow the inclusion of neutral molecules like the two methanol or acetonitrile molecules that are included from the solvothermal synthesis. So far it seems not to have been possible to remove the guest species from the cavity, though.

3.3
Phosphates and Phosphonates

3.3.1
Metalalkylphosphonates

These porous materials are closely related to the well known porous aluminophosphates, which are covered in Chapter 6 of this volume, if aluminum is used as the metal center. However, instead of phosphate groups linking the AlO_4 tetrahedra, phosphonates are used here which results in materials with quite different properties.

There are a few examples of crystalline open framework structures with accessible pore systems where the inner surface of the pores is coated with organic groups [35–37].

AlMepO-α [37] is probably the most interesting one of these structures (Fig. 4). The material can be synthesized under hydrothermal conditions from pseudoboehmite and methylphosphonic acid at 220 °C. It is built up from alternating aluminate and methylphosphonate groups forming one-dimensional 18-ring channels with openings of about 0.7 nm, not dissimilar to the channel structure of $AlPO_4$-5. However, while the $AlPO_4$s contain tetrahedrally co-

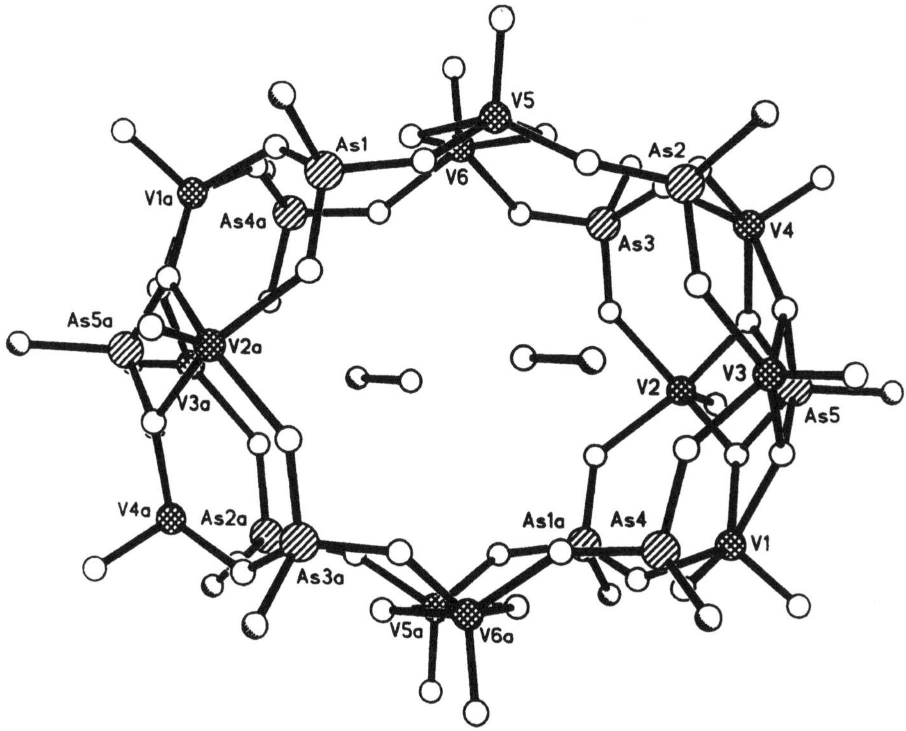

Fig. 3. Structure of the oxovanadiumorganoarsonateanion in the crystal encapsulating two molecules of methanol

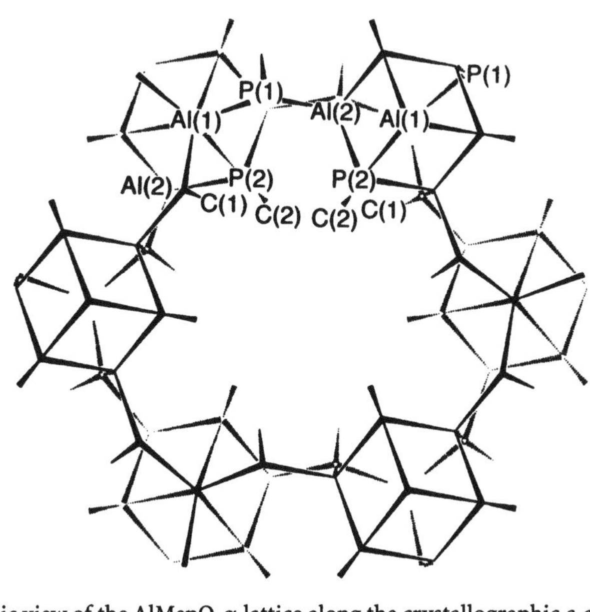

Fig. 4. Schematic view of the AlMepO-α lattice along the crystallographic c-axis

ordinated framework atoms only, there are octahedral aluminum atoms in the AlMepO-α. The channel of this structure is organically coated, since the methyl groups of the phosphonate are exposed at the channel surface. The compound exhibits a fairly high thermal stability up to temperatures of 500 °C in vacuum, and crystals outgassed under vacuum readily adsorb 2,2-dimethylpropane with a kinetic diameter of about 0.6 nm. Other sorptive properties of this material are not known as yet. Since normally $AlPO_4$s are rather hydrophilic, while the phosphonate is hydrophobic, quite different sorptive properties are expected as compared to $AlPO_4$s.

The other examples, an aluminummethylphosphonate with a different structure [36] and a coppermethylphosphonate [35], have smaller pore openings, so that they are probably not useful as sorbents. However, the synthetic approach might be exploited further, and materials with other pore sizes and pore connectivities might be developed.

3.3.2
Transition Metal Phosphates

A wide variety of zeolite-like inorganic structures can be synthesized using transition metals, a phosphate or related phosphorus compound and optionally a template, mostly an amine. One of the most important groups, the molybdenumphosphates, are covered in an excellent review by Haushalter and Mundi [38]. It is again interesting to see that corresponding structures can be found in nature. The most impressive example is certainly the natural iron phosphate cacoxenite [39] which occurs as secondary mineral, commonly associated with other phosphates and iron ores, mostly limonite. Its structure was not solved until 1983 [40] when Moore and Shen obtained large single crystals from Avant's Claim in Arkansas as suitable material for single crystal X-ray diffraction studies (Fig. 5). The most astonishing feature of the structure of this mineral is the enormous free pore diameter of 1.4 nm along the crystallographic c-axis of the hexagonal crystals. Many of the building units of this mineral can be found in other phosphates or related structures. The $(PFe_6^{3+}O_{28})$-unit has a strong resemblance to the central girdle in the Keggin structure discussed in the previous section. The iron in this structure is octahedrally coordinated, the aluminum lies in the center of a trigonal bipyramid, and the phosphorus is in the expected tetrahedral coordination. Adsorption data show that the channels are well accessible to oxygen and water. A substantially lower uptake has been observed for organic molecules which was attributed to the hydrophilicity of the samples and the problems encountered in fully removing the water during the activation of the samples. The lack of thermal stability – the X-ray pattern already substantially changes after heating to 200 °C – prevents proper activation [41].

So far it has not been possible to synthesize cacoxenite or other microporous iron phosphates, although many attempts to obtain new open framework iron phosphate compounds have been made. However, in most cases the organic templates are not removable [42, 43]. Moore and Shen [40] already pointed out that other open framework phases might form under hydrothermal conditions, i. e., at moderate pressures and low temperature.

Fig. 5. View in the (001) direction of the natural mineral cacoxenite. For clarity, oxygen atoms are omitted

Years later, Haushalter and Mundi [38] took this "mineralogical wisdom" and formulated their synthetic concept. Hydrothermal conditions and the combination of early transition elements that are able to form M–O–M bonds as well as M–O–P bonds at low pH and the combination with H_3PO_4 as a phosphate source were the most promising prerequisites to form octahedral-tetrahedral three-dimensional frameworks with pore structures. This is still mainly the concept which chemists try to follow if they intend to synthesize crystalline open framework transition metal phosphates. The first synthesis of an open framework phosphate was, however, reported by Leclaire et al. [44] in 1983, who prepared the molybdenumphosphate $K_4Mo_8P_{12}O_{52}$. What seems even more surprising is that they did not even use hydrothermal techniques. $K_4Mo_8P_{12}O_{52}$, a tunnel structure with cavities in which the potassium ions are located (center to center distance across the tunnel 0.5 to 0.6 nm), can be synthesized by simply heating P_2O_5, MoO_3, Mo and K_2O in appropriate ratios in a fused silica tube under vacuum at 800 °C. One fact the authors pointed out is the unusual +V-valency of molybdenum. One of the bonds of the central molybdenum to oxygen in the oxygen octahedron is supposed to be a Mo=O double bond which shifts the molybdenum well off-center of the octahedron. Based on this work the authors predicted other tunnel structures for compounds containing phosphorus, molybdenum and large monovalent ions like Rb^+, Tl^+ or Cs^+.

Mainly Haushalter et al. followed that synthetic route [45–47] and subsequently obtained similar tunnel structures as were predicted by Leclaire et al. Certainly one of the more astonishing compounds synthesized by this route is $Cs_3Mo_4P_3O_{16}$ [48] (Fig. 6) In this compound eight "Mo_4O_4-cubes" capped with a PO_4 group bridge to an identical cube in the next unit cell to form a "supercube" (Fig. 6). The Cs^+-ions lie at the centers of the windows which are formed by the oxygen atoms of the phosphate groups, i.e., on the face centers of the supercube. It is very surprising to find the large cage of the supercube empty, especially if the fact is considered that this compound is synthesized at 950 °C where most framework structures tend to densify. The variety of structures that could be obtained was appreciably broadened as soon as the researchers focused on the hydrothermal route. However, there are some requirements with respect to the synthetic conditions. In order to suppress the formation of polyoxometallate compounds the oxidation state of molybdenum has to be kept at +5 or lower. This can be done in two ways: either a molybdenum source with an oxidation state of +6 is combined with metallic molybdenum, which has to be very finely divided to allow symproportionation, or the hydrolysis products of $MoCl_5$ and $MoCl_4$ can be used as starting materials [38]. The clear aim of this work was to incorporate large removable cations to get accessible tunnel systems [38] and either exchange them against smaller ones or bring in cations that can be

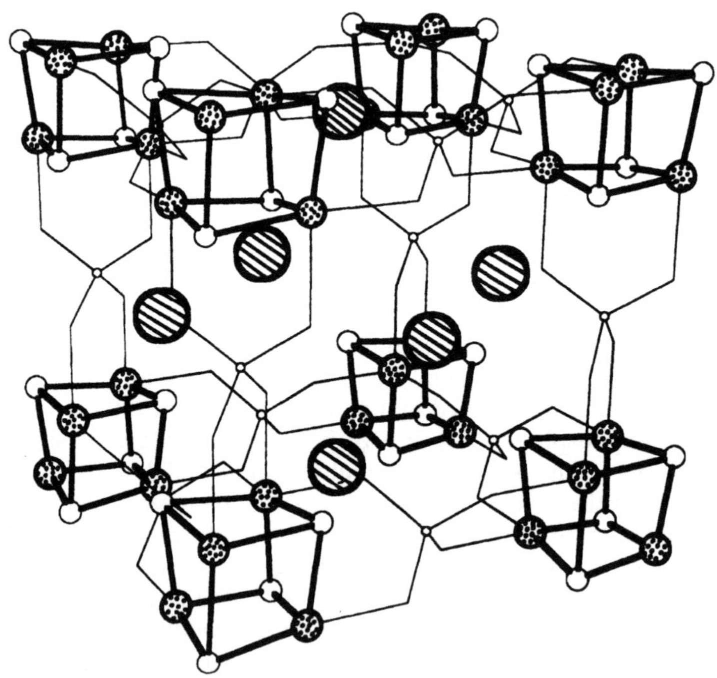

Fig. 6. The supercube of this molybdenum phosphate $Cs_3Mo_4P_3O_{16}$ is formed by four Mo_4O_4 cubes forming the edges of the cell. *Striped circles* represent Cs-atoms, oxygen atoms are omitted for clarity

decomposed which is most easily achieved with protonated amines or tetraalkylammonium ions. Synthesis temperatures have to be kept below 250 °C in this case, though, because of template decomposition during the synthesis. The synthetic efforts resulted in a range of structures, from molecular and ionic aggregates [38] with multinuclear molybdenum clusters from around 30 to 250 atoms, to one-dimensional polymers, formed by Mo_4O_8 cubes connected by phosphate groups [49], and a large range of layered compounds [50–52]. Again, the motif of the layers are cubes connected by phosphate groups, the layers being usually separated by large cations.

The first molybdenum phosphate with real microporosity even in the uncalcined state was $(Me_4N)_{1.3}(H_3O)_{0.7}(Mo_4O_8(PO_4)_2) \cdot 2H_2O$ [53] which can be described in terms of the cubic $Cs_3Mo_4P_3O_{16}$: the removal of one third of the P atoms and oxidation of the Mo3.5+ to Mo5+ ($Mo_4P_3O_{16}^{3-}$ to $Mo_4P_2O_{16}^{2-}$) followed by the removal of one fourth of the $Mo_4P_2O_{16}$ units describe the structure if the voids are ordered in a proper manner. The cavities of the latter structure have a diameter of about 0.7 nm, and the diagonal distance of the cavities is even larger, yet the windows of these cavities measure only 0.28 nm. This is a very strict limitation for this microporous solid as it only allows the penetration of small molecules like water, ammonium, helium and hydrogen, while larger ones like nitrogen or hydrocarbons are excluded. Still the structure has a micropore volume of about 15 vol.-% as can be estimated from water adsorption data [38, 53]. A molybdenum phosphate with a neutral framework of interconnected pores and a micropore volume of more than 35 vol.-% could also be prepared from MoO_3, Mo, H_3PO_4 and H_2O by Haushalter et al. [54]. Again Mo_4 units composed of MoO_6 octahedra are the main building blocks of the framework generating tunnels that run along all three directions parallel to the cell edges. Water adsorption on this compound shows that not only the water in the pores can be desorbed but also the water from the ligand sphere of the molybdenum. This process is reversible as can be shown by X-ray diffraction and from the adsorption isotherm [54]. Again the framework does not adsorb large hydrocarbons but only small molecules.

Another class of molybdenum phosphates is very promising as they can be prepared with organic or inorganic cations varying in size and shape. The $(Mo_2O_2(PO_4)_2(H_2PO_4)^-)$ structure consists mainly of layers formed by corner sharing octahedra of MoO_6 and tetrahedra of PO_4. These layers are connected by phosphate groups forming pores of appreciable size. The ability of the framework to absorb different cations is due to the ability of the phosphate groups to move flexibly around their bond axis [38] altering the angle and rotating and shifting the layers relative to one another. Together with H_3O^+ and Na^+, a body centered tetragonal lattice with a = 0.64 nm and c = 1.6 nm is formed, where the small a results from the 4-fold disorder of the phosphate groups and the large c from the two layer repeat. $CH_3NH_3^+$ in the same compound gives a C-centered monoclinic framework with a = b = 0.91 nm and c = 0.86 nm. Here the unit cell of the compound is rotated by 45° and expanded by the factor of $\sqrt{2}$ compared to the tetragonal lattice mentioned above. The interlamellar phosphate groups all point in the same direction resulting in a polar framework. For Cs^+ and H_3O^+ (Fig. 7a) again a monoclinic framework of a = b = 0.91 nm and c = 1.6 nm

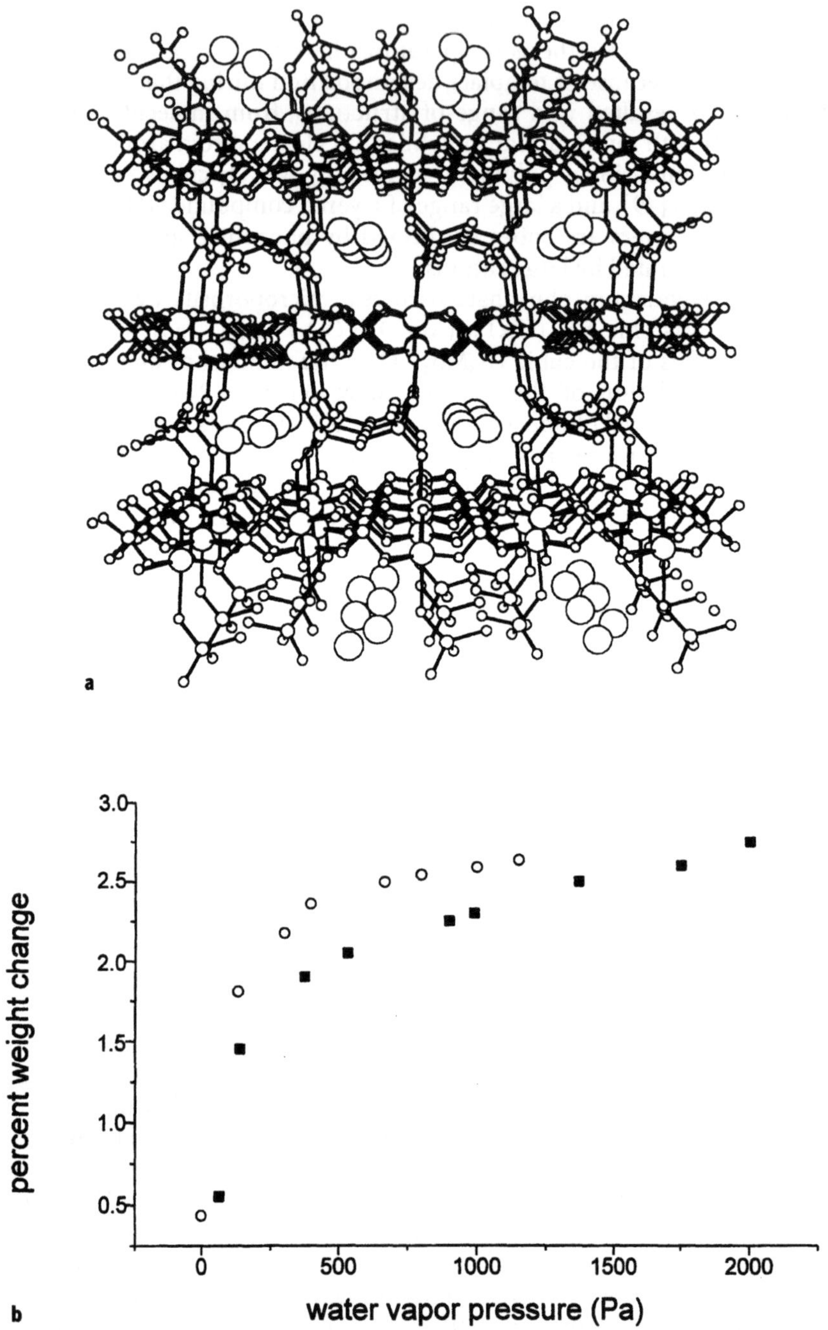

Fig. 7. a Characteristic for the molybdenum phosphate $Cs(H_3O)[Mo_2O_2(PO_4)_2(HPO_4)]$ is its sheet-like structure where phosphate groups connect the layers. **b** Water sorption isotherm for the structure shown in **a** shows an internal void volume of approximately 9%. *Squares*: adsorption branch, *circles*: desorption branch

results, and here the space in between the layers is divided into regions with cations and those with phosphate groups. All these frameworks can be rendered microporous toward water (Fig. 7b) and thermally removing the amine from the framework gives a micropore volume of approximately 30 vol.-% as estimated from the water adsorption data.

The second biggest family of transition metal phosphates that is known to form open framework structures is the vanadium phosphate group. Again many examples of layered [55–59] or polymeric compounds [60] are known, incorporating either organic or inorganic cations. Recent experiments even resulted in the synthesis of layered templated vanadium oxides [61–63]. An outstanding example of the layered vanadium phosphate structures is certainly the layered organophosphonate $((C_2H_5)_2NH_2)((CH_3)_2NH_2)(V_4O_4(OH)_2(C_6H_5PO_3)_4)$ [58] which forms a trilayer of $(C_2H_5)_2NH_2^+$ cations which sandwich between the V/P/O-layers. These are in turn separated from the next layer by an organic bilayer of structurally bound $C_6H_5PO_3$. The V/P/O layers themselves have large cavities formed by rings of four vanadium octahedra and two phosphonate tetrahedra which share corners and form a twelve-membered ring. This structural compound is very interesting because of its possible sorptive properties and other applications. These structures have similarities to the zirconium-phosphonates which will be discussed later. Looking at the polymeric species we find a strong correspondence to molybdenum phosphate structural chemistry.

$((CH_3)_2NH_2)K_4(V_{10}O_{10}(H_2O)_2(OH)_4(PO_4)_7) \cdot 4\,H_2O$ forms chiral double helices of interpenetrating spirals of vanadiumoxo pentamers which are connected by phosphate groups [60]. These double helices are intertwinned with one another and arranged in a way that they generate tunnels and cavities which are filled with the cations employed in the hydrothermal synthesis, $(CH_3)_2NH_2^+$ and K^+. What is most fascinating about this dark blue material is the fact that due to the chirality of its unit cell the single crystals are also chiral and crystallize in enantiomorphs. This gives room for speculations about possible applications of this vanadium phosphate. Firstly, the sorptive properties could render this material interesting. So far no sorptive data are available but the cavities should be accessible to organic molecules. Since the compound is thermally fairly stable, enantioselective sorption or even catalysis could be envisioned. The second possible application lies in the field of optoelectronics. Due to the chirality and the strong paramagnetism at high and antiferromagnetism at low temperature, this vanadium phosphate could be an interesting linear or non-linear optical element.

The variety of materials with pore systems in the vanadium phosphate system is very wide and ranges from solids with small pores consisting of three vanadium octahedra and three phosphate tetrahedra [64] via openings formed by four vanadium octahedra and four phosphate tetrahedra [65–68] to pores consisting of six vanadium octahedra and six phosphate tetrahedra [67, 69]. There are also materials with giant elliptical openings of eight vanadium octahedra and eight phosphate tetrahedra [70, 71], which measure about 1.8 nm in the diagonal of the pore, or alternating VO_6/PO_4 16-ring windows with 0.9 nm openings being the aperture to huge cages of alternating VO_6/PO_4 32-ring cavities [72]. The latter structures are probably the most interesting latest developments in the field of open framework vanadium phosphates (Figs. 8a and 8b).

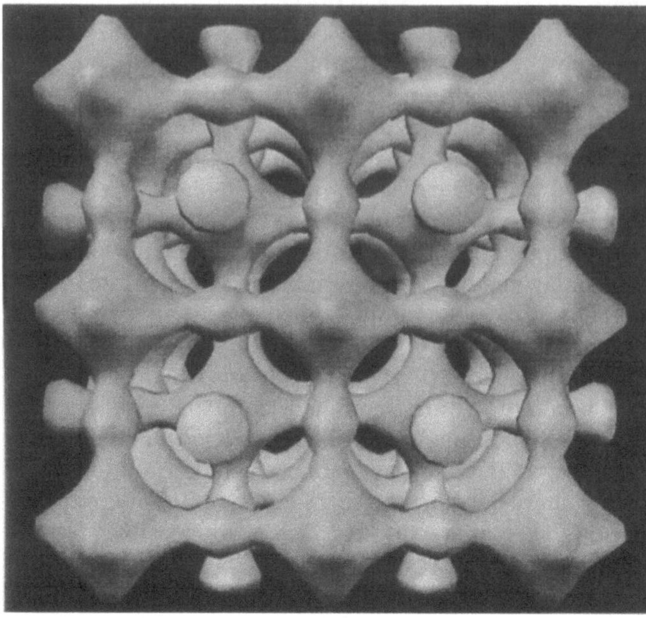

a

b

Fig. 8. **a** Representation of the isosurface of the supercages of $(HN(CH_2CH_2)_3NH)\ K_{1.35}[V_5O_9(PO_4)_2] \cdot x\ H_2O$ forming an interpenetrating network. **b** Isosurface of $(Cs_3[V_5O_9\ (PO_4)_2] \cdot x\ H_2O$ forming a distinct tunnel structure of two crystallographically identical sets of tunnels

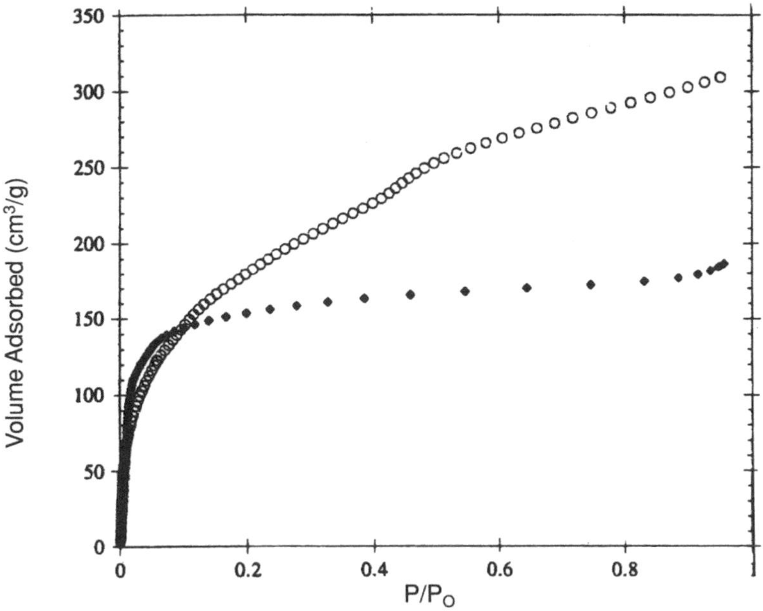

Fig. 8. c Water adsorption isotherms for $(HN(CH_2CH_2)_3NH)K_{1.35}[V_5O_9(PO_4)_2] \cdot x\ H_2O$ (Fig. 8a) (*diamonds*) and barium exchanged $(Cs_3[V_5O_9(PO_4)_2] \cdot x\ H_2O$ (Fig. 8b)) (*circles*). The volume of water vapor is normalized to STP (1.01325×10^5 Pa and 273 K)

$(HN(CH_2CH_2)_3NH)K_{1.35}(V_5O_9(PO_4)_2) \cdot x\ H_2O$ and $(Cs_3(V_5O_9(PO_4)_2) \cdot x\ H_2O$ form open frameworks with large cages which are three-dimensionally interconnected by 0.9 nm windows. Both structures can be prepared hydrothermally from conventional reactants. The DABCO (diazabicyclooctane, $N(CH_2CH_2)_3N$)-templated material is prepared from KVO_3, DABCO, Et_2NH, phenylphosphonic acid, phosphoric acid and water in molar ratios of 1:2:2:0.75:2:100 hydrothermally over 4 days at 170 °C, after which the product can be isolated in form of greenish black truncated octahedra [72]. For $(HN(CH_2CH_2)_3NH)K_{1.35}(V_5O_9(PO_4)_2) \cdot x\ H_2O$ even other templates like piperidine can be employed in the synthesis [72]. Again the sorption data (Fig. 8c) show that large amounts of water can be adsorbed in the cavities of the framework. The sorptive properties can even be improved for $(Cs_3(V_5O_9(PO_4)_2) \cdot x\ H_2O$ by exchanging Cs^+ against Ba^{2+}, and the sorptive volume corresponding to water can be nearly doubled by this ion exchange process which proceeds easily.

Although the structural chemistry of molybdenum and vanadium phosphates is very diverse these compounds seem to be prototypes for open framework transition and main group metal phosphates. Recent results have shown that zeolite-like compounds like zinc/cobalt phosphates [73], indium phosphates [74], nickel/aluminum phosphates [75], cobalt/boron phosphates [76], manganese and cobalt phosphonates with giant 44-membered rings forming ellipsoidal cavities [77] can be synthesized. All these examples show that the diverse chemistry of octahedral-tetrahedral framework structures is a vast field where many interesting structures can still be discovered. There are certainly limitations for these systems, since for many of them template removal is a big problem and often results in structural breakdown. Even if the templates can be removed or exchanged, the windows of these solids are often so small that only very small sorptives like water can penetrate the structures [38]. These facts do not suggest such materials for catalytic applications at present, but further investigations might lead to more promising results in that respect.

3.3.3
Layered and Pillared Zirconium Phosphates and Phosphonates

We have already mentioned a layered vanadium phosphonate structure [58], and we want to come back to this class of materials which has been best explored for the zirconium phosphate and phosphonate system. The basis or backbone of these layered structures is always a crystalline layer of a transition metal phosphate or phosphonate. In the case of the phosphate, it is separated from the next transition metal phosphate layer by organic spacers. These are either intercalated or bound to Brønsted acid sites of P–OH groups. In the case of the phosphonates the organic spacer is formed by the organic side branches of the phosphonate groups [78]. For zirconium two important layered phosphate phases are known, the α-phase and the γ-phase. Synthetically, those phases can be easily obtained by precipitation reactions in aqueous solutions. The α-phase, $\alpha\text{-}Zr(HPO_4)_2 \cdot 2\,H_2O$, can be prepared by refluxing zirconium salts in concentrated H_3PO_4 for long times up to 300 hours or precipitating zirconium fluoro complexes with phosphoric acid [78]. The obtained crystals are monoclinic, and their layers consist of metal atoms lying slightly above and below the mean plane, being bridged alternately from above and below by tetrahedral phosphate groups. The phosphate groups reach into the interlayer region with one P–OH group, and form six-sided cavities with the neighboring layer which contain water molecules. The P–OH groups are also responsible for the van der Waals forces holding the layers together [78].

The γ-phase instead is constituted by two ideal planes of zirconium atoms bound to tetrahedral PO_4 and H_2PO_4 groups. The phosphate groups in this structure share all their oxygen atoms with the zirconium, whereas the H_2PO_4 groups share only two oxygens, and the remaining P–OH groups dangle into the interlayer region. The γ-compounds have the more rigid layers and show a much stronger acidity and, due to the larger interlayer distance, faster intercalation rates [78].

Another interesting feature of this phase is that, due to the topotactic exchangeability of the dihydrogenphosphates against mono- or diphosphonic acids, they can be used as a precursor material for new phosphonates. Due to the Brønsted acid sites, the layered phases of the zirconium phosphates are inorganic ion exchangers with good thermal and chemical stability. Still, highly hydrated cations can only penetrate slowly, especially in the α-phase, as the expansion of the phase is coupled with a high activation energy [78]. For the γ-phase, ion exchange is facilitated as the steric hindrance of diffusion is much lower due to the larger interlayer distance and the flexibility of the H_2PO_4 groups.

There are principally two ways of opening the layers to ions with a larger coordination sphere. The first one is the intercalation of organic molecules like ethanol or alkylamines [78], whereas guests like long-chain alcohols or other organics with low basic strength cannot be inserted so easily. For monoamines intercalation proceeds readily, and at full intercalation the layered α-zirconium phosphate contains 2 moles amine per mole zirconium which form a bilayer in the interlayer region. For the γ-phase, amine uptake for the bilayer formation is much lower with 1 – 1.3 mol per mol zirconium. Compared to the monoamines, diamines form monolayers where both functional groups interact with the Brønsted acid sites of each layer. The intercalated zirconium phosphates swell readily in water and other solvents, while the untreated compounds do not behave like this, probably due to the high charge density and strong layer-layer interactions. This behavior of the intercalated phases gives way also to the second method of enlarging the interlayer spacing by pillaring procedures with the introduction of a wide range of inorganic pillaring ions like Zr, Cr, Al, Si, Fe, Ga and others [79].

Generally, intercalated compounds are treated with solutions containing the desired ions. The resulting materials show high specific surface areas up to 200 m^2g^{-1} or more and pores in the micropore range up to mesopores, while the surface area and pore size distribution can be adjusted by the calcination process [79]. However, materials obtained following such pathways rather resemble pillared clays than zeolitic compounds. Sometimes, calcination at higher temperatures may even render the materials X-ray amorphous, while the surface area and pore size distribution are retained. By these procedures, ion exchange capacities can be increased compared with the original layered phase, and especially here it has been found that the uptake of cations from solution is better if the materials have been NH_4^+-exchanged, as the ammonium cation is a good leaving group. Apart from their ion exchange capacities, inorganically pillared zirconium phosphates have been tested as catalysts, and they were found to be attractive materials due to their relatively sharp pore size distributions and moderate or high Brønsted and Lewis acidity [79]. One example is the fluorinated alumina-pillared α-zirconium phosphate which shows high catalytic activity for the decomposition of isopropyl alcohol together with a selectivity of nearly 100% for the dehydration process [79]. What makes such materials very promising in several catalytic applications is the possibility to incorporate large amounts of heteroatoms, commonly much more than in zeolites or MCM-41 materials. In addition, the resulting pillared material can be altered by ion exchange. This creates a framework variety and atom concentration relative to the layer composition that is achieved only in few other porous solids.

A broad variety of organic derivatives of α-zirconium phosphonates and α-zirconium phosphonates can be obtained simply by replacing H_3PO_4 with H_2O_3PR where R is an organic group. Generally, for these syntheses HF is employed as complexing agent to facilitate the introduction of the organic derivative [78, 80]. For the α-zirconium phosphonates many different organic groups have been introduced, and it seems as if the nature of the groups which can be incorporated depends only on the abilities of the chemist who synthesizes the phosphonate.

Especially interesting is the introduction of organic groups of different size and shape. In this case it is observed that distinct ratios of the different phosphonic acids are incorporated in the layers. Due to the different packing requirements of the organic branches, special distributions which seem energetically favorable tend to form [78]. These distributions often resemble the simplest two-dimensional packings of balls of different size that can be realized. However, not only phosphonates of the H_2O_3PR type can be introduced, but also phosphonic acids of the H_2O_3P–R–PO_3H_2-type have been employed in the synthesis of α-zirconium phosphonates. These compounds connect the layers, and by designing the organic spacer in such a way that, through steric hindrance, neighboring sites in the layer cannot be occupied by organically substituted species, interlayer microporosity can be introduced in α-zirconium phosphonates. Such a compound has pore sizes of about 0.5 nm and an interlayer micropore surface area of 375 m^2g^{-1}. For γ-zirconium phosphonates it is not even necessary to use organic diphosphonates with organic branches as spacers, here the simple incorporation of biphenyldiphosphonate via an exchange process at 75–80 °C is sufficient for the creation of interlayer microporosity due to the low pillaring percentage of only 25% of replaced H_2PO_4 groups in the γ-layers (Fig. 9). For this exchanged material a micropore volume of 0.12 cm^3g^{-1} and a narrow pore size distribution around 0.58 nm were determined. However, the

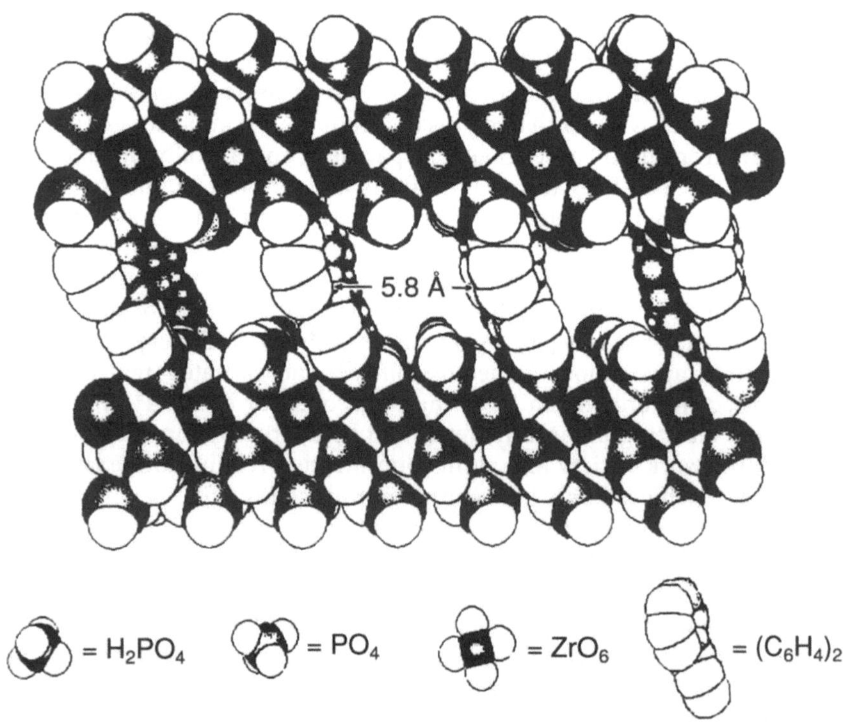

Fig. 9. Lateral view of pillared microporous γ-zirconium phosphonate

thermal stability is relatively low: Decomposition begins around 220–230 °C, which is somewhat lower than for similar non-porous compounds. This effect is attributed to the easier diffusion of gases in the pore system.

Another way of creating porosity was found in the zirconium phosphite-diphosphonate system, where a large excess of phosphorous acid may induce high specific surface areas of 300 to 500 m^2g^{-1} [78]. The main fraction of this surface area is due to intercrystalline mesoporosity, and the size of the pores can be varied from 1.5 to 5 nm with a fairly narrow distribution, simply by changing the ratio and adjusting the concentration of the reagents. Even the interlayer microporosity of the pillared layers can be suppressed by using very short organic spacer groups, and this way purely mesoporous materials can be obtained. The unique feature of these mesoporous solids is the possibility to functionalize the surface of the mesopores by topotactic replacement of the phosphite groups with phosphonic groups possessing catalytically interesting substituents, like for instance -SO_3H. The applications for these porous materials with good chemical and thermal stability, their high surface areas and narrow distributions of inter- and intracrystalline pore sizes lie in the field of ion exchange, protonic conduction and, of course, they are suitable agents for acid catalysis.

3.3.4
Antimonyoxyselenides: Cetineite Type Materials

Unique in their composition as antimonyoxyselenides, in their open framework topology and their optoelectrical behavior are the structural analogues of the natural mineral cetineite which has been synthesized by Liebau and Wang [81–83]. $K_3Sb_7O_9Se_3 \cdot 3\,H_2O$ is a typical example for this class of materials. It can be obtained as transparent dark red hexagonal prisms of approximately 1 mm length and a width of 0.1 mm from elemental antimony and selenium in an aqueous solution of potassium hydroxide under hydrothermal conditions. In the hexagonal structure of $K_3Sb_7O_9Se_3 \cdot 3\,H_2O$ pyramidal SbO_3 groups are connected to channel-like $Sb_{12}O_{18}$ units stretching along the crystallographic c-axis. These channels have an open diameter of approximately 0.7 nm, and the potassium ions are arranged circularly on the inner surface of the channels, while the center of the channel is occupied by a chain of stretched $(H_2O)_6$ octahedra. On their outside the channels are connected by $SbSe_3$ pyramids through weaker secondary bonds. Evidently, potassium and the strongly oriented water molecules act as templating, void filling agents in this structure, but the framework is accessible as the water molecules can easily be removed from large single crystals within a few seconds at ambient temperature under high vacuum. UV-Vis spectra recorded on single crystals and conductivity measurements showed that the framework is semiconducting with the ability of photoconduction [84]. The cetineite-analogous phases are the first examples of porous solids with such properties which suggests uses of the accessible cetineite frameworks in the fields of chemosensoring, optoelectronics or photovoltaic applications.

4 Non-Oxidic Frameworks

All materials mentioned so far are based on oxygen that acts as a bridge between main group or transition elements. In the following we will present materials that are oxygen-free. The first group are the frameworks with the higher chalcogenides as the bridging framework constituents.

4.1 Open Framework Sulfide Structures

The standard approaches for the synthesis of such compounds were for a long time molten flux methods of metal sulfide fluxes [9]. Although large cavities can be obtained by flux methods, the charge compensating cations employed during synthesis are always inorganic and can be only topotactically exchanged but never be totally removed from the framework so that no real open framework structures result. The clue to an open framework is in most cases a void filling cation that can be decomposed like the organic cations in zeolites. One of the earliest reports presenting several new zeolite-like sulfide structures templated with amines was published in 1989 by UOP researchers [85]. Their synthetic concept which proved to be successful was to employ elements which show a preference for tetrahedral coordination, stable thiometallate chemistry in aqueous solutions and which possibly give a minimum framework charge in the resulting hypothetical MeS_2 or $MeMe'S_4$ lattice. As mineralizers they employed either hydroxide ions or HCO_3^- which both proved to be efficient in mobilizing the elemental building units without hydrolyzing the products or the sulfide precursors, mostly the freshly precipitated metal sulfides. As structure directing agents, alkylammonium and amine templates were used in analogy with zeolite and aluminophosphate syntheses. Many tin and germanium sulfides and metal thiogermanate or stannate structures can be synthesized following this concept and these procedures.

As an outstanding example we want to discuss the structure of a cobalt/manganese germanium sulfide [85] which displays important building units and principles of the open framework sulfide chemistry. The basic building unit of this lattice is the adamantine-like Me_4S_4 unit (Fig. 10a), a structural element which is certainly one of the most important ones and which can be found in many lattices [86]. Three adamantine units are connected together by one central MeS_3SH tetrahedron forming a trimer (Fig. 10b). These trimers are linked together in alternating configuration to chains which run along the principle axis of the unit cell (Fig. 10c). The rhombohedral unit cell itself contains six adamantine units, eight three-connected tetrahedral units and twelve tetramethylammonium cations. Each adamantine T-atom is part of three fused 3-rings, while each MeS_3SH tetrahedron is part of six 12-rings. The 3-ring building units seem to be a typical feature of sulfides due to the restricted T–S–T bonding angles of 103 to 108 degrees. Due to this rigidity, more open structures can be found in the sulfide system, containing also a larger amount of template, as compared to microporous oxides. This large amount of template in the

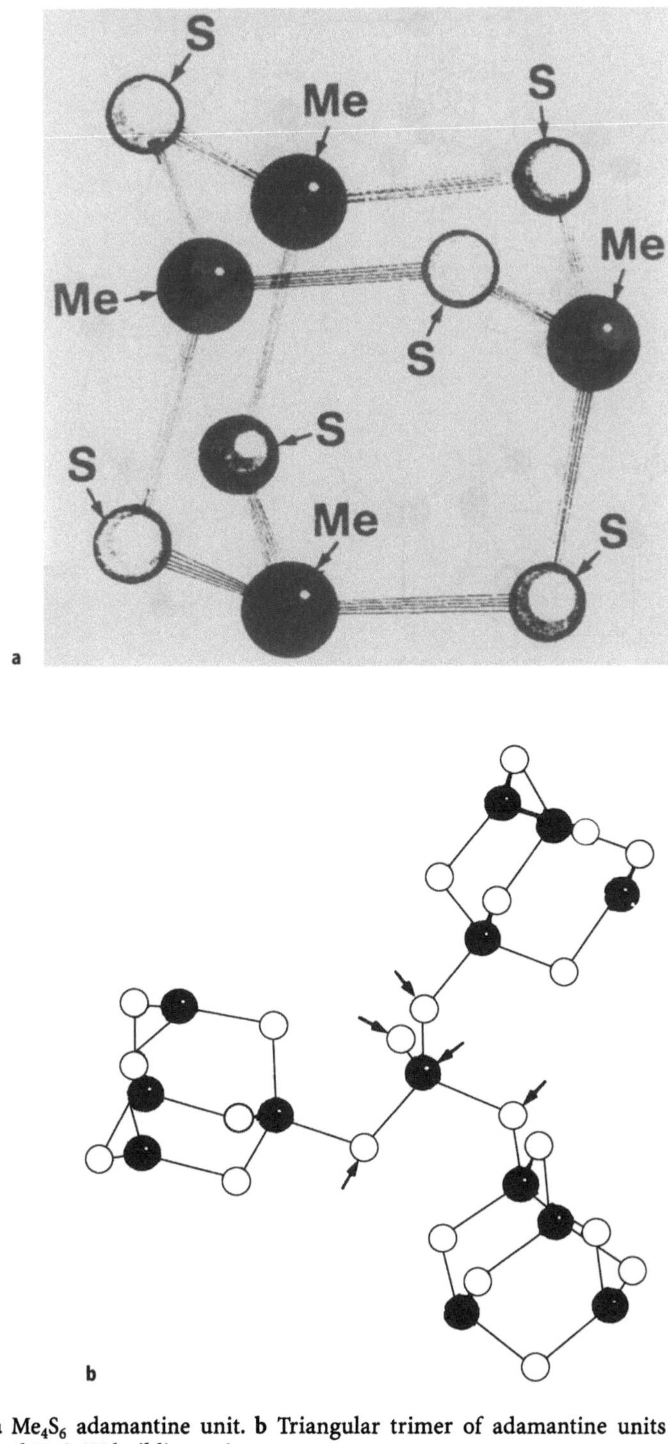

Fig. 10. **a** Me_4S_6 adamantine unit. **b** Triangular trimer of adamantine units linked by the tetrahedral MeS_3SH building unit

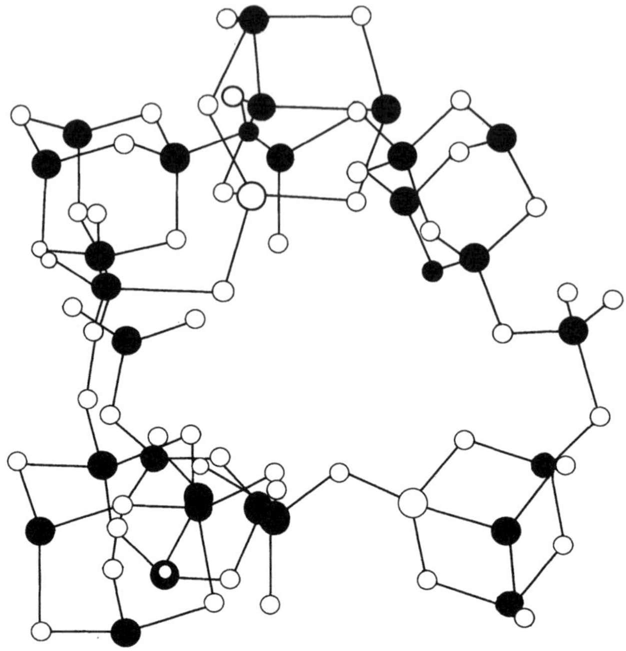

Fig. 10. c Pinwheels formed by the linkage of adamantine trimers

crystals, however, makes the voids also inaccessible to bigger molecules. The situation is complicated further by the fact that sulfides are thermally not as stable as oxides and burnout of the template is mostly accompanied by structural breakdown [85]. However, some structures, like for instance a tetramethylammonium tin sulfide, can be made accessible by moderate activation procedures (150 °C), after which small molecules like H_2O and CO_2 can penetrate the pore systems [85]. Subsequently, several groups presented a variety of new sulfide open framework structures, many of which are well reviewed in [86]. Sulfide frameworks can be synthesized where the structure is dominated by large windows from 16 [87] up to 32-atom rings [88], which are strongly ellipsoidal, but have no open porosity. Arrangements of small sulfide clusters forming cavities are known [86], as well as a large variety of layered materials connected by tunnels [89, 90].

Pillaring of metal sulfides has been achieved [91], and even metal-rich framework structures can be prepared [92]. Even for these materials template removal is not possible, they are still an interesting approach to nano-sized semiconductors, namely so-called anti-dot lattices where the semiconducting lattice surrounds template-filled voids of lower conductivity [86]. Template removal is probably not the essential problem for the design of a nanostructured material, since often the goals of unusual size-dependent optical or electronic properties might be achieved with the composite of template and framework.

4.2 The Second Example of Oxygen-Free Framewoks: Nitridosodalites and Related Phases

The chemistry of phosphonitrides is very different from all the compounds we have described before. It has for a long time been quite a mystery, as the resulting products are often impure, ill-defined and can mostly not be obtained as single crystals. The clue to a series of new structures of phosphorus nitrides was the synthesis of pure crystalline P_3N_5 [93] with which Li_7PN_4 and $LiPN_2$, both good lithium ion conductors, could be obtained and structurally characterized [94,95]. In $LiPN_2$ phosphorus and nitrogen form a β-cristobalite-like framework and show the highest degree of condensation that can be reached by connecting PN_4 tetrahedra over the edges, whereas in Li_7PN_4 isolated PN_4^{7-} ions can be found. The first compound exhibiting a complex anion $P_4N_{10}^{10-}$ was $Li_{10}P_4N_{10}$, which lies with respect to its composition in between the two extremes quoted above and shows also a degree of condensation of PN_4 tetrahedra between the two other structures [93]. The compound can be obtained as colorless crystals by the solid-state reaction of Li_3N and P_3N_5, by reaction of $LiPN_2$ with Li_3N, or by reacting Li_7PN_4 with P_3N_5 stoichiometrically around 700 °C under nitrogen in a tungsten crucible. $Li_{10}P_4N_{10}$ has an ionic structure, and the building unit of the $P_4N_{10}^{10-}$ anion is isoelectronic with P_4O_{10} in a zincblende-like arrangement.

This complex anion has already very small cavities and, like the β-cristobalite-like structure of $LiPN_2$, suggests some similarities to silicate framework chemistry with the prospects of more open structures. The final clue to the synthesis of the sodalite-like structure $Zn_7(P_{12}N_{24})Cl_2$ was reacting Zn_3N_2 and ammonium chloride, producing ammonia in a separated compartment in a fused silica ampoule at 400 °C and keeping P_3N_5 in the other half of the ampoule at 580 °C and 30 bars for 6 days [96]. Later, a simpler route to the nitridosodalite was discovered, using the simple reaction of $ZnCl_2$, NH_4Cl and $(PNCl_2)_3$ in an autoclave reaction at 700 °C. The obtained products are insoluble in water, organic solvents, bases and mineral acids, and only by autoclave treatment in diluted sulfuric acid could the extremely stable framework be hydrolyzed. Other metals like iron, cobalt, nickel and manganese or even hydrogen can be introduced in the cages, and hydrogen which is covalently bound to nitrogen can be replaced by zinc in an autoclaved exchange process at high temperatures [97].

The chloride anions are situated in the center of the sodalite cages and tetrahedrally surrounded by Zn^{2+} ions, chloride can be replaced by chalcogenide anions by reaction of HPN_2 with the corresponding zinc chalcogenide, and the chalcogenide takes the place of the chloride anion forming the center of the charged XZn_4^{6+} cluster in the β-cage [98]. These clusters can be considered as sections of II/IV semiconductors and show clear blue shifts in their absorption spectrum in comparison with the bulk material due to size quantization effects. This is a complete analogy to oxosodalites where the same results were obtained for dispersed semiconductors. Future experiments will show whether more open zeolite-like lattices can be obtained. It will be a great challenge to find removable void-filling and framework-stabilizing agents for such structures. However, even nitridosodalites with empty cages, $Zn_8[P_{12}N_{24}]$, already have the ability to absorb small gas molecules like hydrogen at elevated pressures and temperatures [99].

A different class of nitrides also discovered by Schnick et al. are the open framework silicon nitrides $Ce_3Si_6N_{11}$ and $Pr_3Si_6N_{11}$ [100, 101]. Here the voids of the covalent framework formed by SiN_4 tetrahedra are filled with the rare earth cations. The biggest channels that can be found in these structures are formed by six-membered rings. From the point of view of the zeolite chemist this does not seem too large. However, silicon nitrides offer a variety of features that find several applications, as they are very inert and thermally stable.

Another application with growing interest is the use of open-framework materials like zeolites or layered silicates as ceramic precursors. Certainly, silicon or phosphorus nitrides could be an approach to a similar precursor chemistry for non-oxidic ceramic materials.

4.3 Super Prussian Blue Compounds

In prussian blue, iron ions at the corners of cubes are linked by CN-groups on the edges of the cube. The idea to expand these compounds to porous materials is based on substituting the CN-bridges with longer units. First steps in this direction have been taken by Fischer's group [102–104]. They increased the length of the spacer by inserting a trimethyltin-group, thus creating a [C-N-$Sn(Me_3)$–N–C] bridge, and stabilized the voids by guest species like metallocenes or quaternary ammonium salts. The synthesis of such compounds is surprisingly simple: The cobaltocenium derivative, for instance, precipitates, when an aqueous solution of cobaltocenium perchlorate and trimethyltinchloride is added dropwise to a saturated solution of $K_4[Fe(CN)_6]$.

Related compounds can be prepared using different metal atoms, like iron, ruthenium, copper or tin, and with differing guest species like ferrocene, cobaltocenium, tetra-*n*-butylammonium, tetraethylammonium, and methylviologen. The list of guest species suggests a strong similarity to aluminosilicate or aluminophosphate molecular sieves, since several of these molecules are typical templates for zeolite synthesis.

This similarity is also suggested by the structure of such super prussian blue compounds. Figure 11 shows a view along the c-direction of $(nBu_4N)_{0.5}(Me_3Sn)_{3.5}Fe(CN)_6 \cdot H_2O$. One can clearly identify the channels running along this axis, which are occupied with the alkyl ammonium cation. No sorption data for these materials have been reported yet, but one might expect appreciable adsorption of small molecules already in the template-containing samples. If the template could be removed without destroying the structure, materials with a high porosity should result. Thermal pathways are probably not feasible in this case, since this could also result in destruction of the trimethyltin units. However, it might be possible to exchange the bulky tetra-*n*-butylammonium ion with smaller cations without destruction of the framework.

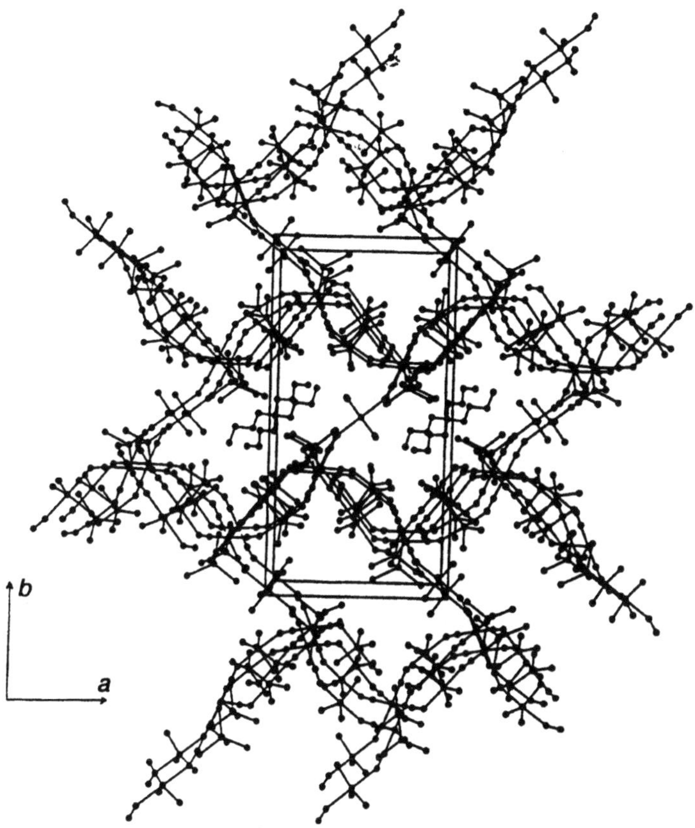

Fig. 11. View of $((nBu_4N)_{0.5}(Me_3Sn)_{3.5}Fe(CN)_6xH_2O)$ in direction of the channels (crystallographic c-axis). nBu_4N^+ ions are omitted in all except two channels

5 Other Approaches to Zeolite-Like Inorganic Materials

In the last section we want to present a variety of other materials that resemble zeolitic arrangements, but often, due to their chemical composition, the less defined pore structure or to preparation methods lack popularity.

A new class of siliceous microporous materials with varying metal contents and high porosities has recently been presented by Maier et al. [105, 106]. These materials which are X-ray amorphous and where no crystalline structure can be seen in the transmission electron microscope are synthesized via a sol-gel-route under hydrothermal conditions without the use of template molecules. The simple reaction of silicon alcoholates and titanium alcoholates which are mixed in ethanol and the condensation of which is induced by the addition of hydrochloric acid, produces a gel that becomes solid after a few days. The following careful calcination process under inert atmosphere and milling of the resulting solid gives the porous material. Detailed analysis of the pore structure parameters, however, have not been given yet, since the base pressure in the adsorption experiments was not sufficiently low to analyze pores in the range below 1 nm.

The synthetic method described offers two important parameters for variation of product properties. The first concerns mainly the polarity of the resulting oxide. By introducing trimethoxymethylsilane in the reaction mixture, oxides with residual methyl groups of varying content can be prepared. The second synthetic variable is the incorporation of a number of metals like titanium, aluminum, zirconium, tin or vanadium in the oxides with ratios of Si/M ranging from 50 down to 1. If these ratios are compared with common zeolites used in catalysis, much more of the catalytically interesting metal can be incorporated in these composite oxides. What seems even more fascinating is the fact that by the introduction of methyl groups the polarity of the resulting amorphous oxide can be adjusted over a wide range, which might be interesting as well for catalytic applications. This has been shown in selective oxidation reactions with tert.-butylhydroperoxide as oxidizing agent. Here it was found that the deactivation of the solid catalyst which is usually brought about by enrichment of polar components in the reaction mixture does not take place if hydrophobic silicon/titanium oxides are used in this process [105].

Another group of materials which are more interesting with respect to host-guest systems but also are not crystalline like zeolites, are intercalative composite materials [107] obtained by chemical modification of layered materials. Such composite materials can possess electrical, optical and mechanical properties which may not be achieved with each of the components separately. Especially layered inorganic hosts with organic polymers as guests can resemble zeolite-like arrangements, and sometimes their structures can be considered more or less open to certain species, meaning that often guests like ions can be inserted in these systems. Good examples for such a system are the polyethyleneoxide/V_2O_5 intercalation compounds. V_2O_5 xerogels are able to encapsulate many water soluble polymers [108], and many monomers can be polymerized oxidatively in the intralamellar space of the host. The PEO/V_2O_5 can simply be prepared by stirring an aqueous solution of a V_2O_5 xerogel together with the polyethyleneoxide, which is readily encapsulated between the layers. Large amounts can be taken up while the interlayer distance appreciably swells, and it can be shown that the PEO chains arrange in a zigzag conformation parallel to the sheets of V_2O_5 [108]. Lithium ions can be inserted via a redox process by exposing the composite to a solution of lithium iodide in acetonitrile, and iodine is produced as a by-product of this exchange procedure. It is interesting to see that the exchange and introduction of the lithium is possible via a redox reaction in which the vanadium oxide is reduced and iodide oxidized. The lithium ions are mobile in between the layers, and this results in good ionic conductivity in the composite material.

A final, really unexpected example of a zeolite-like material is the cluster compound $Cu_{146}Se_{73}(PPh_3)_{30}$ which forms trigonal prismatically shaped triphenylphosphonium coated copper selenide clusters of 4 nm length and a thickness of about 2 nm [109]. This material crystallizes in a most amazing structure where the cluster molecules arrange in a way that they form unidimensional channels in the hexagonal crystals along the crystallographic c-axis (Fig. 12). These channels are normally filled with the solvent molecules from the crystallization which is tetrahydrofuran, but upon drying the organic molecules can be removed without destruction of the material. Sorptive properties are not known so far, but they might be similar to the adsorption behavior of MCM-41 type materials. Future research will show whether this is just a singular example or whether other clusters with appropriate geometry tend to crystallize in open structures.

While this material lies rather in the pore size range of the ordered mesoporous oxides, structures which are described as "skeletal" by biologists might form ordered pore systems on an even larger-size scale. These structures often form the solid backbones of marine algae or other organisms which live in marine environments. Fore several years Mann's group has been researching the biomimetic growth of inorganic materials. Mesoporous or macroporous aragonite in coral-like arrangements can be obtained from microemulsions of oil, supersaturated calcium bicarbonate solutions and surfactant [110]. The pore sizes of these hollow aggregates can be controlled by the relative concentration of water and oil over the size of the emulsion droplets which seem to be embedded by the rapidly crystallizing aragonite. These porous calcite arrangements can be obtained as small spheres of approximately 1 μm diameter or they can be deposited on metal substrates. Even hollow spheres can be produced by covering uniform polystyrene spheres with the microemulsions with subsequent

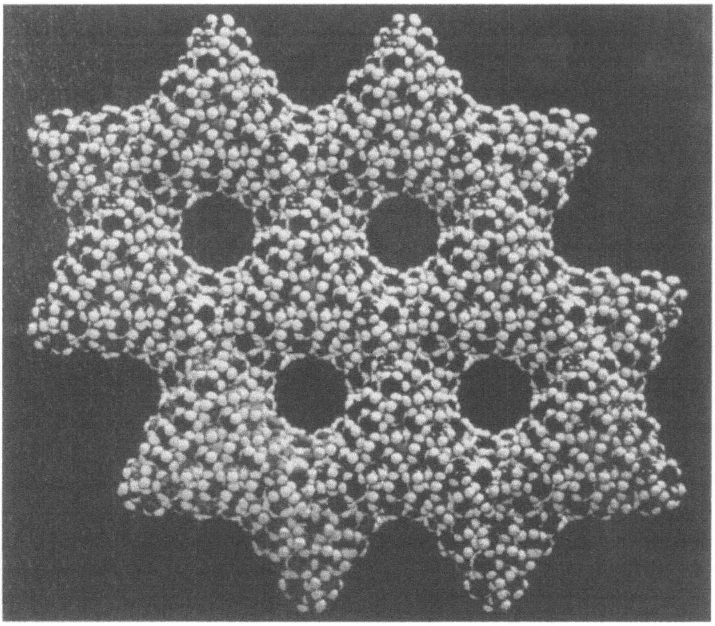

Fig. 12. Hexagonal arrangement of cluster molecules of $Cu_{146}Se_{73}(PPh_3)_{30}$. View in direction of the channels

removal of the emulsion phase after aragonite crystallization and finally dissolving the polystyrene sphere in an appropriate organic solvent.

This could be a totally new route to produce light-weight ceramics, catalytic supports or membranes on the basis of these cellular frameworks, provided that this technique is applicable to other materials than calcium carbonate.

6 Conclusions

We have presented a wide variety of zeolite-like materials which can be obtained by various synthetic techniques. Many of the resulting solids are not porous so far because the guest species included in these structures have not been removed. For other materials it is not known, whether they have structural porosity, since sorption analysis, which belongs to the standard tools of zeolite scientists, is not routinely being used by solid-state chemists working on the non-conventional zeolite-like compounds. If contacts between the different communities intensified, much progress could certainly be made in the future.

The compounds described here enrich the world of advanced materials and are suitable for many special applications. Yet there is one great obstacle in the synthesis of these compounds that has still not been overcome: All the approaches for the synthesis are mainly based on a combination of practical knowledge, good estimations, chemical intuition and synthetic luck. However, a deep understanding of the processes that lead to the desired open frameworks, is still lacking [111]. What is often called a "designed material" is frequently the result of years of synthetic efforts, screening synthesis conditions and characterizing the products. Together with the suitable application it might seem as if the final material has been designed, even if it was not. We certainly do not want to underrate the achievements of the chemists who were able to make this large variety of materials available. However, the future challenges for materials scientists working in this field will be the development of models to understand the underlying chemical pathways leading to the formation of such materials which will then eventually lead to the a priori design of materials that are suitable for the desired application.

7 References

1. Zhang J, Burnham CW (1994) Am Miner 79:168
2. Feng Q, Kanoh H, Miyai Y, Ooi K (1995) Chem Mater 7:148
3. Feng Q, Kanoh H, Ooi K, Tani M, Nakacho Y (1994) J Elektrochem Soc 141:135
4. RM Barrer (1982) Hydrothermal Chemistry of Zeolites. Academic, New York
5. Latroche M, Brohan L, Marchand R, Tournoux, M (1989) Solid State Chem J 81:82
6. Wanatabe M, Mori T, Yamauchi S, Yamamura H (1995) Solid State Ion 79:376
7. Wanatabe M, Komatsu Y, Sasaki T, Fujiki Y (1991) J Solid State Chem 92:80
8. Gareh J, Boucher F, Evain M, Connor C J O', Li S (1996) J Solid State Chem 122:41
9. Schöllhorn R (1980) Angew Chem 92:1015
10. Ozuku T, Kitagawa M, Sawai K, Hirai T (1991) J Electrochem Soc 138:360
11. Rossouw MH, Liles DC, Thackeray MM, David WIF, Hull S (1992) Mater Res Bull 27:221

12. Burns RG,.Burns VM, Stockmann HW (1983) Am Mineral 68:972
13. Giovanolli R (1985) Am Mineral 70:202
14. Post JE, Bish DL (1988) Am Mineral 73:861
15. Golden DC, Chen CC, Dixen JB (1986) Science 231:717
16. Shen YF, Zerger RP, DeGuzman RN, Suib SL, McCurdy L, Potter DI, O'Young CL (1993) Science 260:511
17. DeGuzmann RN, Shen YN, Neth EJ, Suib SL, O'Young CL, Levine Newsam SJM (1994) Chem Mater 6:815
18. Yin G, Xu WQ, Shen YN, Suib SL, O'Young CL (1994) Chem Mater 6:1803
19. Bach S, Henry M, Baffier N, Livage J (1990) J Solid State Chem 88:325
20. Suib SL (1996) Annu Rev Mater Sci 26:135
21. Turner S, Busek PR (1979) Science 203:456
22. Nenoff TM, Harrison WTA, Stucky GD (1994) Chem Mater 4:388
23. Behrens EA, Poojary DM, Clearfield A (1996) Chem Mater 8:1236
24. Rocha J, Brandao P, Lin Z, Esculcas AP, Ferreira A, Anderson MW (1996) J Phys Chem 100:14978
25. Feng S, Tsai M, Greenblatt M (1992) Chem Mater 4:388
26. Harrison WT, Gier TE, Stucky GD (1995) Zeolites 15:408
27. Tytko KH, Glemser O (1976) Adv Inorg Chem Radiochem 19:239
28. Pope M, Müller A (1991) Angew Chem 103:56
29. Pope M (1983) Heteropoly and Isopoly Oxometallates. Springer, Berlin Heidelberg
30. Kozhenikov I (1995) Catal Rev Sci Eng 37(2):311
31. Saha K, Ali M, Banerjee P (1993) Coord Chem Rev 122:41
32. Hölscher M, Englert U, Zibrowius B, Hölderich WF (1994) Angew Chem 106:2552
33. Guo J, Zavalij P, Whittigham MS (1994) Chem Mater 6:357
34. Khan MI, Zubieta J (1994) Angew Chem 106:784
35. LeBideau J, Payen C, Palvadeau P, Bujoli B (1994) Inorg Chem 43:4885
36. Maeda K, Kiyozumi Y, Mizukami F (1994) Angew Chem 106:2427
37. Maeda K, Akimoto J, Kiyozumi Y, Mizukami F (1995) Angew Chem 107:1313
38. Haushalter RC, Mundi LA (1992) Chem Mater 4:31
39. Fischer J (1966) Amer Min 51:1811
40. Moore PB, Shen J (1983) Nature 306:356
41. Szostak R, Kuvadia R, Brown J, Thomas TL (1989) In: Jacobs PA, Santen RA van (eds) Zeolites: Facts, figures, future. Elsevier Science, Amsterdam, p 439. Stud Surf Sci Catal 49A
42. Li KH, Wu LS (1994) J Chem Soc Dalton Trans 1577
43. Cavellec M, Riou D, Ninclaus C, Greneche J-M, Ferey G (1996) Zeolites 17:250
44. Leclaire A, Monier J-C, Raveau B (1983) J Solid State Chem 48:147
45. Lii KH, Haushalter RC (1987) J Solid State Chem 69:320
46. Haushalter RC, Lai FW (1988) J Solid State Chem 76:218
47. Haushalter RC (1990) J Solid State Chem 89:215
48. Haushalter RC (1987) J Chem Soc Chem Commun 1566
49. Mundi L, Strohmaier KG, Goshorn DP, Haushalter RC (1990) J Am Chem Soc 112:8182
50. Mundi LA, Haushalter RC (1990) Inorg Chem 29:2881
51. Mundi LA, Haushalter RC (1992) Inorg Chem 31:3050
52. Peascoe R, Clearfield A (1991) J Solid State Chem 95:289
53. Haushalter RC, Strohmaier KG, Lai FW (1989) Science 246:1289
54. Mundi LA, Strohmaier KG, Goshorn DP, Haushalter RC (1990) J Am Chem Soc 122:8182
55. Benhamada L, Grandin A, Borel MM, Leclaire A, Leblanc M, Raveau B (1992) J Solid State Chem 96:390
56. Haushalter RC, Wang Z, Tompson ME, Zubieta J (1993) Inorg Chem 32:3700
57. Haushalter RC, Wang Z, Tompson ME, Zubieta J, O'Connor CJ (1993) Inorg Chem 32:3966
58. Khan MI, Lee Y-S, O'Connor CJ, Haushalter RC, Zubieta J (1994) J Am Chem Soc 116:4525
59. Harrison WTA, Hsu K, Jacobson AJ (1995) Chem Mater 7:2004
60. Soghomonian V, Chen Q, Haushalter RC, Zubieta J, O'Connor CJ (1993) Science 259:1596
61. Zhang Y, Haushalter RC, Clearfield A (1996) J Chem Soc Chem Commun 1055

62. Zhang Y, DeBord JRD, O'Connor CJ, Haushalter RC, Clearfield A, Zubieta J (1996) Angew Chem 108:1067
63. Zhang Y, Haushalter RC, Clearfield A (1996) Inorg Chem 35:4950
64. Benhamada L, Grandin A, Borel MM, Leclaire A, Raveau B (1992) J Solid State Chem 97:131
65. Haushalter RC, Wang Z, Tompson ME, Zubieta J (1993) Inorg Chem 32:3700
66. Bonavia G, DeBord J, Haushalter RC, Rose D, Zubieta J (1995) Chem Mater 7:1995
67. Haushalter RC, Wang Z, Tompson ME, Zubieta J (1995) Inorg Chim Acta 232:83
68. Pingyun XB, Feng P, Stucky, GD (1995) J Chem Soc Chem Commun 1337
69. Crespoa P, Grandin A, Borel MM, Leclaire A, Raveau B (1993) J Solid State Chem 105:307
70. Soghomonian V, Chen Q, Haushalter RC, Zubieta J (1993) Angew Chem 105:601
71. Zhang Y, Clearfield A, Haushalter RC (1995) Chem Mater 7:1221
72. Khan MI, Meyer LM, Haushalter RC, Schweitzer AL, Zubieta J, Dye J (1996) Chem Mater 8:43
73. Whang D, Hur N-H, Kim K (1995) Inorg Chem 34:3363
74. Dhingra SS, Haushalter RC (1994) J Solid State Chem 112:96
75. Meyer LM, Haushalter RC (1994) Chem Mater 6:349
76. Sevov SC (1996) Angew Chem 108:2814
77. LaDuca R, Rose D, JRD DeBord, Haushalter RC, O'Connor CJ, Zubieta J (1996) J Solid State Chem 123:408
78. Alberti G, Casciola M, Constantino U, Vivani R (1996) Adv Mater 8:291
79. Olivera-Pastor P, Maireles-Torres P, Rodriguez-Castellon E, Jimenez-Lopez A, Cassagneau T, Jones DJ, Roziere J (1996) J Chem Mater 8:1758
80. Clearfield A, Wang JD, Tian Y, Stein E, Bhardwaj C (1995) J Solid State Chem 117: 275
81. Liebau F, Wang X (1992) American Crystallographic Association, Annual Meeting Abstract Book, 128
82. Liebau F, Wang X (1995) suppl Eur J Mineral 7:152
83. Liebau F, Wang X (1995) Z Kristallogr 210:693
84. Simon U, Schüth F, Schunk S, Wang X, Liebau F (1997) Angew Chem 109:1138
85. Bedard RL, Wilson ST, Vail LD,.Bennett JM, Flanigen EM (1989) In: Jacobs PA, Santen RA van (eds) Zeolites: Facts, figures, future. Elsevier Science, Amsterdam, p 375. Stud Surf Sci Catal 49A
86. Bowes CL, Ozin GA (1996) Adv Mater 8:13
87. Axtell EA, Kanatzidis MG (1996) Chem Mater 8:1350
88. Jiang T, Lough AJ, Ozin GA, Young D, Bedard RL (1995) Chem Mater 7:245
89. Bensch W, Schur M, Angew Chem, submitted
90. Marking GA, Kanatzidis MG (1995) Chem Mater 7:1915
91. Bissessur R, Heising J, Hirpo W, Kanatzidis MG (1996) Chem Mater 8:318
92. Li J, Guo H-Y, Zhang X, Kanatzidis MG (1995) J Alloys and Comp 218:1
93. Schnick W, Berger U (1991) Angew Chem 103:857
94. Schnick W, Lücke J (1990) J Solid State Chem 87:101
95. Schnick W, Lücke J (1990) Z Anorg Allg Chem 588:19
96. Schnick W, Lücke J (1992) Angew Chem 104:208
97. Schnick W, Lücke J (1994) Z Anorg Allg Chem 620:2014
98. Wester F, Schnick W (1996) Z Anorg Allg Chem 622:1281
99. Weitkamp J, Ernst S, Cubero F, Wester F, Schnick W (1997) Adv Mater 9:247
100. Schlieper T, Schnick W (1995) Z Anorg Allg Chem 621:1535
101. Schlieper T, Schnick W (1996) Z Kristallogr 211:254
102. Schwarz P, Siebel E, Fischer RD, Apperley DC, Davies NA, Harris RK (1995) Angew Chem 107:1311
103. Brimah AK, Siebel E, Fischer RD, Davies NA, Apperley DC, Harris RK (1994) J Organomet Chem 475:85
104. Schwarz P, Eller S, Siebel E, Soliman TM, Fischer RD, Apperley DC, Davies NA, Harris RK (1996) Angew Chem 108:1611

105. Maier WF, Martens JA, Klein S, Heilmann J, Parton R Vercruysse K, Jacobs PA (1996) Angew Chem 108:222
106. Klein S, Maier WF (1996) Angew Chem 108:2376
107. Ozin GA (1992) Adv Mater 4:612
108. Liu Y-L, Schindler JL, DeGroot DC, Kannewurf CR, Hirpo W, Kanatzidis MG (1996) Chem Mater 8:525
109. Krautscheid H, Fenske D, Baum G, Semmelbaum M (1993) Angew Chem 105:1364
110. Walsh D, Mann S (1995) Nature 377:320
111. Schöllhorn R (1996) Angew Chem 108:2493

Pillared Clays: Preparation, Characterization and Applications

P. Cool · E. F. Vansant

Laboratory of Inorganic Chemistry, Department of Chemistry, University of Antwerp (UIA), Universiteitsplein 1, B-2610 Wilrijk, Belgium.
E-mail: cool@uia.ua.ac.be; evansant@uia.ua.ac.be

1 Introduction

Pillared Clays (PILCs) are an interesting class of 2-dimensional microporous materials. Due to their high surface area and permanent porosity they are very attractive solids for adsorption and catalysis purposes. Since their porosity can be localized in the larger micropore region, these substrates form a bridge between the microporous zeolites on one hand and the inorganic meso- and macroporous solids (e.g. silica, alumina) on the other. The history of PILCs started in 1955, but the first extensive studies appeared only around 1980. During this pioneering work, mainly organic cations and organometal pillars were used. Now, the inorganic polyoxycations are of greatest interest because of their high thermal stability. By changing the nature and, hence, the size of the pillars, different pore-sizes are obtained. In this way, it becomes possible to tune the porosity. This porosity combined with the properties of both pillar and host are very important for certain applications, such as selective gas adsorptions, catalytic reactions, etc.

In this review on the principle of pillaring, various layered structures with different pillaring species will be discussed.

2
The Principle of Pillaring

Although clays are very useful for many applications in the field of catalysis, adsorption and ion exchange, they have one main disadvantage: their lack of permanent porosity. Smectites swell upon hydration, but upon severe dehydration (heating) the layers collapse and the interlayer surface becomes no longer accessible for chemical processes.

To avoid this problem, researchers found a way to prop open the clay layers by the introduction of stable pillars in the interlayer region. By doing so, a high pore volume is created. The pillared interlayered clay (PILC) maintains its porosity during the hydration or dehydration process. The idea is schematically represented in Fig. 1.

The principle of pillaring can be traced back to the early work of Barrer and McLeod, wherein they showed that permanent porosity in montmorillonite

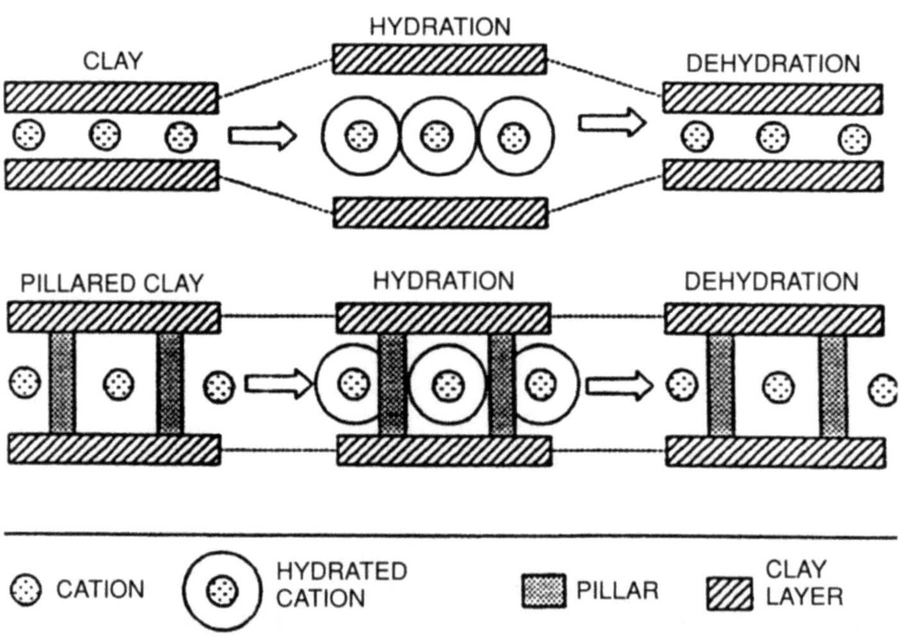

Fig. 1. Hydration-dehydration behavior of a clay and a pillared clay

could be introduced by replacing the interlayer alkali metal ions with tetraalkylammonium ions [1]. During the oil crisis (1973), these pillared solids caught the attention of catalysis researchers since they had pores with the appropriate size for cracking. In spite of their thermal instability, other pioneers like Brindley and Sempels [2] and Vaughan et al. [3] started to develop the inorganic pillared clays. The first fundamental study on an inorganic PILC appeared in the late 1970s. These types of PILCs are still of great interest since they are stable up to temperatures exceeding 773 K.

The concept of pillaring is very straightforward and consists of two main steps: first, the interlamellar small cations are exchanged for other, bulky ions. A second or calcination step converts the inorganic polyoxycation precursors into rigid, stable metal oxide pillars, tightly bonded to the clay layers (Fig. 2).

It has become clear that both the parent clay and the pillars must possess several properties in order to allow the pillaring process to be successful. The clay should have a moderate cation exchange capacity (CEC) to obtain the best swelling and alkali or alkaline earth ions in the interlayer space which can easily

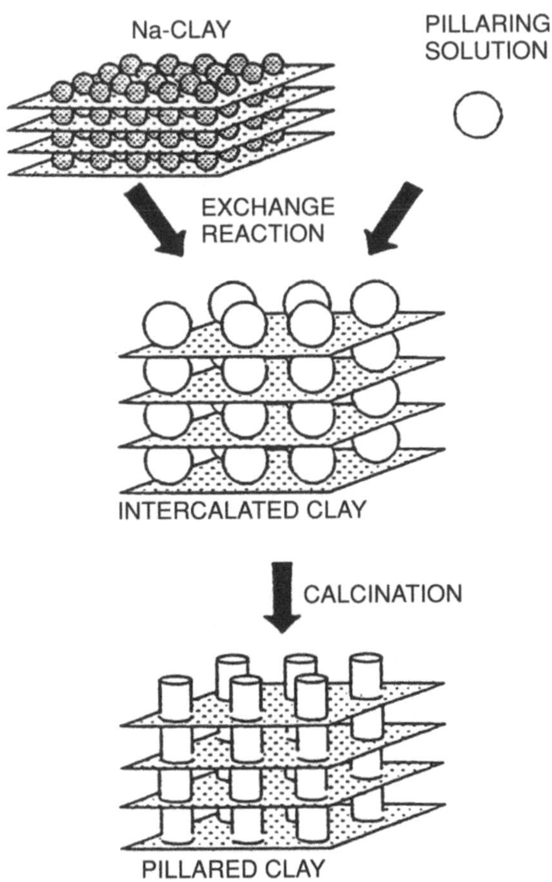

Fig. 2. The principle of pillaring

be exchanged. From this point of view, smectites are the ideal host. The pillaring cation requires a high positive charge and must be dissolvable in the polar solvent used to swell the clay.

Nowadays, the pillaring of clays is a well established technique in the preparation of new porous solids. It is an easy and controllable way to introduce porosity in a rather inexpensive substrate. A variety of pillaring species is available, all with specific sizes and compositions. In this way, it is possible to prepare PILCs with different interlayer free spacings which determine the pore size. The interpillar distance can be tuned by varying the charge of both host and pillars.

Depending on the final purpose of the material, the appropriate pillar can be chosen. This flexibility in the PILC synthesis is one of the main advantages compared to other porous substrates, such as zeolites, which have one definite pore size. The technique not only focuses on clays, but other layered structures serve as host materials as well. Examples are layered double hydroxides (anionic clays), metal(IV)- phosphates and phosphonates, layered silicic acids, etc. [4, 5].

3 The Variety of Pillaring Species

A brief overview of the different well known pillaring species is given in Table 1.

The organic cations were the first pillaring species under investigation. Intercalation of alkylammonium and dialkylammonium ions of various chain length by Barrer resulted in interlayer distances of 2.2–4.0 Å [1]. Surface areas (SA) of up to 300 m^2/g ($CH_3NH_3^+$) and micropore volumes of 0.098 cm^3/g ($^+NH_3(CH_2)_4NH_3^+$) were observed. Although the cations become easily intercalated between the layers, these organo-PILCs have some disadvantages. There is no strong interaction between the amine pillars and the clay sheets so that they remain exchangeable for other cations. Second, their thermal stability is limited to about 523 K.

Other species, like organometals, were used for pillaring in an attempt to increase the thermal stability and to introduce catalytically active elements into the PILC. Examples are the use of tris(ethylene-diamine)cobalt(III) or shortly $Co(en)_3^{3+}$, M(2,2-bipyridine)-complexes and M(o-phenanthroline)-complexes. However, these pillared materials continue to be thermally unstable (above 350–450 °C) due to the presence of C-C bonds. The first stable pillars were obtained by the intercalation of organo-metal complex precursors, e.g., tris (acetylacetonato)silicon(IV) or $Si(acac)_3^{3+}$ and the Fe(III)-acetato complex $Fe_3O(CH_3COO)_6(CH_3COOH) \cdot 2H_2O^+$. The in-situ hydrolysis of the Si-precursor and the intercalation of partially hydrolyzed Fe-complexes followed by a thermal treatment lead to stable oxide pillars. The interlayer free spacing (IFS) of the Si-PILC was approximately 3 Å with a surface area of 200 m^2/g. Due to the fast and uncontrollable hydrolysis of the Fe(III)-acetato precursor, broad pore-size distributions (PSD) with basal spacings between 6 and 10 Å were found, combined with a low porosity. Up to 773 K, the pillared solids remained unaffected.

Metal cluster complexes were proposed to introduce mainly transition metals between the clay layers. Niobium, tantalum and molybdenum chlorides were exchanged on the clay surface and oxidized at 513 K under vacuum to transform

Table 1. Overview of the most common pillaring species

Class	Examples	Ref.
Organic cations	alkylammonium,	[6]
	dialkylammonium	[7]
	DABCO	[8]
Organometallic cations	$Co(en)_3^{3+}$	[9]
	M(2,2-bipyridine)-complexes	[10]
	M(*o*-phenanthroline)-complexes	[11]
	$Si(acac)_3^{3+}$	[12]
	$Fe_3O(OCOCH_3)_6CH_3COOH^+$	[13–16]
Metal Clusters	$Nb_6Cl_{12}^{n+}$, $Ta_6Cl_{12}^{n+}$, $Mo_8Cl_8^{4+}$	[17]
Polyoxycations	$Al_{13}O_4(OH)_{24}(H_2O)_{12}^{7+}$	[18–20]
	$Zr_4(OH)_8(H_2O)_{16}^{8+}$	[21–23]
	$(TiO)_8(OH)_{12}^{4+}$	[21, 24]
	$Cr_n(OH)_m^{(3n-m)+}$	[25, 26]
	Hydrolyzed Fe-salts	[27, 28]
Oxide Sols	TiO_2–, TiO_2–SiO_2 sol	[21, 29, 30]
	$Si_2Al_4O_6(OH)_8$ or imogolite	[31]
Mixed Oxide Pillars	Fe/Al	[32]
	Fe/Cr, Fe/Zr	[33]
	La/Al	[34]
	$GaAl_{12}O_4(OH)_{24}(H_2O)_{12}^{7+}$	[35]
	Cr/Al	[36]
	$LaNiO_x$	[37]

into the metal-oxide form. An advantage here is the well known size of the complexes (no polymerization), but low surface areas resulted. Intercalation of $M_6Cl_{12}^{n+}$ gave a surface area of only 64 m^2/g.

The direct intercalation of metal oxide sols (DIMOS) according to Pinnavaia is a good technique to achieve stable oxide pillars [31]. Aqueous oxide sols with uniform particle sizes of 20–80 Å can be used to create mesoporous pillared clays. The materials are also called supergallery PILCs, a term covering all substrates with an interlayer distance substantially larger than the thickness of the host layers. Ti-sol prepared from Ti-alkoxides results in Ti-PILCs with interlayer free spacings of 13 Å and surface areas of 300 m^2/g.

Mixed sols of SiO_2–TiO_2 create interlayer free spacings of 25–35 Å. The pore size is, however, not of the same dimension, the spherical sol particles pack irregularly between the layers and form pores of non-uniform size in the microporous range. A schematic representation of this mixed supergallery PILC is given in Fig. 3.

Using template molecules like OTMA (octadecyltrimethylammonium) orients the sol particles, and large pores (30–40 Å) will be formed (see Fig. 4). Extremely high surface areas and pore volumes amounting to 600 m^2/g and 0.8 cm^3/g, respectively, are created.

A sol solution containing a tubular aluminum silicate called imogolite or $Si_2Al_4O_6(OH)_8$ gives interlayer distances of about 34 Å and a surface area of

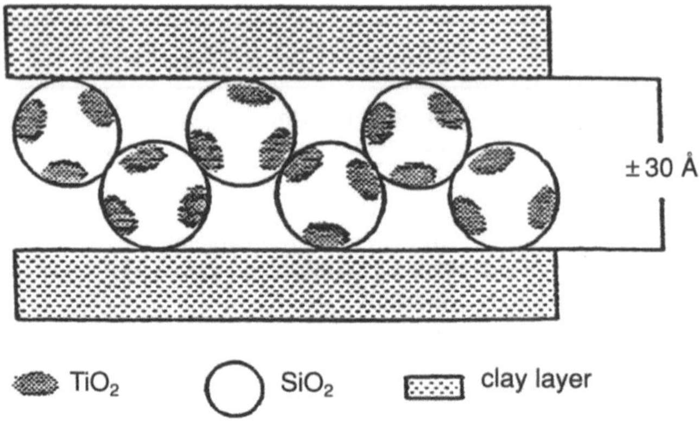

Fig. 3. Mixed SiO_2-TiO_2 PILC

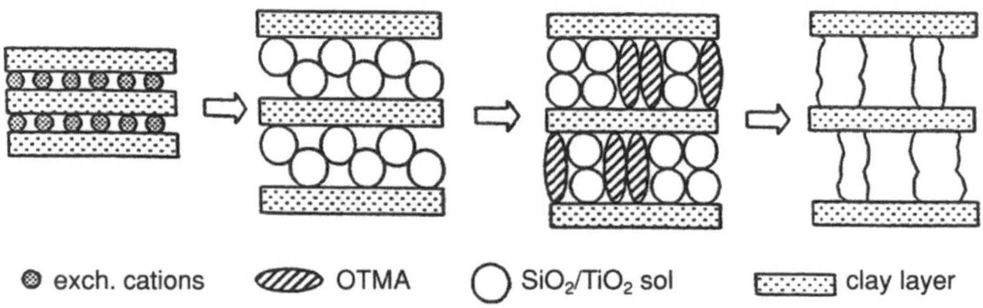

Fig. 4. SiO_2-TiO_2 pillared clay synthesized using OTMA as template

460 m^2/g. Both the intra- and intertubular channels contribute to the final micropore volume. The structure of these Tubular Silicates Layered Silicates (TSLS) is presented in Fig. 5.

By far the most popular and most useful for pillaring are the polyoxycations. They are synthesized by a controlled hydrolysis of metal salts to give complexes with the desired properties for pillaring: high charge, bulky character, and convertable to stable oxide pillars upon heating. Table 2 gives an overview of the most common PILCs together with their characteristics. Figure 6 shows some typical pillaring precursors.

Among them, only the Al-PILC exhibits a narrow pore size distribution [16]. The hydrolysis is the most easy to control, and the pillar dimensions are not very sensitive to differences in the hydrolysis conditions. Interlayer free spacings between 7 and 10 Å are always obtained.

The Zr-tetramers polymerize to form sheets which link together to form a 3-dimensional polymerized pillar. The height of this pillar is determined by the number of constituting sheets, depending on the hydrolysis conditions. Careful control of the polymerization reaction should result in PILCs with different

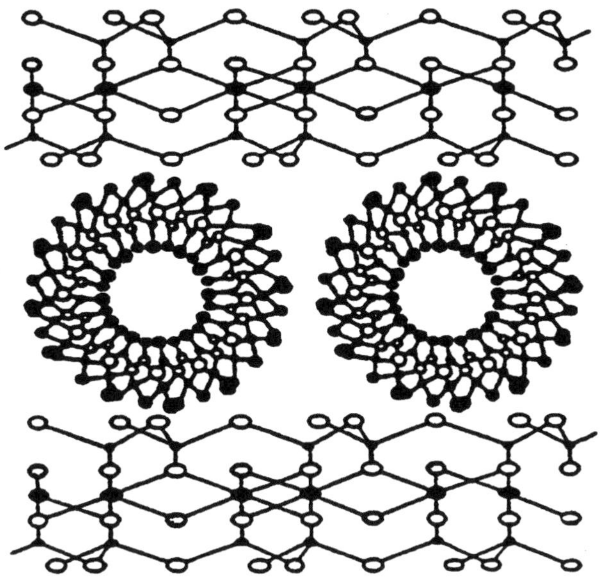

Fig. 5. Front view of a tubular imogolite-PILC

Table 2. Properties of pillared clays intercalated with polyoxycation pillars

Pillar species	Surface area (m^2/g)	Interlayer free spacing (A)
Al_2O_3	200–400	7–10
ZrO_2	200–300	4–14
Cr_2O_3	±300	±12
Fe_2O_3	100–300	±15+mesopores
TiO_2	±300	±14–18+mesopores

basal spacings, but in most cases broad pore size distributions (PSD) are obtained [16]. The formation of this pillar precursor will be discussed in more detail.

Pillaring with hydrolyzed Cr-polyoxycations $[Cr_n(OH)_m]^{(3n-m)+}$ was first reported by Yamanaka and Brindley [38]. The obtained IFS of 16.8 Å collapsed at temperatures above 473 K. Tzou and Pinnavaia succeeded in the synthesis of a more stable PILC, with a surface area of 350 m^2/g and an interlayer free spacing of 11.7 Å even after heat activation at 773 K [26]. Both the basal spacing and the surface area strongly depend on the hydrolysis and the calcination conditions.

Iron salts, e.g. Fe(III)nitrate, chloride or perchlorate, polymerize rapidly in the presence of Na_2CO_3 to form large molecules of very low charge. As a result of this low charge, many Fe-pillars need to be intercalated to neutralize the negative charge density on the clay sheets. The stuffing of the interlayer region with pillars explains the rather low surface area and microporosity of these solids.

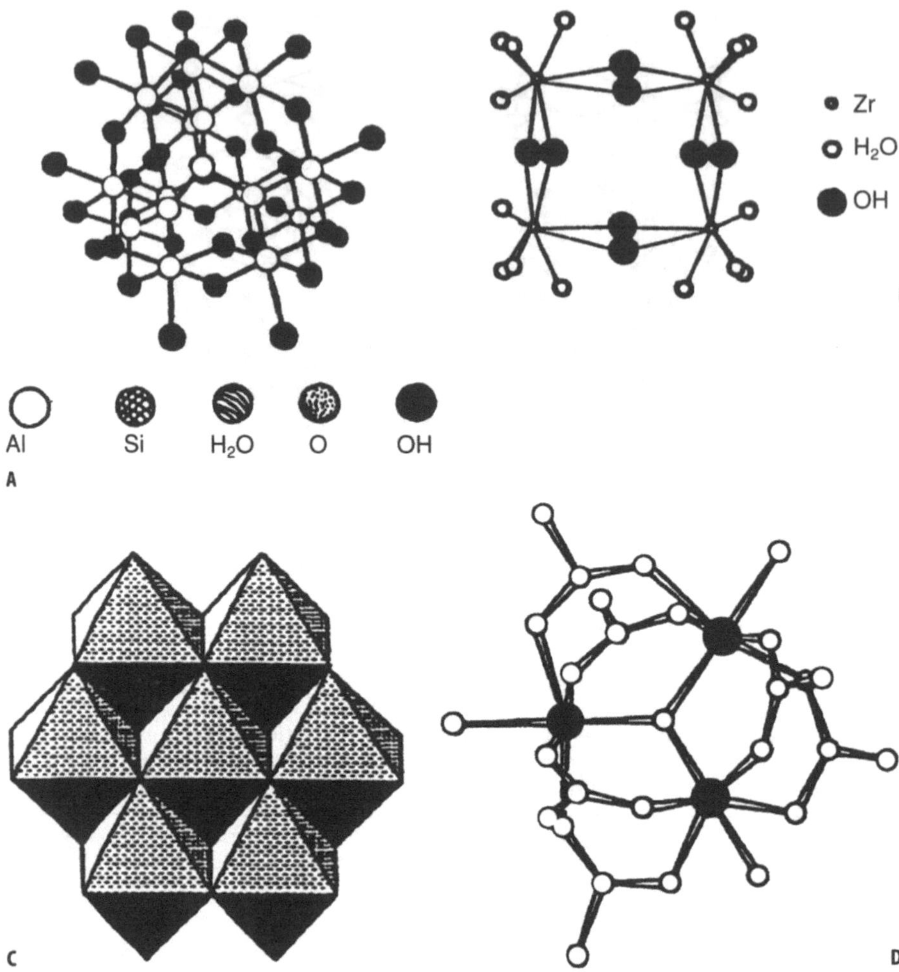

Fig. 6. Illustration of some pillaring precursors for Al-PILC (**A**), Zr-PILC (**B**), Ti-PILC (**C**) and Fe-PILC (**D**)

$TiCl_4$ or Ti-alkoxides rapidly hydrolyze in HCl and polymerize to a sol which results in large polynuclear species with the formula $[TiO(OH)_2]_n$. Broad pore size distributions with a maximum interlayer free spacing of 30 Å and surface areas up to 300 m^2/g have been reported.

Mixed oxide pillared clays are a new form of pillared clays. They can be obtained by a combined hydrolysis of metal salts: La/Al, Al/Fe, Fe/Cr, Fe/Zr, Cr/Al. Another route is the synthesis of complexes containing different metal cations: $GaAl_{12}$-Keggin structures or binary oxides. These mixed pillared clays will have other, more specific catalytic and adsorption properties, while the size and charge of the pillaring species can be changed. The main drawback of these pillared materials is the characterization of the pillars.

As an example, Heylen et al. showed that the incorporation of chromium in iron oxide pillars has a positive effect on the adsorption affinity for cyclohexane, CCl_4 ($P_{eq}=4.7\times10^3$ Pa), and CO_2 ($P_{eq}=4.7\times10^4$ Pa) at 273 K [33]. An improvement of the selectivity as well as the capacity was noticed for the mixed Fe/Cr-PILC in comparison to the parent Fe-PILC.

It becomes clear that a broad pore size range can be covered by the different polyoxycations, though the pore size distribution of PILCs is not uniform as is the case for zeolites. One must keep in mind that careful control of the hydrolysis conditions is necessary in order to obtain the narrowest pore size distribution. Variations in the pillar composition (size, charge, etc.) due to hydrolysis phenomena result in variations of both the slit width and the interpillar distance.

3.1 The Al-Pillaring Precursor

In the literature [39], four different methods are known for preparing Al-pillaring solutions:

1. hydrolysis of Al^{3+}-salts (usually $AlCl_3$) with NaOH,
2. hydrolysis of Al^{3+}-salts (usually $AlCl_3$) with Na_2CO_3 (CO_3^{2-} reacts with H_3O^+ to form OH^- ions and CO_2 gas),
3. electrolysis of $AlCl_3$, and
4. dissolution of Al metal in HCl (with liberation of H_2).

Route 1 is a well established technique and most often utilized in order to prepare fresh Al-solutions on a laboratory scale. On the industrial scale, route 4 is preferred for the preparation of hydrolyzed Al-solutions, such as Locron L (Hoechst) and Chlorhydrol (Reheis Chemical Co.). These stock solutions are very concentrated and need to be diluted prior to pillaring. The commercially available Locron L can be used as an alternative to preparation method 1.

The pillaring specie of interest formed in this way is the Al^{3+}-oligomer $[Al_{13}O_4(OH)_{24+x}(H_2O)_{12-x}]^{(7-x)+}$ (briefly Al_{13}) or Keggin ion [40]. Its structure is described as a prolate spheroid, consisting of one central AlO_4 tetrahedron surrounded by twelve octahedra of aluminum hydroxide. The three edge-sharing octahedra have free corners that are occupied by oxygen, bridging hydroxy groups, or bridging water molecules. Each octahedral Al^{3+} ion is at an equal distance from its neighbors. A representation of this Keggin structure is given in Fig. 7. The Al_{13} dimensions were proposed by Clearfield and Roberts to be 9.5 and 7 Å, for the long and the short axis of the spheroid, respectively [41].

Besides Al_{13} several other species might be present in the pillaring solution, depending on the preparation conditions: $[Al(OH)_x(H_2O)_{6-x}]^{(3-x)+}$ monomers, $[Al_2(OH)_x(H_2O)_{10-x}]^{(6-x)+}$ dimers and also higher polymeric forms (Al_{poly}) [42]. The most critical parameters which determine the composition of the Al-pillaring solution and which need to be closely controlled are the temperature, the aluminum concentration, the degree of hydrolysis OH/Al, the pH and the aging time. A distribution curve of the various Al^{3+} species vs the pH and the OH/Al ratio for a solution of $AlCl_3$ hydrolyzed with NaOH is given in Fig. 8 [42, 43].

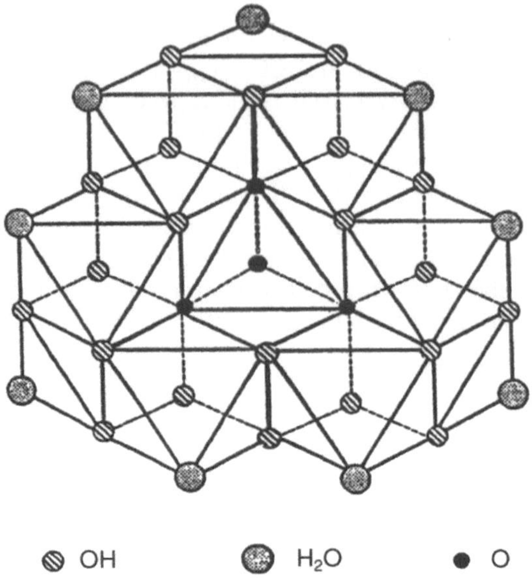

Fig. 7. Schematic representation of the Al-Keggin ion (Al_{13})

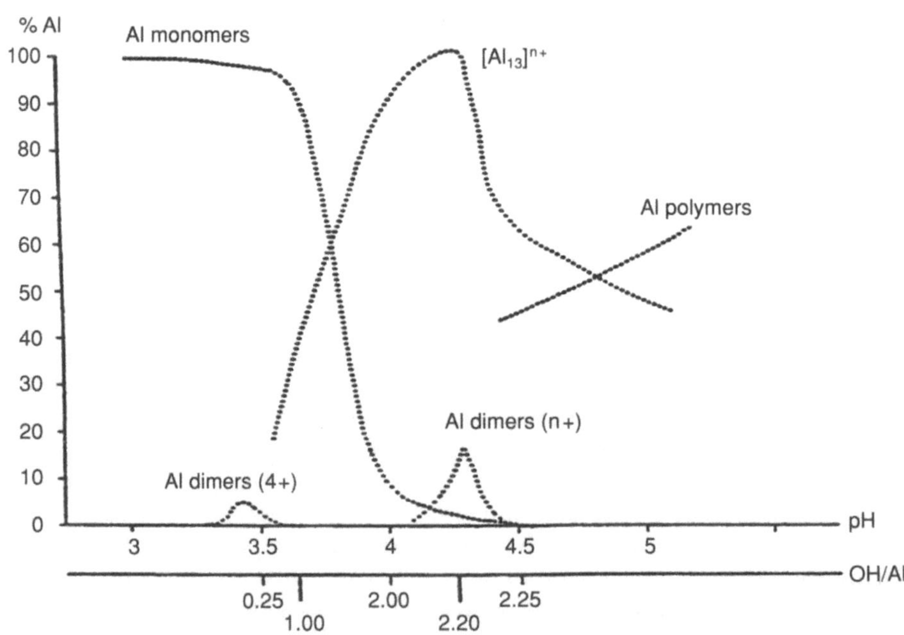

Fig. 8. Distribution curve of the various Al^{3+} species present in the $AlCl_3$ solution hydrolyzed with NaOH

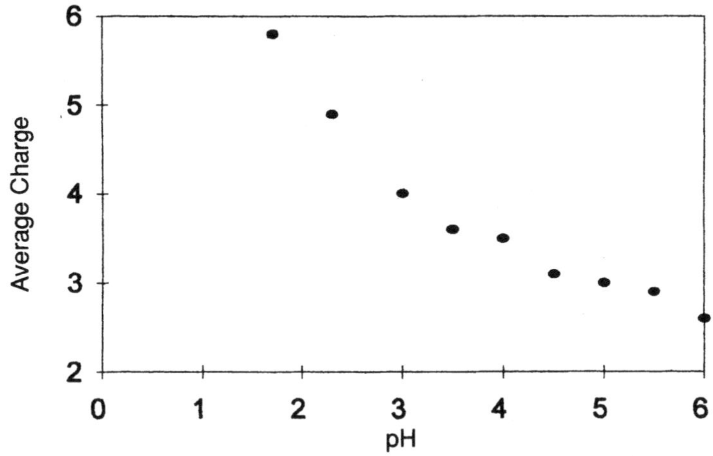

Fig. 9. Influence of the pH on the charge of the Al_{13} Keggin ion

At OH/Al < 1, the monomeric forms are dominant, while for 1.5 < OH/Al < 2.3, Al_{13} becomes the main species in solution. By a further increase of the hydrolysis degree the formation of polymers is promoted. It must be said that the distribution curve is also dependent on the preparation method. For the commercial Al_{13} solution, the percentage of Al_{13} is more or less constant over a wider range (1 < OH/Al < 2.5), but never exceeds 40 % of the relative occurrence, regardless of the hydrolysis degree.

As the OH/Al ratio is directly related to the pH of the solution, the acidity is also a very important factor to obtain good control of the charge of the Al_{13} polyoxycations. Vaughan was the first to report a charge variation with the pH [44]. The curve in Fig. 9 shows this charge dependence on the pH of the solution utilized.

Though, in the original formula, the Keggin ion carries a charge of + 7, the decrease in this charge with increasing basicity can be described as a sequence of deprotonation steps:

$$Al_{13}O_4(OH)_{24}(H_2O)_{12}{}^{7+} + H_2O \rightarrow Al_{13}O_4(OH)_{25}(H_2O)_{11}^{6+} + H_3O^+ \rightarrow \rightarrow$$

Even after intercalation between the clay sheets, the hydrolysis of Keggin ions proceeds, and the degree of hydrolysis depends on the type of host [45, 46].

It is worth mentioning that the Keggin polyoxycations can be separated from the pillaring solution to obtain a pure Al_{13} solution. The cations are precipitated as the sulfate salt, and subsequently the crystals are dissolved with $Ba(NO_3)_2$. The final solution contains only the Al-Keggin complexes, as evidenced by ^{27}Al-NMR spectroscopy [47].

3.2 The Zr-Pillaring Precursor

Compared to the Al-pillaring, the Zr-pillaring of clays is less investigated and documented in the literature. Despite the many experiments, the nature of the zirconium solutions, especially the polymerized species on hydrolytic polymerization, remain unclear, and the clay intercalates derived from these species are not definitely identified. Zirconium oxychloride solutions are usually employed as pillaring solution. In these studies the starting material is $ZrOCl_2 \cdot 8H_2O$. Although the solution chemistry of Zr is complex, it is known that the zirconyl ion is present as a tetrameric cation with the general formula $[Zr_4(OH)_8(H_2O)_{16}]^{8+}$ [48, 23]. A representation of this type of pillaring precursor has been given previously in Fig. 6. The four Zr-ions are located at the corners of a slightly distorted square and are linked together by OH bridges above and below the plane of the square.

When dissolved in water, a rapid polymerization of the tetramers occurs. Clearfield suggested the following sequence of events during the aging or hydrolysis process of Zr-pillaring solutions [49]:

$$[Zr_4(OH)_8(H_2O)_{16}]^{8+} \rightarrow [Zr_4(OH)_{8+x}(H_2O)_{16-x}]^{(8-x)+} + x\,H^+ \quad (1)$$

$$\rightarrow Zr(OH)_4 \cdot x\,H_2O \quad (2)$$

$$\rightarrow ZrO_2 \quad (3)$$

The solution becomes highly acidic due to hydrolysis of the tetramers, as depicted in Eq. (1). Subsequent polymerization results in the formation of colloidal zirconia in which the particles are crystallized in the monoclinic phase, Eqs. (2) and (3).

The hydrolysis and polymerization reactions starting from the tetrameric Zr-cation, as given in Eq. (1), are believed to occur in the following way: As a first step, rapid hydrolysis leads to cationic species with the formula $[Zr_4(OH)_{14}(H_2O)_{10}]^{2+}$. Here, two of the Zr-ions have a single positive charge, and the other two are neutral. Figure 10 shows one corner of a tetramer undergoing this hydrolysis reaction.

Upon aging of the solution, larger species are formed, which are believed to be due to an olation reaction, in which a neutral site on one tetramer reacts with a singly charged site on another tetramer. This results in the formation of large „rafts" or two-dimensional hydroxy polymers [23]. The olation reaction is depicted in Fig. 11. The tetramers are joined by hydroxy bridges (or -ol bridges) to form edge-to-edge bonding.

Three-dimensional polymerization proceeds by oxolation between the two-dimensional layers: the layers are joined by oxo (-O-) bridges to form face-to-

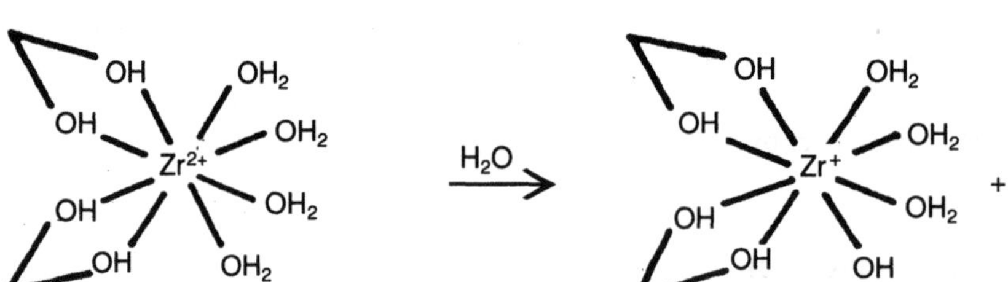

Fig. 10. Hydrolysis reaction of $Zr_4(OH)_8(H_2O)_{16}^{8+}$. Only one corner of the tetramer undergoing the reaction is shown

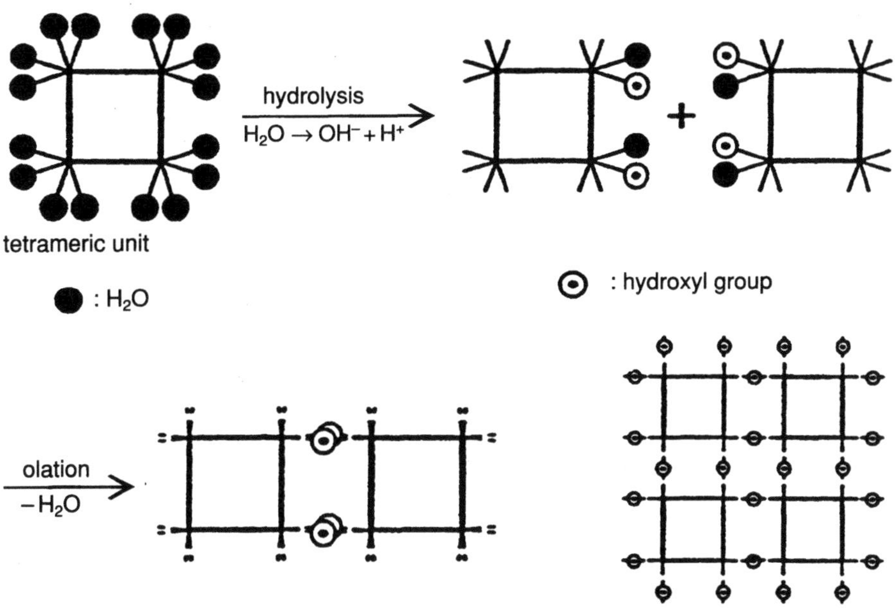

Fig. 11. The olation reaction. The solid lined squares indicate the original tetrameric units, and the dashed lines denote new ol bridges

face bonding with elimination of water by condensation of hydroxyl groups (Fig. 12a). In this face-face bonding, the tetramer layers stack together in such a way that the fluorite-type lattice of tetragonal ZrO_2 can be formed, as illustrated in Fig. 12b. When the three-dimensional assembly approaches the size of 12–24 tetramers, precipitation occurs of crystallites with a size of 30 Å. Then, on prolonged refluxing, crystal growth and crystal perfection with transformation to the monoclinic form take place.

As in the case of the Al-pillaring precursor, several experimental parameters are known to affect the degree of polymerization in the solution here as well [48]:

Influence of the pH. An increasing pH favors the hydrolysis reaction, and this in turn leads to an increase in the polymerization degree. One must note the precipitation of $Zr(OH)_4$ at a pH of 3.

Influence of the concentration. The concentration of the zirconyl chloride solution affects the degree of polymerization in as much as it affects the pH of the solution. A solution of lower concentration will have a higher pH, and so the degree of polymerization will be higher.

Influence of temperature and time. Heating leads to an enhanced polymerization, as demonstrated by Clearfield and Vaughan, simply due to an increase in the rate of hydrolysis with temperature [50]. Increasing the aging time has a similar effect.

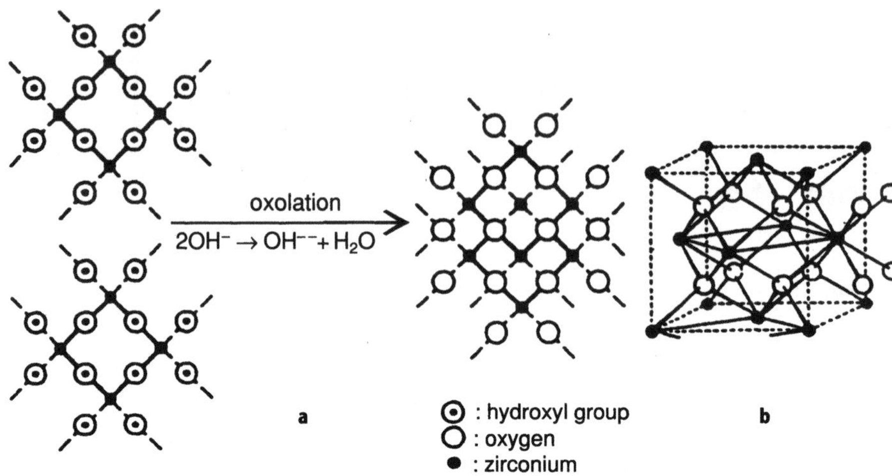

Fig. 12. **a** The oxolation reaction, **b** the fluorite structure of ZrO_2

The major specie present in the Zr-pillaring solution at room temperature is the zirconium tetramer. Its dimensions are 8.9 Å in width and length and 5.8 Å in thickness. Ohtsuka et al. reported that the intercalation leads to products with an interlayer free spacing (IFS) of ca. 7 Å [23]. Aging or refluxing the Zr-solution leads to pillared solids with higher interlayer spacings. The polynuclear specie with the upper limit of size before precipitation, which can be formed in zirconium oxychloride solutions, consists of 12 tetramers. Here, three two-dimensional layers, each composed of four tetramers, are linked. The thickness has been calculated as follows (assuming 2.5 Å as the oxygen anion diameter): 3×5.8 Å -2×2.5 Å $=12.4$ Å. In the literature, interlayer spacings after pillaring ranging from 4 to 14 Å have been reported, depending on the preparation conditions [22, 38, 51, 52]. Broad pore size distributions exist, since the fast hydrolysis and polymerization reactions are very difficult to control.

4
The Preparation of Pillared Interlayered Clays

The general synthesis procedure, as depicted in Fig. 13, consists of four main steps:

1. Purification and saturation of the clay into the Na^+-form,
2. Preparation of the pillaring solution,
3. Exchange reaction between the interlayer Na^+-ions and the polyoxycations present in the pillaring solution, and
4. Calcination with formation of a stable PILC.

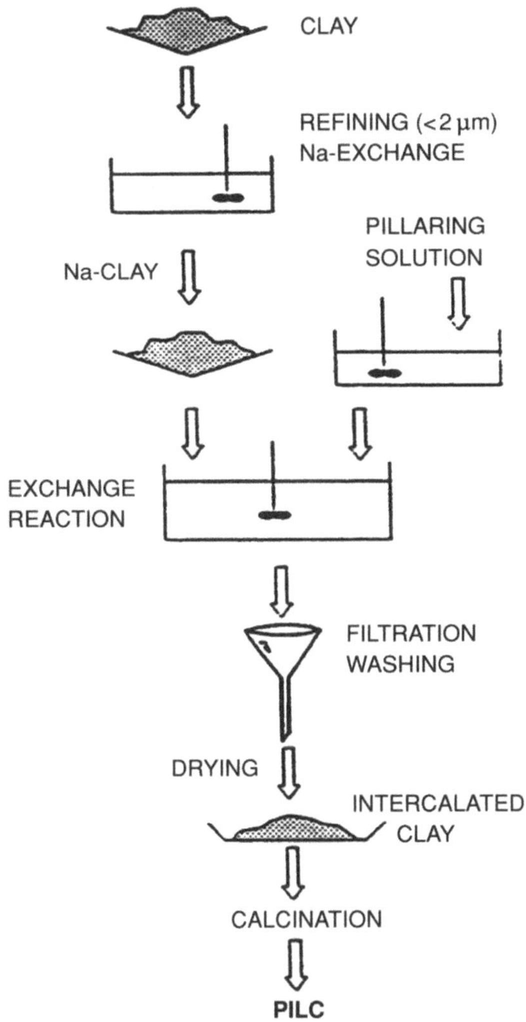

Fig. 13. General preparation procedure for pillared interlayered clays

4.1
The Parent Clays

The parent clays are usually in the Na^+-form when applied as basic substrates for pillaring. As discussed earlier, Na^+ as the charge balancing ion allows the best hydration which, in turn, facilitates the intercalation of the pillaring precursors. For the natural clays, a size fraction < 2 μm is small enough to obtain a homogeneous swelling. Since plates of this size still tend to orient in a face-to-face stacking, natural clays need to be refined, purified and saturated with sodium prior to use. The fraction > 2 μm containing all the clay impurities, can be separated by centrifugation.

The natural smectite hectorite, when obtained from the Clay Repository of the Clay Minerals Society, still contains a lot of carbonate impurities. In order to eliminate these impurities, hectorite has to be pretreated with a sodium acetate/acetic acid buffer of pH 4 [53]. In this way, all carbonates are transformed into H_2CO_3, whereupon H_2O and CO_2 (liberated from the solution) are formed. Subsequent exchange in a NaCl solution and removal of the excess chloride give Na^+-hectorite. The synthetic laponite, supplied by Laporte Inorganics, is already in the Na^+-form and free of any impurity.

4.2 The Pillaring Solutions

Pillaring solutions of Al and Zr have been prepared. In the method of Lahav et al., 0.2 M $AlCl_3$ is hydrolyzed with 0.2 M NaOH, leading to an OH/Al ratio of 2.33 and a pH of 4 [54]. The final concentration of the Al-solution is 0.07 M. The aging process is performed under reflux conditions for 24 h. An alternative and less complicated synthesis simply includes the dilution of the commercially available Locron L solution. This highly concentrated Al-solution (4.6 M) is diluted with demineralized water to obtain the same concentration, OH/Al and pH as in the first method.

For the Zr-pillaring solution 0.1 M $ZrOCl_2 \cdot 8H_2O$ is used. The aging process is also performed under reflux. The pH of the final solution is close to 1, i.e. very acidic due to polymerization.

4.3 The Ion Exchange Reaction

The intercalation is performed by adding the clay, as a powder or in suspension, to the pillaring solution. The mechanism is based on the ion exchange process between the interlayer Na^+ ions and the Al- or Zr-pillaring precursors. The total amount of pillars added can be changed by varying the amount of aluminum or zirconium per mass of clay. These amounts are always in excess of the CEC of the clay used. Now, the amount of pillar precursors, which becomes intercalated, depends on the CEC of the clay and the charge of the pillar. However, the real situation is more complex. The presence of other species with different charge in the solution will interfere with the stoichiometric exchange of the Na^+-ions by the Al-Keggin ions or the Zr-tetramers. Besides, as mentioned previously, a further hydrolysis of intercalated polyoxycations occurs upon exchange on the clay, leading to different charges. Usually, 1–2% clay suspensions in water are employed to pre-swell the clay, which facilitates subsequent intercalation. However, Molinard clearly proved that highly porous, homogeneously intercalated clays can be prepared using the clay as a powder [18]. In this way, the handling of large quantities of water is avoided, which is more economic.

4.4
The Washing and Drying Procedure

After the ion exchange reaction, the pillared clay is separated from the solution by centrifugation and washed with demineralized water to remove the excess pillaring solution and Cl^--ions. Very importantly, washing seems to improve the quality of the prepared PILC. It promotes a homogeneous distribution of the pillars between the layers, creating larger interlayer spacings (from 12 Å without washing to approximately 18 Å after washing for the Al-PILC) [45, 55, 56]. Still at this stage, hydrolysis and pH variations determine the nature and charge of the intercalated species.

Drying is also a critical parameter in the preparation of PILCs. Slow drying (or air-drying) allows the clay layers to settle down in an ordered, parallel way (face-to-face stacking). This favors the microporosity and crystallinity of the final PILC. Fast techniques, like freeze-drying, fix the random orientation of pillared clay plates or aggregates to form a card-house structure. This structure also exists for the laponite clay. PILCs dried in this way exhibit a larger mesoporosity, but are less crystalline.

4.5
The Calcination Step

A calcination step at 573–773 K in air converts the Al- and Zr-polyoxycation precursors into rigid alumina and zirconia oxide pillars. This heating process is necessary to obtain a stable PILC with a permanent microporosity, not subject to swelling or hydrolysis phenomena. During calcination, dehydration and dehydroxylation reactions of the charged pillar precursors occur to give neutral oxide particles. The equilibrium in electrical balance is maintained by the release of protons during the conversion at elevated temperature:

$$[Al_{13}O_4(OH)_{24}(H_2O)_{12}]^{7+} \rightarrow 6.5\ Al_2O_3 + 20.5\ H_2O + 7\ H^+$$
$$[Zr_4(OH)_8(H_2O)_{16}]^{8+} \rightarrow 4\ ZrO_2 + 16\ H_2O + 8\ H^+$$

Since all conditions for the Hofmann-Klemen effect are fulfilled, the liberated protons will migrate into the empty octahedral positions of dioctahedral clays. Some problems are encountered for this type of clays, since a large part of the CEC is now lost. When the trioctahedral hectorite and laponite serve as host for pillaring, the problem of proton migration can be avoided. The protons remain present in the interlayer region of the PILCs and are still available for further ion exchange.

Pinnavaia et al. described the bonding of the pillars to the hectorite clay sheets, upon calcination, as a layer cross-linking mechanism [57]. ^{29}Si and ^{27}Al MAS-NMR data on Al-pillared smectite clays indicated the existence of two mechanisms for the linking of the pillars to the clay sheets. It was shown that the layer composition of the host clay plays a very important role in determining the layer reactivity upon heating.

In the case of natural fluorohectorite and beidellite structurally significant transformations were detected during the calcination. For beidellite, in which the isomorphic substitution took place in the tetrahedral layer, a covalent bond between Al^{3+} or Si^{4+} from the T-layer and Al^{3+} from the pillar was proposed [58]. The protons released during calcination are able to migrate to the negative charges located in the T-sheet. A subsequent proton attack causes the formation of Si-OH silanol groups and protonated Si-(OH)-Al bridges. Now, a reaction with the pillars is possible and leads to Si_{clay}-O-Al_{pillar} or Al_{clay}-O-Al_{pillar} bonds. From ^{27}Al and ^{29}Si MAS-NMR experiments, the latter mechanism was found to be more likely. This model of cross-linking to the pillars induces the inversion of the AlO_4 clay units.

At that time, Plee et al. stated that the layer reactivity is solely related to the origin of the layer charge, and only occurs in case of tetrahedrally charged smectites [58]. Later, however, Pinnavaia et al. proved that the mechanism of cross-linking was also possible for the octahedrally substituted fluorohectorite clay [57]. F^- present in the former clay, was found to be responsible for the labilization of the Si-O bonds of the tetrahedral layer (40). This promotes the linking between the host layers and the pillars to form Si_{clay}-O-Al_{pillar} covalent bonds by an inversion of the SiO_4-tetrahedrons. This mechanism of cross-linking for fluorohectorite is represented in Fig. 14.

This model of covalent oxygen cross-linking of the tetrahedral clay layer and the pillars does not hold for other types of clay. ^{29}Si and ^{27}Al MAS NMR studies

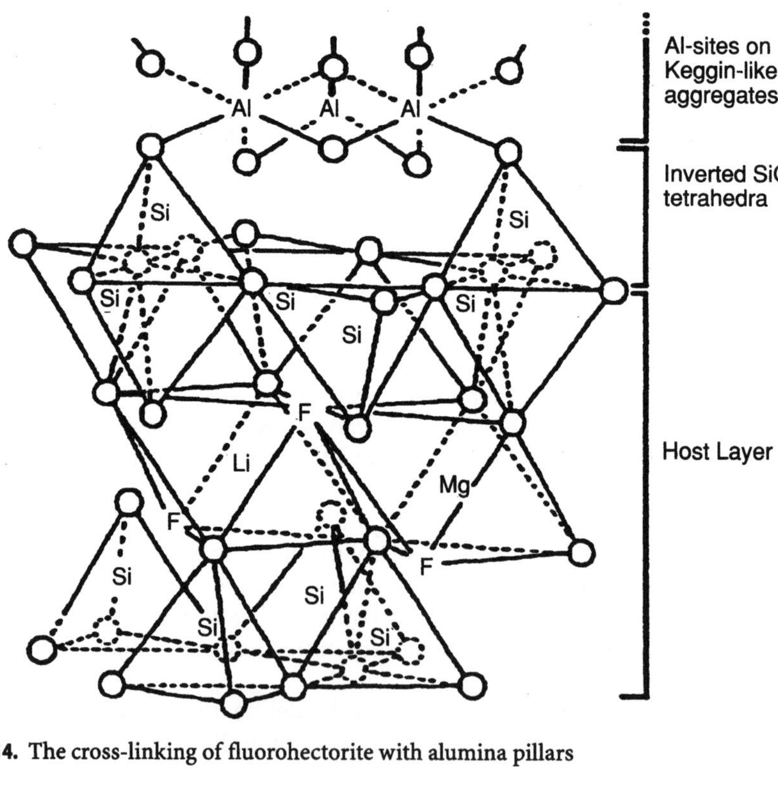

Fig. 14. The cross-linking of fluorohectorite with alumina pillars

showed that pillared montmorillonite and laponite retained their layer constitution after calcination, so a different bonding must be assumed. The clay layers are unreactive and pillaring presumably involves van der Waals interactions or, more likely, simple dative bond formation between the layer oxygens and coordinatively unsaturated sites on the Al-pillars.

5 Modification of Pillared Clays

In order to achieve a porous material with the appropriate properties for adsorption and separation purposes, additional modifications of pillared clays are sometimes necessary. These applications require a high adsorption capacity, a high selectivity towards certain gas molecules, a high adsorption strength or other properties. The modifications can be performed either during the synthesis or after the synthesis. A short discussion of the different modification techniques will be given below.

One way to increase the porosity of pillared clays is the pre-adsorption of template molecules prior to the ion exchange with the pillaring precursors. *n*-Alkylammonium ions are pre-exchanged on the Na^+-clay in an amount which is lower than the CEC. As a result, the pillar density decreases since part of the interlayer space is occupied by the template. During calcination, the organic template molecules are removed, and a homogeneous distribution of pillars results.

Heylen et al. reported a surface area and a micropore volume which was 2.5 times higher for an Fe-PILC synthesized with butylammonium as template, in comparison to the unmodified Fe-PILC [27]. An important increase in the adsorption capacity towards N_2, O_2 and CO at 194 K ($P_{eq}=4.5\times10^4$ Pa) has been observed after the modification. Adsorption capacities of 0.23 mmol/g, 0.17 mmol/g and 0.30 mmol/g for, respectively, N_2, O_2 and CO were reached for the modified BuA-Fe-PILC. When compared to the adsorption on the unmodified Fe-PILC, the following capacities were obtained: 0.00 mmol/g N_2, 0.03 mmol/g O_2 and 0.27 mmol/g CO. Also, it is proven that ammonium templates can be used to optimize the Zr-pillaring of laponite leading to enhanced pore volumes and surface areas.

A control of the porosity might still be possible by a secondary modification of the PILC with carbon. Adsorption of polymers, like polyvinylalcohol (PVA) and hydrocarbons (toluene), followed by a carbonization in an inert atmosphere, leads to C-deposition on the porous PILC-structure, influencing the adsorption properties. The aim of this technique is the achievement of pore-size narrowing to such an extent that PILCs with molecular sieving properties are obtained.

The coke deposition on Ti- and Al-PILC was investigated by Maes and Vansant [59, 60]. The reported techniques were in fact unsuccessful in achieving a controlled pore size modification, but the obtained results can function as a basis for further investigations and optimization. The pillared clay layers were found to be too rigid, preventing the bulky PVA molecules to penetrate the pores. Therefore, PVA adsorption mainly takes place on the external surface of the clay layers. Subsequent carbonization converts the PVA molecules to a carbon coating which results in pore-blocking effects. In this way, a complete pore-blocking is reached for Al-PILC. This phenomenon is not so effective for the Ti-pillared clay, due to the existence of an external TiO_2-phase creating defects in the carbon coating. A narrowing of the PSD for the latter PILC was found, without achieving a controlled pore-narrowing. Toluene is able to enter the Ti-PILC pores and can be situated at the acid sites which are responsible for cracking and coking reactions upon heating. The irregular deposition of coke reduces the micropore volume (pore-filling effect) without leading to a controlled narrowing. In case of the smaller pore-sized Al-PILC, deposition of coke occurs at the slit-pore entrances where pore-entrance narrowing and blocking take place.

The incorporation of metals into the pillars, when performed during the synthesis, results in the formation of mixed oxide-pillars (examples are given in Table 1). In this way specific adsorption sites are created in the PILC, exerting a positive influence on the adsorption capacity and selectivity towards gases [33].

A secondary ion modification involves the direct exchange in the PILC of both cations and anions from an alkaline or acidified salt solution, respectively. This is possible because of the amphoteric character of the hydroxy groups, present on the pillars and on the clay layer edges. At low pH, the -OH groups protonate and act as anion exchangers, while at high pH, they deprotonate and become cation exchangers.

Molinard pointed out that the ions introduced can serve as specific adsorption sites and that this type of modification allows fine-tuning of the PILC substrate for certain gas adsorption applications [18]. By introducing Sr^{2+} in Al-PILC, the gas adsorption isotherms at 273 K and 5×10^4 Pa equilibrium pressure showed that the N_2 capacity doubled from 0.06 mmol/g on Al-PILC to 0.12 mmol/g on Sr-Al-PILC. The amount of cations in the PILC influences the adsorption properties. A higher cation loading results in a lower capacity but a higher N_2/O_2 selectivity on Ca-Al-PILC, as was proven by adsorption measurements. For the anion exchanged Cl-Al-PILC, more O_2 than N_2 was adsorbed (0.09 mmol/g and 0.07 mmol/g, respectively) at 273 K and 5×10^4 Pa. This affinity for O_2 should result from a specific interaction with the Cl^--ions. The CO_2 capacity decreased in comparison with the unmodified substrate from 0.5 mmol/g on Al-PILC to 0.2 mmol/g on Cl-Al-PILC at 273 K and 0.5 bar. After modification with other anions, like F^- and PO_4^{3-}, the same effect was observed.

6
Pillared Clay vs Delaminated Clay: Structure and Applications

When the lateral dimensions of the clay layers are small (≤ 0.05 µm) and the layer morphology is lath-like, the flocculation of smectite clays by polyoxycations can lead to delaminated aggregates [45, 61]. Under these conditions, the previously discussed "card-house" structure has been proposed for delaminated clays. It differs dramatically from the well-ordered F-to-F lamellar structures formed by pillared clays when the layer size is large (≤ 2 µm) and pancake-like in morphology. Schematic drawings of the pillared and delaminated clay

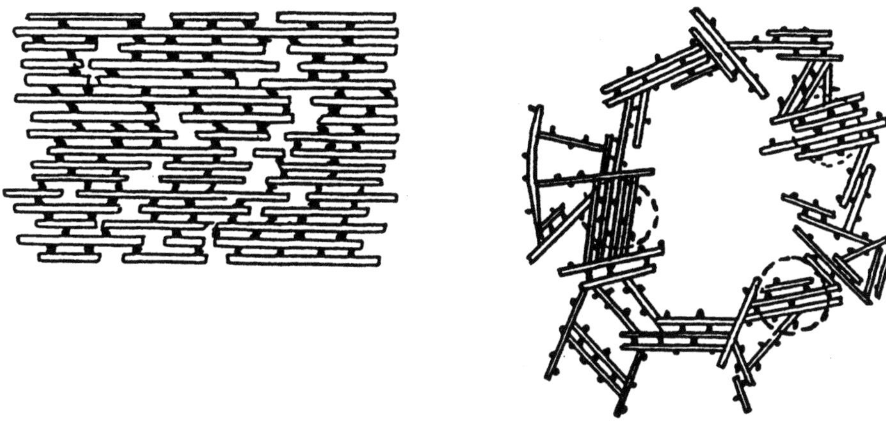

Fig. 15. Face-to-face pillared clay (*left*) and delaminated clay (*right*). The *circled regions* indicate the contribution of microporosity

structures are given in Fig. 15. Delaminated structures are favored by the small particle size fraction of the synthetic laponite. The more commonly observed lamellar structures are typified by pillared derivatives of the natural minerals hectorite, montmorillonite, beidellite and others. Delaminated laponite is formed when it is pillared with Al_{13}-Keggin ions. This delaminated clay structure has a high surface area and is mainly mesoporous.

The different porosity characteristics of delaminated clays compared to pillared clays will influence their catalytic and adsorption properties.

Occelli et al. determined the gas oil cracking ability of a delaminated Al-laponite and found some distinct advantages for this type of material over pillared clays [62]. The activity of the mesoporous delaminated catalyst for gas oil cracking was shown to be higher than for PILC and similar to the activity of an amorphous silica-alumina catalyst. However, its selectivity more nearly resembles the one of a typical commercial zeolite-promoted fluidized catalytic cracking (FCC) catalyst or a pillared clay. This suggests the existence of F-to-F stacked clay sheets, generating microporosity besides the large mesoporosity in the card-house structure. Due to their much greater mesoporosity, the desorption of high-molecular-weight hydrocarbons is favored, which would be retained as coke in the PILC structure. The enhanced mesoporosity can be an advantage in catalytic reaction systems where mass transfer is an important kinetic consideration. Also in gas adsorption applications, the diffusion of gases into the pores will be facilitated. However, delaminated clays lack the molecular sieving properties of well-ordered microporous pillared clays.

Also for other applications, including the adsorption of large sized environmental toxicants (e.g. chlorophenols), delaminated clays seem to be very appropriate adsorbents [63]. Al-delaminated laponite (ADL) is more effective in adsorbing pentachlorophenol (PCP) than the Al-pillared montmorillonite (APM). So delamination facilitates the adsorption process. The binding of PCP onto the substrate is attributed to interactions between the PCP molecule and the immobilized Al_2O_3 species. Now, the greater adsorption capacity observed for ADL relative to APM arises from the greater dispersion and availability of the Al-oxide aggregates in the clay having a delaminated structure.

7 Other Applications of Pillared Clays

The main applications of PILCs are situated in the field of catalysis and adsorption. The acidity of PILCs is a very important feature in controlling catalytic reactions [64, 65a, 65b]. Pillared clays exhibit Lewis (L) as well as Brønsted (B) acid sites, and the ratio L/B is depending on the type of clay. The pillars are the major sources of Lewis acidity while the structural OH-groups of the clay contribute to the Brønsted acidity. For those clays containing 3-coördinated Al^{3+} substituted for Si^{4+} in the T-layer, the Al^{3+} is a Lewis acid site. Also coördinatively unsaturated Al^{3+} located at the edges of the O-sheet is a Lewis acid, however, upon hydration the sites are converted into an octahedrally coördinated Al with Brønsted acidity. Most structural Brønsted acid sites are located therefore at the edges of the clay sheets. The high L/B ratio of a delaminated pillared laponite compared to a pillared clay confirms an alumina-covered card-house structure.

In the literature, the performance of several acid-catalyzed reactions on PILC have been reported, of which some will be mentioned below [66]:

- Cumene cracking was carried out as a test reaction for Brønsted acidity.
- Propylene oligomerization is catalyzed by Lewis sites on Al-pillared montmorillonite.
- The disproportionation of trimethylbenzene is a possible route to produce durene (1,2,4,5-tetramethylbenzene). PILCs have been reported to be catalysts for this reaction, and the shape selective effect was discovered in this reaction from the product distribution with durene being the major product. This disproportionation reaction is catalyzed by the PILC Lewis sites, while the isomerization of trimethylbenzene (side reaction) is well correlated to the Brønsted acidity.

The commercial separation of air into N_2 and O_2, an industrially very important process, is achieved by either cryogenic distillation or pressure swing adsorption (PSA). The use of pillared clays forms an interesting alternative for the carbon molecular sieves and zeolites currently applied as adsorbents in PSA techniques. Both the capacity and the selectivity towards air components are very important features in gas adsorption applications.

8 References

1. Barrer RM, McLeod DM (1955) Trans Faraday Soc 51:1290
2. Brindley GW, Sempels RE (1977) Clay Minerals 12:229
3. Vaughan DEW, Lussier RJ, Magee JS Jr (1979) US Patent 4,176,090
4. Mitchell IV (1990) Pillared layered structures: current trends and applications. Elsevier, London

5. Occelli ML, Robson HE (1992) Synthesis of microporous solids, vol II: Expanded clays and other microporous solids. Van Nostrand Reinhold, New York
6. Barrer RM (1978) Zeolites and clay minerals as sorbents and molecular sieves. Academic Press, London
7. Barrer RM (1990) In: Mitchell IV (ed), Pillared layered structures: current trends and applications. Elsevier Science Publishers, London
8. Van Leemput L, Stul MS, Maes A, Uytterhoeven JB, Cremers A (1983) Clays and Clay Miner 31:261
9. Knudson MJ Jr, McAtee JL Jr (1974) Clays and Clay Miner 22:59
10. Berkheiser VE, Mortland MM (1977) Clays and Clay Miner 25:105
11. Loeppert RH Jr, Mortland MM, Pinnavaia TJ (1979) Clays and Clay Miner 27:201
12. Endo T, Mortland MM, Pinnavaia TJ (1981) Clays and Clay Miner 29:153
13. Yamanaka S, Doi T, Sako S, Hattori M (1984) Mat Res Bull 19:161
14. Martin-Luengo MA, Martins-Carvalho H, Ladriere J, Grange P (1989) Clay Minerals 24:495
15. Doff DH, Gangas NHJ, Allan JEM, Coey JMD (1988) Clay Miner 23:367
16. Maes N, Vansant EF (1995) Microporous Mater 4:43
17. Pinnavaia TJ (1986) In: Setton R (ed) Chemical reactions in organic and inorganic constrained systems. Reidel, Boston
18. Molinard A (1994) Physicochemical and gas adsorption properties of ion exchanged alumina pillared clays. PhD thesis, University of Antwerp
19. Schoonheydt RA, Van Den Eynde J, Tubbax H, Leeman H, Stuyckens M, Lenotte I, Stone WEE (1993) Clays and Clay Miner 41:598
20. Schoonheydt RA, Leeman H, Scorpion A, Lenotte I, Grobet PJ (1994) Clays and Clay Miner 42:518
21. Yamanaka S, Nishihara T, Hattori M (1987) Mater Chem Phys 17:87
22. Burch R, Warburton CI (1986) J of Catalysis 97:503
23. Ohtsuka K, Hayashi Y, Suda M (1993) Chem Mater 5:1823
24. Sterte J (1986) Clays and Clay Miner 34:658
25. Brindley GW, Yamanaka S (1979) Amer Miner 64:830
26. Tzou MS, Pinnavaia TJ (1988) In: Burch R (ed) Pillared Clays, Catalysis Today 2:243
27. Heylen I, Vanhoof C, Vansant EF (1995) Microporous Mater 5:53
28. Burch R, Warburton CI (1978) J Chem Soc Chem Commun 117
29. Yamanaka S, Nishihara T, Hattori M (1988) Mater Res Soc Symp Proc 111, Pittsburgh 283
30. Yamanaka S, Takahama K (1993) In: Ishizaki K, Sheppard L, Okada S, Hamasaki T, Huybrechts B(eds) Porous Materials, Ceramic Transactions 31
31. Pinnavaia TJ (1992) In: Occelli ML, Robson HE (eds) Expanded clays and other microporous solids. Van Nostrand Reinhold, New York
32. Bergaya F, Hassoun N, Barrault J, Gatineay L (1993) Clay Miner 28:109
33. Heylen I, Maes N, Molinard A, Vansant EF (1994) In: Vansant EF (ed) Separation technology. Elsevier, Amsterdam
34. Sterte J (1991) Clays and Clay Miner 39:167
35. Bradley SM, Kydd RA, Yamdagni R, Fyfe CA (1992) In: Occelli ML, Robson HE(eds) Expanded clays and other microporous materials. Synthesis of microporous materials, vol II. Van Nostrand Reinhold, New York
36. Zhao D, Yang Y, Guo X (1995) Zeolites 15:58
37. Skaribas SP, Pomonis PJ, Grange P, Delmon B (1992) J Chem Soc Faraday Trans 88:3217
38. Yamanaka S, Brindley GW (1979) Clays and Clay Miner 27:119
39. Jones SL (1988) In: Burch R (ed) Pillared Clays, Catalysis Today 2:2–3
40. Clearfield A, Kuchenmeister M (1992) In: Bein T (ed) Supramolecular architecture, ACS Symp Ser, 499. Am Chem Soc, Washington DC
41. Clearfield A, Roberts BD (1988) Inorg Chem 27:3237
42. Bottero JY, Cases JM, Fiessinger F, Poirier JE (1980) J Phys Chem 84:2933
43. Bottero JY, Tchoubar D, Cases JM, Fiessinger F (1982) J Phys Chem 86:3667
44. Vaughan DEW (1988) In: Burch R (ed) Pillared Clays, Catalysis Today 2:187

45. Pinnavaia TJ, Zhou MS, Landau SD, Raythatha RH (1984) J of Molecular Catalysis 27:195
46. Jones JR, Purnell JH (1993) Catalysis Letters 18:137
47. Furrer G, Ludwig C, Schindler PW (1992) Journal of Colloid and Interface Science 149:56
48. Bartley GJJ (1988) In: Burch R (ed) Pillared Clays, Catalysis Today 2:233
49. Clearfield A (1990) J Mater Res 5:161
50. Clearfield A, Vaughan PA (1956) Acta Cryst 9:555
51. Baksh MSA, Kikkinides ES, Yang RT (1992) Ind Eng Chem Res 31:2181
52. Bartley GJJ, Burch R(1985) Applied Catalysis 19:175
53. Jaynes WF, Traina SJ, Bigham JM, Johnston CT (1992) Clays and Clay Miner 40:397
54. Lahav N, Shani U, Shabtai I (1978) Clays and Clay Miner 26:107
55. Figueras F, Klapyta Z, Massiani P, Mountassir Z, Tichit D, Fajula F, Gueguen C, Bousquet J, Auroux A (1990) Clays and Clay Miner 38:257
56. Schoonheydt RA, Leeman H (1992) Clay Miner 27:249
57. Pinnavaia TJ, Landau SD, Tzou MS, Johnson ID (1985) J Am Chem Soc 107:7222
58. Plee D, Borg F, Gatineau L, Fripiat JJ (1985) J Am Chem Soc 107:2362
59. Maes N, Vansant EF (1996) J Porous Materials 3:257
60. Maes N, Vansant EF (1997) J Porous Materials 4:5
61. Occelli ML, Landau SD, Pinnavaia TJ (1987) J of Catalysis 104:331
62. Occelli ML, Landau SD, Pinnavaia TJ (1984) J of Catalysis 90:256
63. Zielke RC, Pinnavaia TJ (1988) Clays and Clay Miner 36:403
64. Kikuchi E, Matsuda T (1988) In: Burch R (ed) Pillared Clays, Catalysis Today 2:2–3
65. a. Occelli ML (1986) J of Molecular Catalysis 35:377
65. b. Occelli ML, Finseth DH(1986) J of Catalysis 99:316
66. Ming-Yuan H, Zhonghui L, Enze M (1988) In: Burch R (ed) Pillared Clays, Catalysis Today 2:2–3

Subject Index

Springer and the environment

At Springer we firmly believe that an international science publisher has a special obligation to the environment, and our corporate policies consistently reflect this conviction.
We also expect our business partners – paper mills, printers, packaging manufacturers, etc. – to commit themselves to using materials and production processes that do not harm the environment. The paper in this book is made from low- or no-chlorine pulp and is acid free, in conformance with international standards for paper permanency.

MIX
Papier aus verantwortungsvollen Quellen
Paper from responsible sources
FSC® C105338

If you have any concerns about our products,
you can contact us on
ProductSafety@springernature.com

In case Publisher is established outside the EU,
the EU authorized representative is:
Springer Nature Customer Service Center GmbH
Europaplatz 3, 69115 Heidelberg, Germany

Printed by Libri Plureos GmbH
in Hamburg, Germany